Unifying Scientific Theories

This book is about the methods used for unifying different scientific theories under one all-embracing theory. The process has characterized much of the history of science and is prominent in contemporary physics; the search for a "theory of everything" involves the same attempt at unification.

Margaret Morrison argues that contrary to popular philosophical views, unification and explanation often have little to do with each other. The mechanisms that facilitate unification are not those that enable us to explain how or why phenomena behave as they do.

A feature of this book is an account of many case studies of theory unification in nineteenth- and twentieth-century physics and of how evolution by natural selection and Mendelian genetics were unified into what we now term evolutionary genetics.

The book emphasizes the importance of mathematical structures in unification and claims that despite this common feature, theory unification is a multifaceted process for which no general account can be offered.

Margaret Morrison is Professor of Philosophy at the University of Toronto.

T0275630

Unifying Scientific Theories

Physical Concepts and Mathematical Structures

MARGARET MORRISON

University of Toronto

CAMBRIDGE
UNIVERSITY PRESS

CAMBRIDGE UNIVERSITY PRESS
Cambridge, New York, Melbourne, Madrid, Cape Town, Singapore, São Paulo

Cambridge University Press
The Edinburgh Building, Cambridge CB2 8RU, UK

Published in the United States of America by Cambridge University Press, New York

www.cambridge.org
Information on this title: www.cambridge.org/9780521652162

First published 2000
This digitally printed version 2007

A catalogue record for this publication is available from the British Library

Library of Congress Cataloguing in Publication data
Morrison, Margaret, 1954–
Unifying scientific theories : physical concepts and mathematical
structures / Margaret Morrison.
p. cm.
Includes index.
ISBN 0-521-65216-2 (hbk.)
1. Unified field theories. I. Title.
QC794.6.G7M67   1999
530.14′2 –dc21   99-20874
                                CIP

ISBN 978-0-521-65216-2 hardback
ISBN 978-0-521-03760-0 paperback

Contents

Acknowledgements   *page* vii

Introduction   1

1   The Many Faces of Unity   7
   1.1   Kepler: Unity as Mathematical Metaphysics   7
   1.2   Kant: Unity as a Heuristic and Logical Principle   12
   1.3   Whewell: Unity as Consilience and Certainty   16
   1.4   Logical Empiricism: Unity as Method and Integration   22
   1.5   Unity as Explanation   25

2   Unification, Realism and Inference   35
   2.1   The Friedman Model   38
   2.2   The Importance of Conjunction   39
   2.3   Reduction versus Representation   43
      2.3.1   Is Reduction a Viable Approach?   43
      2.3.2   The Problem of Many Models   47
   2.4   Consilience and Unification   52
   2.5   Unification as an Evidential or Epistemic Virtue   57
   Appendix   Derivation of the van der Waals Law: Historical Details   59

3   Maxwell's Unification of Electromagnetism and Optics   62
   3.1   Development of Electromagnetic Theory: The Early Stages   64
      3.1.1   Maxwell circa 1856: "On Faraday's Lines of Force"   64
      3.1.2   "On Physical Lines of Force"   68
   3.2   Unification and Realism: Some Problems for the
       Electromagnetic Theory   79
   3.3   The Electromagnetic Theory: Later Developments   81
      3.3.1   The Dynamical Theory   81
      3.3.2   The Treatise on Electricity and Magnetism   90
   3.4   Realism and Dynamical Explanation   99
      3.4.1   Philosophical Conclusions   105

4   Gauges, Symmetries and Forces: The Electroweak Unification   109
   4.1   Fermi and Beta Decay   110
   4.2   Symmetries, Objects and Laws   114

4.3   The First Steps to Unity: Symmetry and Gauge Theory   118
4.4   Unity through Symmetry-Breaking   121
4.5   Renormalization: The Final Constraint   127
4.6   When Unity Isn't Enough   130
4.7   Unified Theories and Disparate Things   135
Appendix 4.1   Weinberg's Lepton Model   140
Appendix 4.2   Renormalization   143

5   Special Relativity and the Unity of Physics   147
5.1   Electrodynamics circa 1892–1904: Lorentz and the Origins of
      Relativity   149
      5.1.1   1895: *Versuch* and the Theory of Corresponding States   156
5.2   Einstein circa 1905: Lorentz Transformed   162
      5.2.1   Synchronizing Clocks: The Here and Now   165
      5.2.2   Deriving the Transformation Equations: The
              Interconnection of Space and Time   169
      5.2.3   Synthetic Unity: From Form to Content   176
      5.2.4   The Most Famous Equation in the World   179
5.3   From Fields to Tensors: Minkowski Space-Time   183

6   Darwin and Natural Selection: Unification versus Explanation 192
6.1   On the Nature of Darwin's Theory   193
6.2   Kitcher's Argument Patterns: Explanation as Derivation
      and Systematization   196
6.3   Methods and Causes: History and Influence   202
6.4   Does Unity Produce Explanation?   206

7   Structural Unity and the Biological Synthesis   210
7.1   Synthesis and Unity   210
7.2   Mathematics and Theory   213
      7.2.1   Mendelian Ratios   213
      7.2.2   Fisher's Idealized Populations   214
      7.2.3   Wright's Path Coefficients   224
7.3   Disunity and Explanation   227

Conclusions   232

Notes   238
References   257
Index   267

Acknowledgements

As is usually the case with large projects like books, the completion of the work is in no small part due to the generosity of others – friends, colleagues and institutions. Perhaps my greatest debt is to the Social Sciences and Humanities Research Council of Canada for generously supporting my research for many years. Without their assistance this work might never have been finished, or even undertaken. I am also grateful to the Wissenschaftskolleg zu Berlin for a fellowship in 1995–96 and to the Connaught Fund at the University of Toronto for financial support at the beginning stages of the project. Several colleagues and friends have offered comments and advice at one stage or another that greatly improved both the content and structure of the book, particularly Jed Buchwald, Francis Everitt, Paul Forster, Stephan Hartmann, Colin Howson, Elie Zahar and an anonymous referee for Cambridge University Press. Paul Teller gave me extensive comments on the entire manuscript and offered valuable suggestions that forced me to present my ideas in a much clearer way. The main arguments of Chapters 4, 6 and 7 were presented as lectures at the Centre for the Philosophy of the Natural and Social Sciences at the London School of Economics. I would like to thank Nancy Cartwright and the Centre for their support and hospitality and for allowing me the opportunity to discuss my work with the other fellows and students. Michael Ashooh and Nick Oweyssi helped in the final stages of preparing the manuscript. Sheldon Glashow kindly answered correspondence about the electroweak theory, and Steven Weinberg graciously took time out to talk to me about his own views on unity and his early work on the lepton model. Thanks also go to Terence Moore for his interest in the project and his encouragement in its early stages and to Jim Mobley for his careful and helpful editorial assistance. Finally, I am grateful to the philosophy department at the University of Toronto for providing the kind of flexible environment that makes doing research possible and enjoyable.

Chapters 4 and 7 contain material from two papers delivered at the Philosophy of Science Association (PSA) meetings in 1994 and 1996, respectively. I wish to thank the association for permission to cite from "Unified Theories and Disparate Things" (PSA 1994) and "Physical Models and Biological Contexts" (PSA 1996). A good deal of the material in Chapter 3 first appeared in *Studies in the History and Philosophy of Science* (1990) as a paper entitled "A Study in Theory Unification: The Case of Maxwell's Electromagnetic Theory". I also wish to thank Elsevier for allowing me to reprint that material here. Chapter 2 draws on my paper "Unification, Reduction and Realism", which first appeared in the *British Journal for the Philosophy of Science* in 1990.

Introduction

"Unity" has become a much-maligned word in history and philosophy of science circles, the subject of criticism that is both normative and descriptive. The concepts of "unity of science" and "unity of method" and even the notion of a "unified theory" have been criticized for being either politically undesirable (Dupré 1996) or metaphysically undesirable (Galison and Stump 1996), or else simply nonexistent – the products of a misrepresentation of scientific practice. Critics of unity claim that when we look at scientific practice we see overwhelming evidence for disunity, rather than the coherent structure we have been led to believe characterizes science. Although some of these arguments are extremely persuasive, the desire to banish unity altogether has resulted, I believe, in a distortion of the facts and a misunderstanding of how unity actually functions in science. It is simply a mistake to deny that science has produced unified theories. So where does the evidence for disunity come from? In order to answer this question, we need to look to theory structure as a way of clarifying the nature of that unity. The task then, as I see it, is not so much one of defending a strong version of unity at all costs, but rather of providing an analysis of how it is achieved and how it functions. To that end I have chosen to focus on theory unification as the basis for my discussion of unity. Not only do unified theories provide the foundation for a more general notion of scientific unity, but also there has been a great deal of attention paid to theory unification in the philosophical literature (e.g., Friedman 1983; Glymour 1980; Kitcher 1989).

Although there are undeniable instances of theory unification, to ignore instances of disunity in science would also be to disregard the facts. So instead of trying to counter examples of disunity with ones of unity, I want to show that once we have some understanding of (1) how unity is produced, (2) its implications for a metaphysics of nature and (3) its role in theory construction and confirmation, it will cease to occupy the undesirable role attributed to it by the advocates of disunity. This book addresses, primarily, these three issues, but not by providing a general "theory" of unification, because no such account is, I think, possible. Instead, I shall draw attention to some general features of unified theories, thereby providing the reader with some insight into the complex nature of theory unification and the philosophical consequences that result from a better understanding of the process. One such consequence is the decoupling of unification and

1

explanation. Rather than analysing unification as a special case of explanatory power, as is commonly done in the literature, I claim that they frequently have little to do with each other and in many cases are actually at odds.

If one were asked to list the most successful scientific theories of the modern era, two obvious entries would be Newtonian mechanics and Maxwell's electrodynamics. The feature common to both is that each encompasses phenomena from different domains under the umbrella of a single overarching theory. Theories that do this are typically thought to have "unifying power"; they unify, under a single framework, laws, phenomena or classes of facts originally thought to be theoretically independent of one another. Newton's theory unified celestial and terrestrial mechanics by showing how the motions of both kinds of objects could be described using the law of universal gravitation. Maxwell's theory *initially* brought together electromagnetism and optics by demonstrating that the calculated velocity for electromagnetic waves travelling through a material medium (an aether) was in fact equal to the velocity of light – that light waves and electromagnetic waves were in fact motions of one and the same medium. Later versions of the theory did not rely on the aetherial medium to achieve this result; the value for V was derived from the field equations expressed in the abstract mechanics of Lagrange. In both Newton's theory and Maxwell's theory the unification consists, partly, in showing that two different processes or phenomena can be identified, in some way, with each other – that they belong to the same class or are the same kind of thing. Celestial and terrestrial objects are both subject to the same gravitational-force law, and optical and electromagnetic processes are one and the same.

Because each of these theories unifies such a diverse range of phenomena, they have traditionally been thought to possess a great deal of explanatory power. For example, we can "explain" the nature of radio waves by showing that they are simply a type of electromagnetic radiation; and we can explain the tides by demonstrating that they are a manifestation of gravitational force. In fact, much of the literature in the philosophy of science has analysed unification exclusively in terms of explanation; see especially Friedman (1983) and Kitcher (1981, 1989). A unified theory is simply one that explains several different phenomena using the same laws. And, frequently, what it means for a theory to have explanatory power is analysed in terms of its ability to unify. The best explanation then typically will be the one that unifies the greatest number of phenomena. Accounts of explanation, such as the deductive nomological (D-N) model, that focused primarily on explanation as deduction or derivability could also account for unification within those parameters. Although certain formal and material constraints needed to be fulfilled if the explanation was to count as a unification, the point remained – unification was a special case of explanation.

In the case of the D-N model, we explain a particular phenomenon, termed the explanandum, by showing how it is derivable from a set of laws and initial conditions that, taken together, constitute the explanans. We know that the unification produced by Newton's theory involved a synthesis of Galileo's laws of terrestrial mechanics with Kepler's laws for celestial bodies. So if we consider why bodies in free fall move in accordance with Galileo's laws and why planetary motions exhibit the uniformities specified by Kepler's laws, we can explain these facts by showing how these laws are special consequences of universal gravitation and Newton's laws of motion. So, in that sense, Newton's theory unifies the two domains insofar as it can accommodate two separate kinds of phenomena – phenomena that had been governed by two distinct domains of laws. However, because Newton's theory, strictly speaking, does not give the same predictions for the same range of phenomena that, say, Galileo's laws do (i.e., the former implies a steady increase in acceleration as a freely falling body nears the ground), in order to adhere to the proper deductive structure for the D-N model we must add specific auxiliary assumptions: that the earth is a homogeneous sphere with a particular mass and radius and that we are concerned with free fall over short distances near the surface of the earth. We must then modify the explanandum so that it is no longer a question of why freely falling bodies move in accordance with Galileo's laws, but why these bodies obey Galileo's laws to a high degree of approximation. Hence, we don't say that Newton's theory explains Galileo's laws in virtue of there being a straightforward deductive relationship between them; the deductive structure of the explanation is maintained only by reinterpreting the fact to be explained.

This kind of unification is thought to provide explanation and increased understanding because it shows that a wide variety of phenomena actually obey regularities that are manifestations of very few basic laws. And it supposedly can show why the laws that have been superseded by the unifying theory were true in the cases they covered. What the D-N model provides, then, is a kind of explanation by subsumption; when we ask why a particular phenomenon behaves as it does, a law is cited from which the behaviour can be derived. The law of gravitation will tell us everything about the way the moons of Jupiter move around the planet, and, among other things, it tells us why the earth and the other planets move around the sun in the way they do. This is possible because all of these bodies obey the same law of gravitational force. So unification is again defined, as it were, in terms of the explanation of diverse phenomena under one law.

There have been many criticisms of the D-N model, not because of the way it handles unification but because of its failure to produce an "understanding" of the processes to be explained. Explanation by derivation from quantitative laws very often doesn't provide what Richard Feynman calls the "machinery" of a particular system. The machinery is what gives us the mechanism that explains why, but more importantly *how*, a certain process takes place. When we ask about the propagation of electromagnetic waves, we want to know not just their velocity but also how they travel through space and the mechanism responsible for their propagation. Maxwell's first account of electrodynamics explained this in terms of the aether; but later, more abstract formulations of the theory left the mechanism unspecified, thereby sacrificing explanatory power. The point I want to stress here is that simple obedience to the same laws often does not provide the kind of explanation we typically search for in the face of puzzling phenomena. Although it gives some information about how we can expect a system to behave, we frequently want to know more than that; we want to know about the machinery, part of which involves knowledge of the causal behaviour of the system. It is this feature that enables us to understand how certain processes take place.

Admittedly, such an explanation is not always possible; we may simply be unable to determine the material behaviour or conditions that produce a particular event or effect. But even when this type of explanation is not available, the theory in question still may be able to unify a group of phenomena. Indeed, such was the case with later formulations of Maxwell's electrodynamics – formulations that did not include a mechanical account of how electromagnetic waves travelled through space. Another example is the unification of terrestrial and celestial phenomena in Newton's *Principia*. Although influenced by Cartesian mechanics, one of the most striking features of the *Principia* is its move away from explanations of planetary motions in terms of mechanical causes. Instead, the mathematical form of force is highlighted; the planetary ellipses discovered by Kepler are "explained" in terms of a mathematical description of the force that produces those motions. Of course, the inverse-square law of gravitational attraction explains why the planets move in the way they do, but there is no explanation of how this gravitational force acts on bodies (how it is transported), nor is there any account of its causal properties.

What this suggests is that explanation and unification may not be as closely related as has typically been thought; unity is possible without a satisfactory level of explanatory power. Moreover, when unification is analysed in terms of something like the D-N framework, it becomes clear that the account of unification that results provides virtually no understanding of *how* the unifying process takes place. Because unity is understood simply in terms of derivability, there is no sense of how the phenomena become integrated within a theoretical edifice. In fact, as long as unification is seen through the lens of explanatory power, its articulation becomes linked to some prior notion of explanation that may or may not be fulfilled. Hence, rather than connecting unification and explanation by analysing one in terms of the other, a good deal of my analysis will be devoted to showing how the two actually diverge. I want to argue, using specific examples of unified theories, that the mechanisms crucial to the unifying process often supply little or no theoretical explanation of the physical dynamics of the unified theory.

The examples of unified theories we shall consider have two common features: (1) they embody a mathematical structure or mathematization of the phenomena that furnishes an abstract, general framework capable of uniting diverse phenomena under a single theory; (2) that framework typically contains a theoretical parameter, quantity or concept that "represents" the unifying mechanism – that is, a parameter that functions as the necessary piece of theoretical structure that either facilitates or represents the unification of distinct phenomena. For example, in Maxwell's electrodynamics it is electric displacement (a fundamental quantity in the field equations), together with the Lagrangian formalism, that allows Maxwell to furnish a field-theoretic description of the phenomena. This is accomplished without any explanation of how electromagnetic waves are propagated through space. Despite its prominence, electric displacement is given no theoretical explanation in Maxwell's mature theory; that is, he offers no physical dynamics explaining the nature of displacement. And to the extent that unification relies on mathematical structures like Lagrangian mechanics, it becomes easy to see how explanatory detail is sacrificed for the kind of generality and abstraction that facilitate unification.

Given this separation of unification and explanation, together with the role of mathematical structures in unifying theories, it becomes necessary to rethink the impact of theory unification for a metaphysical thesis about unity in nature. Again, whether or not the former provides evidence for the latter cannot be answered in a general way. Their connection will depend on the kind of unity a particular theory exemplifies. In the examples, I distinguish two different types of unity: reductive unity, where two phenomena are identified as being of the same kind (electromagnetic and optical processes), and synthetic unity, which involves the integration of two separate processes or phenomena under one theory (the unification of electromagnetism and the weak force). I claim that in the latter case there is no ontological reduction, and consequently the unification offers little in the way of support for claims about a physical unity in nature. Although reductive unity does seem to involve an ontological component, any conclusions we draw about ontology must ultimately depend on the way in which the unity was achieved. In other words, are there good physical reasons for thinking that two processes are one and the same, or have they simply been brought together with the aid of an abstract mathematical structure or model?

Briefly, then, my thesis is twofold: First, unification should not be understood as a form of explanatory power, for the mechanisms that facilitate the unification of phenomena often are not the ones that could enable us to explain those phenomena. Second, although unification is an important part of the scientific process, an analysis of how it takes place reveals that it can in some instances have very few, if any, implications for a reductionist metaphysics and an ontological unity of nature.

I begin with a discussion of the various ways in which "unity" has been conceived throughout the history of science and philosophy. The historical portion of Chapter 1 focuses on Kepler, Kant and Whewell. Although all three agreed on the importance of theoretical unity as a goal to be pursued, they had different metaphysical views about the source of that unity and how it functioned in an explanatory capacity. I have chosen to highlight Kant and Whewell because both figure prominently in philosophical discussions of unity, and both are often invoked in attempts to justify the search for unified theories as *the* methodology for the sciences. Kepler is similarly important because his search for connections between bodies, together with the system of quantifiable forces that he so diligently sought after, represents what has come to be seen as the paradigm of modern mathematical physics. The remainder of the first chapter presents a brief analysis of "unity" as conceived by the founders of the *International Encyclopedia of Unified Science* (Neurath et al. 1971), writers who were responsible for laying the foundations for a twentieth-century philosophy of science. And finally I shall discuss some of the philosophical arguments that have recently been put forward linking explanation, unification and truth. This issue of whether or not unified theories are more likely to be true will be addressed in greater detail in Chapters 2 and 6, where we shall examine Friedman's and Kitcher's accounts of unification. I shall highlight some difficulties with each approach – problems that arise in the application of these philosophical views to particular instances of theoretical unity.

In addition to addressing certain philosophical arguments regarding the nature and status of unification and explanation, we shall also examine several instances of unification encompassing both the physical and biological sciences. These cases will be presented in Chapters 3–5 and 7. What I hope my investigation will reveal are the ways in which theoretical unification takes on different dimensions in different contexts. What this means is that there is no "unified" account of unity – a trait that makes it immune from general analysis. Nevertheless, there are certain features that all unified theories possess, features that enable us to distinguish the process of unifying from that of simply explaining and conjoining hypotheses. Highlighting these will allow us to free theory unification from the kind of metaphysical speculation that fuels the desire for disunity in science. One of the implications of my analysis is that the unity/disunity debate rests on a false dichotomy. Describing science as either unified or disunified prevents us from understanding its rich and complex structure; in fact, it exhibits elements of both. Acknowledging the role played by each will allow for an appreciation of unity and disunity as essential features of both science and nature.

1

The Many Faces of Unity

1.1. Kepler: Unity as Mathematical Metaphysics

In the *Mysterium cosmographicum* Johannes Kepler claimed that it was his intention to show that the celestial "machine" was not a kind of divine living being,

> but a kind of clockwork insofar as the multiplicity of motions depends on a single, quite simple magnetic and corporeal force, just as all the motions of a clock depend upon a simple weight. And I also show that this physical cause can be determined numerically and geometrically. (Kepler 1938, xv:232)

His research began with a specification of certain astronomical hypotheses based on observation; that was followed by a specification of geometrical hypotheses from which the astronomical ones would follow or could be calculated. Those geometrical hypotheses were grounded in the idea that God created the solar system according to a mathematical pattern. Given that assumption, Kepler attempted to correlate the distances of the planets from the sun with the radii of spherical shells that were inscribed within and circumscribed around a nest of solids. The goal was to find agreement between the observed ratios of the radii of the planets and the ratios calculated from the geometry of the nested solids. Although unsuccessful, Kepler remained convinced that there were underlying mathematical harmonies that could explain the discrepancies between his geometrical theory and ratios calculated from observations.

Part of Kepler's unfaltering reliance on mathematical harmonies or hypotheses was based on their direct relationship to physical bodies. He considered a mathematical hypothesis to be physically true when it corresponded directly to physically real bodies. What "corresponding directly" meant was that it described their motions in the simplest way possible. Hence, according to Kepler, physical reality and simplicity implied one another; and it was because nature loves simplicity and unity that such agreement could exist. (Here unity was thought to be simply a manifestation of nature's ultimate simplicity.) Perhaps his most concise statement of the relationship between truth and simplicity or between the mathematical and the physical can be found in the *Apologia*, where Kepler distinguished between "astronomical" and "geometrical" hypotheses:

If an astronomer says that the path of the moon is an oval, it is an astronomical hypothesis; when he shows by what combination of circular movements such an oval orbit may be brought about, he is using geometrical hypotheses.... In sum, there are three things in astronomy: geometrical hypotheses, astronomical hypotheses, and the apparent motions of the stars themselves; and, consequently, the astronomer has two distinct functions, the first, truly astronomical, to set up such astronomical hypotheses as will yield as consequences the apparent motions; second, geometrical, to set up geometrical hypotheses of whatsoever form (for in geometry there may often be many) such that from them the former astronomical hypotheses, that is, the true motions of the planets, uncorrupted by the variability of the appearances, both follow and can be calculated.[1]

One was able to discover the true motions of the planets by determining their linear distances and using simplicity as the guiding principle in interpreting the observations.

Much of his early work in constructing physical theories (before the development of his laws of planetary motion) was dominated by the desire to provide a unified explanation of the causes of planetary motion. The Neoplatonic sun, to which he added a force that pushed the planets along in their orbits, served as the primary model for his solar hypothesis. But the foundation for that hypothesis was the metaphysical principle that one ought to reduce several explanatory devices to a single source. That principle, in turn, was based on Kepler's ideas about the Trinity. The sun served as the principle that unified and illuminated matter in the way that the Trinity symbolized the indivisible, creative God. Kepler then transformed the theological analogy into a mathematical relation in which solar force, like the light in a plane, was assumed to vary inversely with distance. The idea was that there existed one soul at the centre of all the planetary orbits that was responsible for their motions. God the Father created spirit in the same way that the sun dispersed spirit, and the sun emitted a moving force in the ecliptic in accordance with the same mathematical function as light propagating in a plane.

The important relation here, of course, was between mathematical simplicity and unity and the way in which those notions were used to both construct and justify astronomical hypotheses. As mentioned earlier, there was a direct relation between the symmetry of the mathematical relations used to describe physical bodies and the metaphysical underpinnings of those relations found in the Trinity. In his account of the interspacing of solid figures between planetary spheres, Kepler claimed that it ought to follow perfectly the proportionality of geometrical inscriptions and circumscriptions, and "thereby the conditions of the ratio of the inscribed to the circumscribed spheres. For nothing is more reasonable than that the physical inscription ought exactly to represent the geometrical, as a work of art its pattern" (1938, vi:354). And in analogy with the Trinity, he remarked that

there exists everywhere between point and surface the most absolute equality, the closest unity, the most beautiful harmony, connection, relation, proportion and commensurability.

And, although Centre, Surface and the Interval are manifestly Three, yet they are One, so that no one of them could be even imagined to be absent without destroying the whole. (Kepler 1938, vi:19)

Here we see an explicit statement of how unity and simplicity could be, in some cases, manifestations of the same thing. The unifying axiom that the planets were united by a single force, rather than a multiplicity of planetary "souls" acting in isolation, was, of course, also the simplest hypothesis. Hence, simplicity and unity were represented as oneness. In other contexts, however, unity and simplicity were related to each other via a kind of interconnectedness, the one as a manifestation of the many. For Kepler, the latter was apparent in the notion of the Trinity, but we can perhaps see it more clearly in the idea of a nation-state that embodies many people and perhaps many cultures, all of which are united in one identity – citizens of that state. It was that combination of unity and simplicity as a form of *interconnectedness* that provided the empirical basis on which Kepler's astronomical hypotheses were justified.

Although Kepler saw the truth of a physical or astronomical hypothesis as metaphysically grounded in its simplicity or unity, the latter also had to be revealed empirically. Not only did the phenomena have to be describable using mathematically simple relations, but the interconnectedness among those descriptions had to be manifest at the empirical level in order for the hypothesis to be justified. Such was the case in Kepler's famous argument for the elliptical orbit of Mars. Indeed, it was his belief that "physical" hypotheses regarding the quantifiable forces exerted by the sun on the motions of the planets could, in fact, be proved or demonstrated. And it was the idea that "one thing is frequently the cause of many effects" that served as the criterion for the truth or probability of a hypothesis, particularly in the *Astronomia nova*. The key to the argument in Kepler's famous "war on Mars" was the geometrical relation that facilitated the combination of two quantifiable influences of the sun on the planet, the first being the planet's orbit around the sun, and the second its libratory approach to and recession from the sun. Once those two were combined, Kepler could *justify* not only the elliptical orbit of Mars but also the fact that its motion was in accordance with the area law. The synthesis consisted in showing (1) that although libratory motion obeyed a law of its own, it was exactly because of the motion of libration that the planet described an elliptical orbit, and (2) that the second law or area law was valid only for an elliptical orbit. Kepler saw his argument as producing an integrated unity founded on mathematical simplicity. Let us look briefly at the physical details to see how they fit together.

Kepler's dynamical account of libration was modelled on magnetic attraction and repulsion. In *Astronomia nova*, planetary motion was explained by the joint action of the sun and the planets themselves, whereas in his later work, the *Epitome*, the entire action was attributed to the sun. The motive radii of the sun's species not only led the planets around it but also repelled and attracted them depending on whether a planet displayed its "friendly" or "hostile" side toward the sun, that is, depending on which magnetic side was facing the sun. Kepler hypothesized that the source of that magnetism lay in magnetic fibres that passed through the planets. However, the planets themselves did not "exert" any force; rather, the action of the sun communicated a certain "inclination" to the fibres of the planetary body such that its entire libration derived from the sun. Hence, libration was not the result of any action or motion of the planet itself.[2] In order to give an exact account of the mechanism of orbital motion it would be necessary to determine the variations in the propelling and attracting forces throughout the entire path. That would require that one calculate the angles that the radius vectors of the sun made with the magnetic fibres of the planet. The sines of the complements of those angles (the cosines) would provide a measure of those portions of the forces that acted on the planet.[3]

Once Kepler developed the mechanism responsible for libration and proved that it was measured by the versed sine of the arc traversed by the planet, he was able to formally establish that an elliptical orbit resulted as a consequence of libration.[4] From there Kepler went on to prove his second law, which describes the relationship between the time taken by a planet to travel a particular distance on its orbit and the area swept out by the radius vector. Again, the key to the synthesis was that the law, as Kepler formulated it, was valid *only* for an elliptical orbit, thereby establishing an interconnectedness between the dynamics of libratory motion and the geometry governing the motions of the planets. The relationships were all confirmed empirically, making the argument one that was not based solely on formal geometrical constraints, but one that united the physical and mathematical components of celestial phenomena in a simple coherent way. That unity created a kind of justification that not only applied to Kepler's laws themselves but also extended to the metaphysical thesis regarding the relation between mathematical simplicity and truth. In other words, it was a justification determinable through the agreement of results. His laws governing planetary orbits described the simplest possible paths consistent with libratory motion, and the convergence of those results provided further evidence that his physical hypothesis was true. Kepler began with the belief that nature was founded on or determined by mathematical harmonies, and the correspondence of empirical observations with laws based on those harmonies further reinforced the idea of nature's unity and mathematical simplicity.

It is exactly this kind of context, where one sees a particular law or principle yielding different kinds of interconnected results (i.e., the dynamics of libratory motion facilitating the derivation of Kepler's first and second laws in a way that made each interdependent on the other), that we typically take as exemplifying at least some of the qualities of a unified theory.[5] But in the assessment of such a theory, in the determination of its truth or confirmation value, one must be cautious in locating the truth component in the proper place. In other words, if we look at the unity displayed by Kepler's account of planetary motion, it is tempting to describe its explanatory power as being grounded in the dynamics of libratory motion. And when Kepler succeeded in deducing an elliptical orbit from his physics, instead of just arriving at it through a process of observation coupled with manipulation of mathematical hypotheses designed to fit the facts, one is tempted to say that the physical basis of the theory must be true. Tempting as this may be, the question is whether or not we can, or even should, infer, as Kepler did, on the basis of this type of interconnectedness, the truth of the physical hypothesis.

With hindsight we know that Kepler's physics was mistaken, despite the fact that it yielded laws of planetary motion that were retained as empirically true approximations within Newtonian mechanics. The example, however, raises a number of issues that are important for understanding how one ought to think about unification. First, it seems that given what we know about the history of physics, it becomes obvious that one should not, as a general principle, attribute truth to a unifying hypothesis such as Kepler's theory of libratory motion simply because it yields a convergence of quantitative results. Second, and equally important, is the question of whether unification can be said to consist simply in a mathematical or quantitative convergence of different results or whether there needs to be an appropriate dynamical or causal explanation from which these results issue as consequences. The answer one gives to this latter question is significant not only for the link between explanatory power and unification but also for the connection between unified theories and the broader metaphysical thesis about unity in nature. That is, if unity is typically accompanied by an underlying physical dynamics, then it becomes necessary to determine whether or not the unifying power provides evidence for the physical hypothesis from which it emerges. Finally, it certainly is not an uncommon feature of scientific theories that they display, at least to some degree, the kind of interconnectedness present in Kepler's account of planetary motion. Yet his account is not typically thought to be a truly "unified" theory, in that it does not bring together *different* kinds of phenomena. For example, electromagnetic phenomena were thought to be radically different from optical phenomena until they were unified by Maxwell's theory and shown to obey the same laws. In fact, most instances of what we term "unification" are of this sort. Is it necessary, then, to specify particular conditions that must be satisfied in order for a theory to be truly unified, or is the notion of unification simply one that admits of degrees? In other words, do all theories unify to a greater or lesser extent? Each of these issues will be discussed in later chapters, but by way of contrast to Kepler's metaphysical/mathematical picture of unity and simplicity let us turn to Kant's account, which accords to unity a strictly heuristic role by characterizing it as a regulative ideal that guides our thinking and investigation about experience in general and scientific investigation in particular. What is interesting about Kant's notion of unity is that it carries with it no metaphysical commitments; yet it is indispensable for scientific research and the more general quest for knowledge.

1.2. Kant: Unity as a Heuristic and Logical Principle

Within the Kantian framework it is the faculty of reason that is responsible for synthesizing knowledge of individual objects into systems. An example is Kant's notion of the "order of nature", an entire system of phenomena united under laws that are themselves unified under higher-order laws. This systematic arrangement of knowledge is guided by reason to the extent that the latter directs the search for the ultimate conditions for all experience – conditions that are not, however, to be found within the domain of experience itself. That is, we could never unify all our knowledge, because such a grand unification could never be found in experience. Hence, the quest for unity is one that, by definition, is never fulfilled; it remains simply an ideal or a goal – in Kant's terms, a "problem" for which there is no solution. What reason does, then, is introduce as an ideal or an uncompletable task a set of rational conditions that must be satisfied for all of our knowledge to constitute a unified system. Examples of such conditions are (1) that we act as though nature constitutes a unified whole and (2) that we act as if it is the product of an intelligent designer. Consequently, this ideal *regulates* our search for knowledge and directs us toward a unified end. The fact that we can never achieve this complete unity should not and cannot be an obstacle to our constant striving toward it, for it is only in that striving that we can achieve any scientific knowledge.

To the extent that complete unity is not attainable, reason is said to function in a "hypothetical" way; the conditions referred to earlier take on the role of hypotheses that function as methodological precepts. Consequently, the systematic unity that reason prescribes has a *logical* status designed to secure a measure of coherence in the domain of empirical investigation. Kant specifically remarks that we would have no coherent employment of the understanding – no systematic classifications or scientific knowledge – were it not for this presupposition of systematic unity. But how can something that is in principle unrealizable, that is merely and always hypothetical, function in such a powerful way to determine the structure of empirical knowledge? Part of the answer lies in the fact that the search for unity is an essential logical feature of experience.

The notion of a *logical* principle serves an important function in the Kantian architectonic. Principles of reason are dependent on thought alone. The logical employment of reason involves the attempt to reduce the knowledge obtained through the understanding "to the smallest number of principles (universal conditions) and thereby achieve the highest possible unity" (Kant 1933, A305). Although we are required to bring about this unity in as complete a form as possible, there is nothing about a logical principle that guarantees that nature must subscribe to it. In that sense the logical employment and hypothetical employment of reason describe the same function. The logical aspect refers to the desire for systematic coherence, and the hypothetical component is a reminder that this ultimate unity as it applies to nature always has the status of a hypothesis. The principle that bids us to seek unity is necessary insofar as it is definitive of the role of reason in cognition; without it we would have no intervention on the part of reason and, as a result, no coherent systematization of empirical knowledge. In other words, it is a necessary presupposition for all inquiry. And as a logical principle it specifies an ideal structure for knowledge in the way that first-order logic is thought to provide the structure for natural language.

One of the interesting things about the requirement to seek systematic unity is that it not only encompasses a demand for a unified picture of experience but also involves what Kant classifies as "subjective or logical maxims" – rules that demand that we seek not just homogeneity but also variety and affinity in our scientific investigations and classifications. These maxims are the principles of genera (homogeneity), specification (species) and continuity of form (affinity). Homogeneity requires us to search for unity among different original genera; specification imposes a check on this tendency to unify by requiring us to distinguish certain subspecies; and continuity, the affinity of all concepts, is a combination of the previous two insofar as it demands that we proceed from each species to every other by a gradual increase in diversity. Kant expands on this point in the *Jäsche Logic* (sec. 11), where he discusses the concepts "iron", "metal", "body", "substance" and "thing". In this example we can obtain ever higher genera, because every species can always be considered a genus with respect to a lower concept, in the way iron is a species of the genus metal. We can continue this process until we come to a genus that cannot be considered a species. Kant claims that we must be able to arrive at such a genus because there must be, in the end, a highest concept from which no further abstraction can be made. In contrast, there can be no lowest concept or species in the series, because such a concept would be impossible to determine. Even in the case of concepts applied directly to individuals, there may be differences that we either disregard or fail to notice. Only relative to *use* are there "lowest" concepts; they are determined by convention insofar as one has agreed to limit differentiation.

These logical maxims, which rest entirely on the hypothetical interests of reason, *regulate* scientific activity by dictating *particular* methodological practices. Again, this connection between logic and methodology is a crucial one for Kant. At the core of his view of science as a systematic body of knowledge lies the belief that science must constitute a logical system, a hierarchy of deductively related propositions in ascending order of generality. The act of systematizing the knowledge gained through experience enables us to discover certain logical relations that hold between particular laws of nature. This in turn enables us to unify these laws under more general principles of reason.

This classification process, which includes the unification of dissimilar laws and diversification of various species, exemplifies Kant's *logical* employment of reason. A properly unified system exhibits the characteristics of a logical system displaying coherence as well as deductive relationships among its members. Scientific theories are themselves logical systems that consist of classificatory schemes that unify our knowledge of empirical phenomena. Kant recognizes, however, that reason cannot, simply by means of a logical principle, command us to treat diversity as disguised unity if it does not presuppose that nature is itself unified. Yet he claims that the only conclusion which we are justified in drawing from these considerations is that the systematic unity of the manifold of knowledge of understanding, as prescribed by reason, is a *logical* principle. (Kant 1933, A648/B676)

This leaves us in the rather puzzling position of having logical or subjective maxims whose use is contextually determined, while at the same time upholding an overriding principle of unity in nature as prescribed by reason. In other words, Kant seems to sanction the idea of disunity while at the same time requiring that we seek unity. At A649 he discusses the search for fundamental powers that will enable us to unify seemingly diverse substances. Again the idea of such a power is set as a problem; he does not assert that such a power must actually be met with, but only that we must seek it in the interest of reason. As Kant remarks at A650/B678, "this unity of reason is purely hypothetical". Yet in the discussion of logical maxims the principle of unity seems to take on a more prominent role. His example concerns a chemist who reduces all salts to two main genera: acids and alkalies. Dissatisfied with that classification, the chemist attempts to show that even the difference between these two main genera involves merely a variety or diverse manifestations of one and the same fundamental material; and so the chemist seeks a common principle for earths and salts, thereby reducing them to one genus. Kant goes on to point out that it might be supposed that this kind of unification is merely an economical contrivance, a hypothetical attempt that will impart probability to the unifying principle if the endeavour is successful. However, such a "selfish purpose" can very easily be distinguished from the *idea* that requires us to seek unity. In other words, we don't simply postulate unity in nature and then when we find it claim that our hypothesis is true.

For in conformity with the idea everyone presupposes that this unity of reason accords with nature itself, and that reason – although indeed unable to determine the limits of this unity – does not here beg but command. (Kant 1933, A653/B681)

Put differently, the overall demand of reason to seek unity is the primary goal of all cognition in the attempt to reconstruct nature as a logical system. The mere fact that we engage in cognitive goals implicitly commits us to the search for unity. Within that context there are several different methodological approaches that can be employed for achieving systematic classification of empirical knowledge. Reason presupposes this systematic unity on the ground that we can conjoin certain natural laws under a more general law in the way that we reduce all salts to two main genera. Hence, the logical maxim of parsimony in principles not only is an economical requirement of reason but also is necessary in the sense that it plays a role in defining experience or nature as a systematically organized whole. Hence, what appear to be conflicting research strategies, as outlined by the subjective maxims, are simply different ways that reason can attain its end. For example, the logical principle of genera responsible for postulating identity is balanced by the principle of species, which calls for diversity; the latter may be important in biology, whereas the former is more important for physics. But Kant is no reductionist; the idea of a "unified knowledge" is one that may consist of several different ways of systematizing empirical facts.

The logical maxims are not derived from any empirical considerations, nor are they put forward as merely tentative suggestions. However, when these maxims are confirmed empirically, they yield strong evidence in support of the view that the projected unity postulated by reason is indeed well-grounded. But in contrast to the strategy described earlier, the motivation behind the unifying methodology is not based on utilitarian considerations; it is not employed because we think it will be successful. Nevertheless, when we do employ a particular maxim in view of a desired end and are successful in achieving our goal, be it unity, specification or continuity, we *assume* that nature itself acts in accordance with the maxim we have chosen. On that basis we claim that the principles prescribing parsimony of causes, manifoldness of effects and affinity of the parts of nature accord with both reason and nature itself.

We must keep in mind, however, that although these principles are said to "accord with" nature, what Kant means is that although we must *think* in this way in order to acquire knowledge, there is also some evidence that this way of thinking is correct. The latter, however, can never be known with certainty, because we can never know that nature itself is constituted in this way. From the discussion of the logical employment of reason we know that in order to achieve the systematic unity of knowledge that we call science it is necessary that this unity display the properties of a logical system. In other words, if one agrees with Kant that science is founded on *projected* systematization and that this system is ultimately reducible to logical form (non-contradiction, identity and deductive closure over classification systems), then the principles that best cohere with the demand of systematic unity recommend themselves. Parsimony, manifoldness and affinity are not only methodological principles for organizing nature according to our interests; they are also the most efficient way of realizing the one interest of reason – the systematic unity of all knowledge. Because we empirically verify the extent to which this unity has been achieved, we are thereby supplied with the means to judge the success of the maxims in furthering our ends (Kant 1933, A692/B720), but ends that we, admittedly, never attain. We employ a particular maxim based on what we think will be the most successful approach in achieving systematic unity given the context at hand.

As mentioned earlier, the motivating idea for Kant is the construction of a logical system rather than the realization of a metaphysical ideal regarding the unity of nature. Kant is silent on the question of whether or not this notion of systematization constitutes the basis for scientific explanation. Although it seems clear that classification of phenomena does serve some explanatory function, there is nothing in the Kantian account of unity to suggest that it is in any way coincident with explaining or understanding the nature of phenomena. In essence, the Kantian account of unity constitutes a methodological approach that is grounded in the basic principles of human reason and cognition. The unity has a hypothetical and pre-suppositional status; it is an *assumption* that the world is a unified whole, rather than a metaphysical principle stating how the world is *actually* structured. In that sense it is simply an idealization that is necessary for scientific inquiry.

Kant's views about the role of ideas in producing unity both in and for science were taken up in the nineteenth century by William Whewell. His views about unity as a logical system were also adopted, albeit in a different form, in the twentieth century by Rudolph Carnap. Unlike Kant, Whewell took a more substantive approach by linking his notion of unity (termed the consilience of inductions) to explanation by way of a set of fundamental ideas: Each member in the set of ideas would ground a particular science. Consequently, Whewell also adopted a much stronger epistemological position by claiming that unified or consilient theories would have the mark of certainty and truth.

1.3. Whewell: Unity as Consilience and Certainty

In the *Novum Organon Renovatum* William Whewell discusses various tests of hypotheses that fall into three distinct but seemingly related categories. The first involves the prediction of untried instances; the second concerns what Whewell refers to as the consilience of inductions; the third features the convergence of a theory toward unity and simplicity. Predictive success is relatively straightforward and encompasses facts of a kind previously observed but predicted to occur in new cases. Consilience, on the other hand, involves the explanation and prediction of facts of a kind different from those that were contemplated in the formation of the hypothesis or law in question. What makes consilience so significant is the finding that classes of facts that were thought to be completely different are revealed as belonging to the same group. This "jumping together" of different facts, as Whewell calls it, is thought to belong to only the best-established theories in the history of science, the prime example being Newton's account of universal gravitation. But Whewell wants to claim more than that for consilience; he specifically states that the instances where this "jumping together" has occurred

impress us with a conviction that the truth of our hypothesis is certain. . . . No false suppositions could, after being adjusted to one class of phenomena, exactly represent a different class, where the agreement was unforeseen and uncontemplated. That rules springing from remote and unconnected quarters should thus leap to the same point, can only arise from that being the point where truth resides.[6]

Finally, such a consilience contributes to unity insofar as it demonstrates that facts that once appeared to be of different kinds are in fact the same. This in turn results in simpler theories by reducing the number of hypotheses and laws required to account for natural phenomena. Hence, unity is a step in the direction of the goal of ultimate simplicity in which all knowledge within a particular branch of science will follow from one basic principle.

Part of what is involved in a consilience of inductions is what Whewell refers to as "the colligation of ascertained facts into general propositions". This also takes place through a three-step process that involves (1) selection of the idea, (2) construction of the conception and (3) determination of the magnitudes. These steps have analogues in mathematical investigations that consist in determining the independent variable, the formula and the coefficients. It was Whewell's contention that each science had its own fundamental idea; the study of mathematics was based on the ideas of number and space, whereas mechanics relied on the idea of force to make it intelligible. Similarly, the idea of polarity was predominant in the study of chemical phenomena, and ideas of resemblance and difference were crucial to the study of natural history and the classificatory sciences.

Once the requisite idea was chosen, one could then proceed to the construction of a conception, which was a more precise specification of the idea. For example, a circle or a square is a kind of spatial configuration, and a uniform force is a particular manifestation of the general notion of a force. So if we have a phenomenon like the weather and we are trying to establish some order that will assist in predictions, we must decide whether we wish to select (1) the idea of time, and introduce the conception of a cycle, or (2) the idea of force, accompanied by the conception of the moon's action. One selects the appropriate conception by comparing it with observed facts, that is, determining whether or not the weather really is in fact cyclical by comparing the supposed cycle with a register of the seasons. The idea is the core concept that grounds each field of inquiry. When we achieve a consilience of inductions, the result is that two different conceptions governing different classes of facts are seen in a new way, either as belonging to a totally new conception or as manifestations of one or the other of the original conceptions. The important point is that consilience does not involve a "jumping together" of two distinct ideas from different branches of science. The classes of facts usually are drawn from within an individual science.

There is no definitive method for selecting the right idea, nor the right conception for that matter. The only requirement or rule is that the idea must be tested by the facts. This is done by applying the various conceptions derived from the idea to the facts until one succeeds in uncovering what Whewell refers to as the "law of the phenomena".

Although my intention is not to provide a complete analysis of Whewell's account of induction and unification, I do think it important to discuss briefly the mathematical representation of this procedure, in an attempt clarify how unification takes place. The interesting question is whether the convergence of numerical results that occurs in a consilience of inductions is what ultimately constitutes unity or whether there are further implications for the supposed connection between explanation and unification.

In a section entitled "General Rules for the Construction of a Conception" Whewell describes the process as the construction of a mathematical formula that coincides with the numerical expression of the facts. Although the construction of the formula and the determination of its coefficients have been separated into two steps, Whewell claims that in practice they are almost necessarily simultaneous. Once one selects the independent variable and the formula that connects the observations to form laws, there are particular technical processes whereby the values of the coefficients can be determined, thereby making the formula more accurate. These include the methods of curves, of means, of least squares and of residues. In the case of the method of curves, we have a specific quantity that is undergoing changes that depend on another quantity. This dependence is expressed by a curve. The method enables us to detect regularities and formulate laws based not only on good observations but also on those that are imperfect, because drawing a line among the points given by observations allows us to maintain a regular curve by cutting off the small and irregular sinuosities. When we remove the errors of actual observation by making the curve smooth and regular, we are left with separate facts corrected by what Whewell calls their "general tendency"; hence, we obtain data that are "more true than the individual facts themselves" (Whewell 1847, vol. 2). The obstacles that prove problematic for the method are ignorance of the nature of the quantity on which the changes depend and the presence of several different laws interacting with one another.

The method of curves assumes that errors in observation will balance one another, because we select quantities that are equally distant from the extremes that observation provides. In cases where we have a number of unequal quantities and we choose one equally distant from the greater and smaller, we use the method of means rather than the method of curves. The implicit assumption, again, is that the deviations will balance one another. In fact, the method of means is really just an arithmetical procedure analogous to the method of curves, with one significant difference: In the method of curves, observation usually enables us to detect the law of recurrence in the sinuosities, but when we have a collection of numbers we must divide them into classes using whatever selection procedures we think relevant.

The method of least squares is also similar to the method of means. It allows us to discover the most probable law from a number of quantities obtained from observation. The method assumes that small errors are more probable than larger ones, and it defines the best mean as that which makes the sum of the squares of the errors the least probable sum. Finally, the method of residues involves an analysis of unexplained facts that have been left as residue after the formation of a law governing changes of a variable quantity. The residue is analysed in the same way as the original observations until a law is found that can account for it. This continues until all the facts are accounted for.[7]

The notable feature present in these methods, and what is important for our purposes here, is the level of generality that is introduced in order to assist in the formulation of laws of phenomena. Although it is true that induction is the process by which one formulates a general proposition from a number of particular instances, the difference in these cases is that the general conclusion is not simply the result of a juxtaposition or conjunction of the particulars. In each case a "conception" is introduced that is not contained in the bare facts of observations. The conception is the new fact that has been arrived at through a reinterpretation of the data using the relevant methods. This new element or conception can then be superimposed on existing facts, combining them in a unique way. Such was the case with the ellipse law governing the orbit of Mars. What Whewell describes are methods for data reduction that facilitate the formulation of a conception; but one need not employ all of these methods in order to arrive at a conception. For example, after trying both circular and oval orbits and finding that they did not agree with observations of the observed longitudes of Mars (or the area law), Kepler was led to the ellipse, which, taken together with the area law, gives the best agreement with the available observations. Some methods of data reduction were employed, because the object of the exercise was to find a structure that would fit with the observations. As we saw earlier, there was a convergence of numerical results in establishing the ellipse law, which led Kepler to believe that he had hit on the right formulation. Although we don't have predictions of different *kinds* of data or classes of facts, as in the case of a true consilience, we do have better predictions for not only Mars but also Mercury and the earth. In that sense, then, there is a colligation of facts made possible by the introduction of the conception (i.e., the ellipse) based on the idea of space.

So the induction does not consist in an enumerative process that establishes a general conclusion; rather, the inductive step refers to the suggestion of a general concept that can be applied to particular cases and can thereby unify different phenomena. According to Whewell, this "general conception" is supplied by the mind, rather than the phenomena; in other words, we don't simply "read off" the conception from the data. Rather, it requires a process of conceptualization. The inference that the phenomena instantiate this general conception involves going beyond the particulars of the cases that are immediately present and instead seeing them as exemplifications of some ideal case that provides a standard against which the facts can be measured. Again, the important point is that the standard is constructed by us, rather than being supplied by nature. That the conception presents us with an "idealized" standard is not surprising, because the mathematical methods used to arrive at it embody a great deal of generality – generality that obscures the specific nature of the phenomena by focusing instead on a constructed feature that can be applied across a variety of cases.[8] It is this issue of generality that I want to claim is crucial not only to the unifying process but also to the connection (or lack thereof) between unification and explanation. My focus is not so much the notions of data reduction, as described by Whewell, but more general mathematical techniques used to represent physical theories. The importance of calling attention to Whewell's methods is to emphasize the role of mathematics generally in the formulation of specific hypotheses. The more general the hypothesis one begins with, the more instances or particulars it can, in principle, account for, thereby "unifying" the phenomena under one single law or concept. However, the more general the concept or law, the fewer the details that one can infer about the

phenomena. Hence, the less likely it will be able to "explain" how and why particular phenomena behave as they do. If even part of the practice of giving an explanation involves describing how and why particular processes occur – something that
frequently requires that we know specific details about the phenomena in question – then the case for separating unification and explanation becomes not just
desirable but imperative.

It has been claimed by Robert Butts, and more recently by William Harper, and
even by Whewell himself, that in a consilience there is an *explanation* of one distinct
class of facts by another class from a separate domain.[9] However, it is important
here to see just what that explanation consists in. As Butts has pointed out, we
cannot simply think of the explanatory power of consilience in terms of entailment
relations, because in most cases the deductive relationship between the consilient
theory and the domains that it unifies is less than straightforward. The best-known
example is Newton's theory and its unification of Kepler's and Galileo's laws. That
synthesis required changes in the characterization of the nature of the physical systems involved, as well as changes in the way that the mathematics was used and
understood, all of which combined to produce nothing like a straightforward deduction of the laws for terrestrial and celestial phenomena from the inverse-square
law. Given that consilience cannot be expressed in terms of entailment relations, is
it possible to think of the connection between explanatory power and consilience in
terms of the convergence of numerical results? Such seems the case with Maxwellian
electrodynamics, in which calculation showed that the velocity of electromagnetic
waves propagating through a material medium (supposedly an electromagnetic
aether) had the same value as light waves propagating through the luminiferous
aether. That coincidence of values suggested that light and electromagnetic waves
were in fact different aspects of the same kind of process. However, as we shall see
in Chapter 3, whether or not this kind of convergence constitutes an explanation
depends on whether or not there is a well-established theoretical framework in
place that can "account for" why and how the phenomena are unified. The latter
component was in fact absent from Maxwell's formulation of the theory. Yet the
theory undoubtedly produced a remarkable degree of unity.

A similar problem exists in the Kepler case. Recall, for instance, the way in
which Kepler's first and second laws fit together in a coherent way, given the
physics of libratory motion. Although there was an explanatory story embedded in
Kepler's physics, it was incorrect; hence, contrary to what Whewell would claim,
a coincidence of results by no means guarantees the truth of the explanatory hypothesis. Although the convergence of coefficients may count as a unification of
diverse phenomena, more is needed if one is to count this unification as explanatory. This is especially true given that phenomena are often unified by fitting them
into a very general mathematical framework that can incorporate large bodies of diverse data within a single representational scheme (e.g., gauge theory, Lagrangian
mechanics). And mathematical techniques of the sort described by Whewell are
important for determining a general trait or tendency that is common to the data
while ignoring other important characteristics.

But Whewell himself seems to have recognized that more was needed if consilience was to count as truly explanatory; specifically, one needed a *vera causa* to complete the picture. In the conclusion to the section on methods of induction he remarks that those methods applicable to quantity and resemblance usually lead only to laws of phenomena that represent common patterns, whereas inductions, based on the idea of cause and substance, tend to provide knowledge of the essential nature and real connections among things (Whewell 1967, p. 425). Laws of phenomena were simply formulae that expressed results in terms of ideas such as space and time (i.e., formal laws of motion). Causes, on the other hand, provided an account of that motion in terms of force.

Unfortunately, Whewell was somewhat ambiguous about the relations between causes and explanations and sometimes suggested that the inference to a true cause was the *result* of an explanation of two distinct phenomena; at other times he simply claimed that "when a convergence of two trains of induction point to the same spot, we can no longer suspect that we are wrong. Such an accumulation of proof really persuades us that we have a *vera causa*".[10] Although the force of universal gravitation functioned as just such a true cause by explaining why terrestrial and celestial phenomena obeyed the same laws, gravitation itself was not well understood. That is, there was no real explanatory mechanism that could account for the way that the force operated in nature; and in that sense, I want to claim that as a cause it failed to function in a truly explanatory way. Hence, even though a *why* question may be answered by citing a cause, if there is no accompanying answer to the question of *how* the cause operates, or what it is in itself, we fail to have a complete explanation.

With hindsight, of course, we know that Whewell's notion of unity through consilience could not guarantee the kind of certainty that he claimed for it. Regardless of whether or not one sees Whewell's account of consilience of inductions as a model for current science, it is certainly the case that Whewell's history and philosophy of the inductive sciences provided a unity of method that at the same time respected the integrity and differences that existed within the distinct sciences. It provided not only a way of constructing unified theories but also a way of thinking about the broader issue of unity in science. Each science was grounded on its own fundamental idea; some shared inductive methods (e.g., means, least squares), but only if they seemed appropriate to the kind of inquiry pursued in that particular science. In that sense, Whewell was no champion of the kind of scientific reductionism that has become commonplace in much of the philosophical literature on unity. Consilience of inductions was a goal valued from within the boundaries of a specific domain, rather than a global methodology mistakenly used to try to incorporate the same kinds of forces operant in physics into chemistry (Whewell 1847, p. 99).

Now let us turn to another context, one in which the focus is not on unified theories specifically but more generally on unity in science defined in terms of unity of method. I am referring to the programme outlined in the *International Encyclopedia of Unified Science*, a collection of volumes written largely by the proponents of logical empiricism and first published in 1938. Although there were similarities to Whewell's attempt to retain the independence of particular sciences, the proponents of that version of the unity of science (Neurath) claimed that the localized unity achieved within specific domains carried no obvious epistemic warrant for any metaphysical assumptions about unity in nature. Their desire to banish metaphysics also resembled the Kantian ideal of unity as a methodology. What is especially interesting about that movement, as characterized by each of the contributions to the *Encyclopedia*, is the diversity of ideas about what the unity of science consisted in. Although that may seem the appropriate sort of unity for an encyclopedia, more importantly it enables us to see, in concrete terms, how unity and disunity can coexist – evidence that the dichotomy is in fact a false one.

1.4. Logical Empiricism: Unity as Method and Integration

It has frequently been thought that the unity of science advocated by the logical empiricists had its roots in logical analysis and the development of a common language, a language that would in turn guarantee a kind of unity of method in the articulation of scientific knowledge. In his famous 1938 essay "Logical Foundations of the Unity of Science", published in the *International Encyclopedia of Unified Science* (Neurath et al. 1971), Rudolph Carnap remarks that the question of the unity of science is a problem in the logic of science, not one of ontology. We do not ask "Is the world one?", "Are all events fundamentally of the same kind?". Carnap thought it doubtful that these philosophical questions really had any theoretical content. Instead, when we ask whether or not there is a unity in science we are inquiring into the logical relationships between the terms and the laws of the various branches of science. The goal of the logical empiricists was to reduce all the terms used in particular sciences to a kind of universal language. That language would consist in the class of observable thing-predicates, which would serve as a sufficient reduction basis for the whole of the language of science. Despite the restriction to that very narrow and homogeneous class of terms, no extension to a unified system of laws could be produced; nevertheless, the unity of language was seen as the basis for the practical application of theoretical knowledge.

We can see, then, that the goal of scientific unity, at least as expressed by Carnap, is directly at odds with the notion of unity advocated by Whewell. The kind of reductionist programme suggested by the logical unity of science would, according to Whewell, stand in the way and indeed adversely affect the growth of knowledge in different branches of science. According to him, the diversity and disunity among the sciences were to be retained and even encouraged, while upholding a unity within the confines of the individual branches of science.

But, as with the problem of the unity of science itself, within the logical-empiricist movement there were various ways in which the notion of unity was understood, even among those who contributed to the *International Encyclopedia of*

Unified Science. Views about what constituted the unity of science and how the goal was to be pursued differed markedly from the more traditional account of logical empiricism and the reductionism expressed by Carnap. For example, John Dewey, a contributor to the *Encyclopedia*, saw the unity of science as largely a social problem. In addition to a unification of the results obtained in science there was also the question of unifying the efforts of all those who "exercise in their own affairs the scientific method so that these efforts may gain the force which comes from united effort" (Dewey 1971, p. 32). The goal was to bring about unity in the scientific attitude by bringing those who accepted it and acted upon it into cooperation with one another. Dewey saw this problem as prior to the technical issue of unification with respect to particular scientific results. Unlike Carnap, who concerned himself with more technical problems and the development of methods that would achieve a logical reduction of scientific terms, Dewey believed that the unity-of-science movement need not and should not establish in advance a platform or method for attaining its goal. Because it was a cooperative movement, common ideas ought to arise out of the very process of cooperation. To formulate them in advance would be contrary to the scientific spirit.

Dewey saw the scientific attitude and method as valuable insofar as such practice had brought about an increase in toleration; indeed, in his view that attitude formed the core of a free and effective intelligence. Although the special sciences can reveal what the scientific method is and means, all humans can become scientific in their attitudes (i.e., genuinely intelligent in their ways of thinking and acting), thereby undermining the force of prejudice and dogma.

Yet another account of the unity of science was articulated by Otto Neurath, who saw unified science as a type of encyclopedic integration. It certainly was no accident that the *International Encyclopedia of Unified Science* brought together authors with diverse views on the topic of unity, but views that nevertheless could be integrated together in a way that could achieve a common goal. Hence, the *Encyclopedia* itself stood as the model for a unified science as envisioned by Neurath. Each contribution from a given scientific field was brought together with others that expressed diverse opinions within a wider set of agreements, agreements that lent unity and the spirit of Deweyan cooperation to the project.

But how should one understand unity as encyclopedic integration? At what point do differences begin to obscure the unified core that binds together the diversity of opinion and method? If one adopts a Whewellian approach, the answer is relatively straightforward: Unity existed within each science, and across domains there was a common approach to the discovery of knowledge that had its origin in the doctrine of fundamental ideas. However, for Neurath, as well as some of his fellow contributors, the aim of the *Encyclopedia* was to synthesize scientific activities such as observation, experimentation and reasoning and show how all of those together helped to promote a unified science. Those efforts to synthesize and systematize were not directed toward creating *the* system of science, but rather toward encouraging encyclopedism as both an attitude and a programme. One starts with a certain group of scientific statements that may or may not be axiomatizable; they are then combined with others expressed in similar form. Such an encyclopedic integration of statements stresses the incompleteness of an encyclopedia, rather than the completeness of a system of knowledge.

The idea is perhaps better expressed by analogy with an orchestra, in which the different instruments play not only in harmony but also in discord with one another (Kallen 1948). Orchestration requires diversity in order to sustain itself; even the state of perfect harmony cannot be achieved without the necessary independence of each member of the group. In that sense the orchestration is self-directed from within, rather than imposed from without. Indeed, this notion of an orchestration of the sciences involves the kind of unity of action that forms the core of Neurath's vision of the *Encyclopedia*. Similarly, the unity of language expressed in Neurath's physicalism acts more as a coordinating mechanism than as an imposed structure that must be adhered to.[11] In other words, its task is to facilitate integration, not impose it.

If Neurath's physicalism is in fact non-constraining in that it does not impose a particular structure on how scientific practice/theorizing is carried out, then it begins to have quite radical implications for the philosophy of science. Philosophy ceases to be a kind of meta-science that dictates acceptable principles of rationality and method and becomes instead a form of inquiry whose normative force arises not out of first principles but out of practical cooperation with other disciplines. As a result, the normative component can be only pragmatically motivated; it is fallible and continuously evolving as inquiry itself evolves. Philosophy can provide the tools for analysis and contribute to the synthesis of reason and practice, but it in no way provides a "grounding" for scientific activity that transcends the practice itself.

In essence, then, what emerges from this version of the logical-empiricist or scientific-empiricist picture is a kind of unity through disunity. There is nothing to suggest that the kind of localized systematization of the sort favoured by Whewell would necessarily be ruled out; one could quite legitimately have unity within a particular science, physics for example, that emerges as the result of unifying theories, without any extension of this unity to other domains. But there is also nothing in this view of unity as orchestration to suggest that unity within a science in the form of reduction to a few basic principles is necessarily a goal to be desired or sought after. Clearly, this empirical approach to unity is at odds with the grand systematization desired by Kant, as well as the picture of a mathematically harmonious universe endorsed by Kepler. Yet one sees elements of each of those accounts in contemporary scientific and philosophical attitudes toward unity. The power of gauge theory in the development of quantum field theory pays homage to the mathematical idea of symmetry that formed the core of Kepler's physics. Molecular biology gives credence to the Kantian desire for systematization of knowledge across theoretical boundaries. Fields like astrophysics and cosmology, whose theoretical foundation is based in diverse and sometimes mutually contradictory models, lend support to the image of science as orchestration, whereas some areas in high-energy physics are more accurately depicted by the Whewellian model that stresses unity only within a particular domain. For example, there may be no reason to assume that gravity can be incorporated into the unified structure that describes electromagnetism and the weak force.

Mirroring these various scientific attitudes are current philosophical accounts regarding the nature of unification and its connection to explanation. Aside from the discussion of Whewell, and perhaps Kepler, I have said very little about the connection between explanatory power and unity. As a way of briefly addressing that issue and setting the problem in its philosophical context, let us consider some of the points and questions that have emerged in the debate about the nature of unification and explanation. As discussed briefly in the Introduction, one common way of characterizing unification is through explanation; theoretical unity is thought to exist when, in a manner similar to Whewell's consiliences, a theory can explain a number of diverse facts. Similarly, the explanatory power of a theory is sometimes defined in terms of its ability to unify diverse facts under one law. We shall next look briefly at the merits of these views. More detailed accounts of unification and its connection with explanation, specifically those of Kitcher and Friedman, will be discussed in later chapters.

1.5. Unity as Explanation

Much of the debate about the character of explanation centres on the tension between the search for an objective criterion that can define scientific explanation and a subjective component that seems necessary if we are to link explanation with some notion of understanding. In his frequently quoted article "Explanation and Scientific Understanding", Michael Friedman (1974) tries to establish a link between these two goals and lists three desirable properties that a philosophical theory of explanation should have. First, the conditions that specify what counts as a good explanation should be sufficiently general so that most, if not all, scientific theories that we consider explanatory should come out as such. Second, explanations should be objective – they should not depend on the idiosyncrasies and changing tastes of scientists and historical periods. In other words, there is no room for what Friedman calls non-rational factors, such as which phenomenon one happens to find more natural, intelligible or self-explanatory. To the extent that there is an objective and rational way in which scientific theories explain, a philosophical theory of explanation should tell us what it is. Finally, the theory providing the explanation should connect explanation and understanding by telling us what kind of understanding scientific explanations provide and how they provide it.

Friedman attempts to articulate just such a theory by isolating what he sees as a particular property of the explanation relation. This property should be possessed by most instances of scientific explanation and should be one that is common to scientific theories from various historical periods. The property in question is the "unification" achieved by particular scientific theories – the ability to derive large numbers of diverse phenomena from relatively few basic laws. How exactly is this property to be identified with explanation? If we think of the laws of Newtonian mechanics as allowing us to derive both the fact that the planets obey Kepler's laws and the fact that terrestrial bodies obey Galileo's laws, then, claims Friedman, we have reduced a multiplicity of unexplained independent phenomena to one. Hence, our understanding of the world is increased through the reduction in the total number of independent phenomena that we must accept as given. *Ceteris paribus*, the fewer independent phenomena, the more comprehensible and simple the world is. By replacing one phenomenon or law with a more comprehensive one, we increase our understanding by decreasing the number of independently acceptable consequences.

Friedman's strategy for giving a precise meaning to this notion of reduction of independent phenomena was shown to be technically flawed by Philip Kitcher (1976). However, in addition to the technical difficulties, some of Friedman's more intuitive claims regarding the connection between unification and explanation are by no means unproblematic. In his discussion of the kinetic theory of gases, Friedman claims that the theory explains phenomena involving the behaviour of gases, such as the fact that they approximately obey the Boyle-Charles law, by reference to the behaviour of the molecules that compose the gas. This is important because it allows us to deduce that any collection of molecules of the sort that compose gases will, if they obey the laws of mechanics, also approximately obey the Boyle-Charles law. The kinetic theory also allows us to derive other phenomena involving the behaviour of gases – the fact that they obey Graham's law of diffusion and why they have certain specific-heat capacities – all from the same laws of mechanics. Hence, instead of these three brute facts, we have only one: that molecules obey the laws of mechanics. Consequently, we have a unification that supposedly increases our understanding of how and why gases behave as they do. The unifying power of the mechanical laws further allows us to integrate the behaviour of gases with other phenomena that are similarly explained.

The difficulty with this story is that it seems to violate Friedman's second condition for a theory of explanation: that it be objective and not dependent on the changing tastes of scientists and historical periods. We know that the kind of straightforward mechanical account that he describes is not the technically correct way of explaining the behaviour of gases, given the nature of quantum statistical mechanics. Although it may have been a perfectly acceptable explanation at that time, today we no longer accept it as such. But perhaps what Friedman had in mind was the claim that because the laws of mechanics unify a number of different domains, they themselves can be thought to provide an objective explanation of the phenomena. That is, the search for unifying/explanatory theories is more than simply an objective methodological goal. When we do find a theory that exhibits the kind of powerful structure exemplified by mechanics, we typically want to extend the notion of objectivity beyond the idea of unifying power simpliciter to the more substantive claim of "unification through mechanical laws". Hence, it is the explanatory theory, in this case mechanics, that provides us an objective understanding of the phenomena. And to the extent that mechanics is still used to explain phenomena in certain domains, it should be considered objective. However, as the historical record shows, the importance of mechanical explanation is indeed linked to specific historical periods, and in many cases when we make use of mechanical explanations we do so on the basis of expediency. We know that the accurate "objective" description is either too complicated or not really required for the purposes at hand. Hence the notion of changing tastes of scientists is one that is directly linked to theory change, and to explanation as well. That is, with every theoretical change comes a change regarding what constitutes an "objective explanation". And although this encompasses much more than "matters of taste", decisions to accept an explanation as merely useful are ones that are nevertheless contextual and hence pragmatically determined.

We need only look at the development of physics in the nineteenth century to see how contextually based mechanical explanation had become even at that time. As a mathematical theory, mechanics was able to deal not only with rigid bodies, particle systems and various kinds of fluids but also, in the form of the kinetic theory of gases, with the phenomenon of heat. Unfortunately, however, the kinetic theory was far from unproblematic. Maxwell had shown that the system of particles that obeyed the laws of mechanics was unable to satisfy the relation between the two specific heats of all gases (Maxwell 1965, vol. 2, p. 409). In other words, the equipartition theorem, which was a consequence of the mechanical picture, was incompatible with the experimentally established findings about specific heats. Similarly, the second law of thermodynamics could not be given a strict mechanical interpretation, because additional features, like the large numbers of molecules, needed to be taken into account. A statistical description involving an expression for entropy in terms of a molecular distribution function was thought by some, including Boltzmann, to provide a solution to the problem. Some favoured mechanical explanation, while others opposed it.

The place of mechanical explanation in Maxwellian electrodynamics was also unclear. Although the theory was developed using a series of mechanical models and analogies, the final formulation of the theory was in terms of Lagrange's dynamical equations. The advantage of that method was that it enabled one to proceed without requiring any detailed knowledge of the connections of the parts of the system, that is, no hypotheses about the mechanical structure were necessary. Mechanical concepts were used by Maxwell as a way of showing how electromagnetic phenomena *could* be explained, but they by no means formed the core of the theory. In addition, difficulties in establishing a mechanical account of the behaviour and constitution of material bodies led the theorist Willard Gibbs to exercise extreme caution in his claims about the validity of the theory presented in *Statistical Mechanics* (Gibbs 1902). Too many unresolved difficulties in the theory of radiation and the specific-heats problem made many sceptical of the legitimacy of mechanical descriptions of physical phenomena.

Others, like Ernst Mach, and to some extent Pierre Duhem, disliked mechanical explanations for what were largely philosophical reasons. Because of Mach's unfaltering reliance on empiricism/phenomenalism as the proper method for acquiring

scientific knowledge, mechanism and its hypotheses about the ultimate constituents of matter simply lacked the kind of justification that he thought a proper
scientific account should have. There simply was no legitimate means of knowing whether or not mechanical phenomena could provide the ultimate explanatory
ground. In the end, the mechanical explanation that had proved so powerful since
the time of Newton gave way to quantum mechanics and eventually quantum field
theory, with the old ideal of a mechanically based physics being gradually abandoned. My point, then, is that we can, at most, claim objectivity for only a certain
kind of understanding that arises from within the confines of a specific theory at a
particular time in history. The fact that the theory may provide a unified account
of the phenomena does not eliminate the fact that the acceptability of its explanatory structure is historically situated and may be a matter of what particular individuals see as appropriate. It isn't the case that mechanics wasn't an explanatory
theory; rather, its explanatory power was ultimately linked to the very factors that
Friedman wants to rule out: changing tastes and historical periods.

What content can we give, then, to the claim that the unifying/explanatory
programme provided by the mechanical paradigm satisfied the goal of objectivity? It clearly was the case that preference for mechanical explanations was tied
to the preferences of specific groups and/or individual scientists.[12] However, neither that nor the fact that the mechanical conception of nature was historically
rooted need detract from its objectivity. Decisions either to pursue or to abandon
the search for mechanical explanations were firmly rooted in scientific successes
and failures, as well as in philosophical presuppositions about the correct methodology for science. In that sense the objectivity of mechanical explanation was and
is ultimately linked to the objectivity of the scientific method. Although there can
be no doubt that mechanics was a very powerful foundation for physics, its broad
unifying/explanatory power was not, as we have seen, based on a wholly consistent
foundation. The difficulties with the kinetic theory, together with the historical
contingency that accompanied mechanical explanation, paint a picture of objectivity quite different from the one Friedman suggests; they reveal an objectivity
that was grounded in localized requests for information determined by the particular theories and periods in history.

One might want to argue that the objectivity Friedman claims for his account
is nothing more than a *measure* of the reduction of independent facts through
the use of more basic, comprehensive ones (i.e., it is an objective fact whether
or not such a reduction has occurred). Hence, if this is a constraint on explana-
tion/understanding, it is also an objective fact whether or not this reductivist goal
has been attained. But surely this feature cannot be divorced from the fundamental
worth of the reducing theory. Even though, as Friedman notes, the basic phenomena may themselves be strange or unfamiliar, one nevertheless expects a fundamental coherence between theory and experiment. That coherence was simply absent
from the kinetic theory and its mechanical structure. In that sense, objectivity cannot be merely a procedural feature of an explanatory theory that involves reduction

of facts, but instead it must integrate the philosophical theory of explanation and the scientific explanatory theory. In other words, one's theory of explanation must employ concepts and constraints that are applicable to the ways in which scientific theories themselves evolve. That evolution has had a history that has influenced the kinds of explanations that have been deemed acceptable; it is only by taking account of that history that we can begin to see how "objectivity" has emerged. Instead of characterizing objectivity as something transcendent to which theories and explanations aspire, we must recognize that part of what makes an explanation "objective" is its acceptability in the context in which it is offered, something that will, undoubtedly, have a temporal dimension.

The difficulty at the core of Friedman's account is his identification of explanation and unification. Some of these issues arise again in the context of his more recent account of unification and realism, as discussed in Chapter 2. For now, suffice it to say that it is, and should be, an objective question whether or not a theory has unified a group of phenomena. Moreover, the unification should not depend on historical contingencies. It is simply a fact that Newton's theory unified celestial and terrestrial phenomena and that Maxwell's theory unified electromagnetism and optics. Yet we no longer accept the physical dynamics required to make those theories explanatory. By separating explanation and unification we can retain our intuitions about the context independence of theory unification while recognizing the historical aspects of explanation. Although the broader notion of unity in science may have several different interpretations, there nevertheless seem to be good reasons for thinking that theory unification is more clear-cut. We ought to be able to determine, in a rather straightforward way, the extent to which a particular theory has unified different domains. Indeed, much of this book is dedicated to showing how that can be done.

Another attempt to "objectify" explanation has been proposed by Clark Glymour (1980). He claims that there are two different reasons for belief in a scientific theory: reasons provided by the explanations the theory gives and reasons provided by the tests the theory has survived. The two qualities that explanations have that lend credence to theories are their ability to eliminate contingency and their unifying power. For example, Glymour claims that perhaps the most comprehensive way to explain the ideal-gas law is to show that it simply is not possible for a gas to have pressure, volume and temperature other than as the gas law requires. So instead of demonstrating that a regularity is a necessary consequence of a theory, one shows that the regularities are necessary in and of themselves. One thereby explains the regularity by identifying the properties it governs with other properties "in such a way that the statement of the original regularity is transformed into a logical or mathematical truth" (Glymour 1980, p. 24). Consequently, the statements that identify properties are, if true, necessarily true, and thereby transform the contingent regularity into a necessary truth. A simple example hinges on the identification of gravity with curved space-time. Provided this identification is true, then if general relativity is true, the identification is necessarily true.

Why is this so? It is so because on such a picture the field equation of general relativity states an identity of properties, and hence if it is true, it is necessarily so. As a result, the equation of motion of the theory, because it is a consequence of the field equation, is also necessary. In physics, these identities usually are definitional in form, but are expressed in terms of a mathematical equation. For instance, consider one of the field equations of electrodynamics, div $\mathbf{B} = 0$ where $\mathbf{B}$ is the magnetic-flux intensity; if we introduce the vector potential $\mathbf{A}$ and claim that $\mathbf{B}$ is equivalent to curl $\mathbf{A}$ we get div curl $\mathbf{A} = 0$. Because the divergence of a curl is always zero, we have a mathematical identity that supposedly affords an explanation of the Maxwell field equation. Moreover, because the field equation follows as a necessary consequence of the mathematical identity, it is also necessary.

Although this scheme provides a relatively straightforward and powerful explanatory strategy, it implicitly assumes a direct and unproblematic correspondence between the mathematical structure of our theories and the physical systems represented by the mathematical formalism. Although the nature of this correspondence is one of the most important unanswered questions in philosophical analyses of mathematical physics, there are some partial answers to the question that would seem to caution against taking Glymour's analysis as a general scheme for providing explanations. If we think about the use of mathematical structures like group theory and the Lagrangian formalism, we quickly see that what is established is, at best, a structural similarity between the mathematical framework and a physical system. Although it was Lagrange's intention to provide an account of mechanics, he wished to do so by eliminating the Newtonian idea of force, replacing it with the kinetic potential L (excess of kinetic energy over potential energy). But in modern physics, the uses to which Lagrange's equations are put extend far beyond mechanics, making the Lagrangian formalism a method for framing equations of motion for physical systems in general, rather than providing mechanical explanations of phenomena.

Both the breadth of the Lagrangian method and its weakness as an explanatory structure come from the use of generalized coordinates q_i used in place of rectangular coordinates to fix the position of the particle or extended mass [where $x = x(q_1, q_2, q_3)$, and so on for y and z]. It is important to note that the interpretation of these coordinates can extend well beyond simply position coordinates; for instance, in the Lagrangian formulation of electric circuits given by Maxwell, the q_i terms were interpreted as quantities of electricity with unspecified locations. The q_i terms then are functions of time and need not have either geometrical or physical significance. In modern accounts they are referred to as coordinates in a configuration space, and the $q_i(t)$ terms as equations of a path in configuration space. Hence, because no conclusions about the nature of a physical system (other than its motion) can be reached on the basis of its Lagrangian representation, it seems unreasonable for us to argue from a mathematical identity to a necessary physical truth on the basis of identification of physical and mathematical quantities. Similarly, consider the Fourier series, as used in the study of heat diffusion.

The structure of the mathematical expression $\sum_r a_r \cos rx$ is that of a presumably convergent series of periodic functions. The sum itself represents the temperature of a point in the body, but this doesn't imply that heat is thought to be composed of an infinite series of basic states. Alternatively, this kind of mathematical structure may be more successfully used in a Fourier-series solution to the wave equation, where the periodic functions refer to harmonics in the study of acoustics.

We can see, then, that the use of mathematical identities as a way of eliminating contingency and providing explanations of physical phenomena or their relations can in fact have the *opposite* effect from the one Glymour proposes. Often an identification of a phenomenon with a particular mathematical characterization is highly contingent, and the generality of such frameworks is such that they provide no unique or detailed understanding of the physical systems that they represent. That is to say, we can predict the motions of phenomena from dynamical principles, but we have no understanding of the causes of motion.[13] Hence, there is no guarantee of explanatory power resulting from the mathematical description afforded by our theories. In fact, as we shall see in later chapters, the generality provided by such mathematical structures can actually detract from rather than enhance the theory's overall explanatory power. It is also worth pointing out that in order for the identities Glymour speaks about to attain the status of a necessary truth, one must first assume that the theoretical specification of the identity is itself true. Of course, if one knew that, then the epistemological problems associated with explanation and its link to evidence would simply vanish.

Finally, there is the additional form of understanding and explanation involving what Glymour calls the recognition of a pattern and the demonstration that diverse phenomena are of a kind that exhibit a common pattern. Examples are Newtonian mechanics and Copernican astronomy. In the Newtonian case, unity is achieved by generating diverse regularities from a single scheme that specifies the acceleration of any body in a system of n point particles in terms of their masses and mutual distances, assuming that they are subject only to mutual gravitational attraction. Once the value of n is specified, the scheme furnishes a linear second-order differential equation. Because there are infinite numbers of possible geometrical configurations and velocities of the n particles, each differential equation will have an infinity of solutions. One can apply this scheme to various values of n and classes of initial conditions for n-particle systems to yield the theorems of celestial mechanics. This achieves a unification insofar as the evidence bears on each of the laws connected in the pattern; without this kind of common pattern there is no reason to assume that the evidence for one group of laws has any connection with any other group. Moreover, it is not simply the case that Kepler's laws are logical consequences of Newton's dynamical theory together with universal gravitation. Instead, on Glymour's account, the pieces of evidence about various systems that satisfy Kepler's laws (the primary planets, the satellites of Jupiter and Saturn, and the moon) all lead to *instances* of the law of universal gravitation, and as a result all provide tests of the law itself. In that way the explanatory power of a theory, in this case Newton's theory, can be connected with the other reason for acceptance, namely, an accumulation of successful tests.

Although I have no dispute with this latter characterization of the relation between unification and *evidence*, it isn't immediately clear that it provides an account of explanation. Glymour himself wants to maintain that explanation can be a variety of things: unification, the description of causal connection, or several others. I want to claim that the kind of unity Glymour describes does not constitute an explanatory relation, nor does it allow us to distinguish between theories that are truly unified and phenomena/theories that merely share a number of common patterns. On this account, Kepler's celestial mechanics qualifies as a unified theory, but not to the degree that Newton's mechanics does. Difficulties then ensue about how to measure the level of unity achieved in a particular context. Instead, I want to claim that truly unified theories display a particular feature in virtue of which the phenomena are joined together, enabling diverse phenomena to be combined into a single theoretical framework. It is this *combining* of phenomena through a particular parameter in the theoretical structure that constitutes an important part of the unifying process, a process that is represented in the mathematical framework of the theory.

This process of combining is crucial for differentiating among the various ways of thinking about the unifying process. First, there is the rather straightforward case in which different phenomena obey the same laws, an example being the inverse-square law governing the forces on both electrical and mechanical phenomena. Second, there is the type of unity provided by Newtonian physics, in which phenomena previously thought to be different were shown to be similar in kind (gravitating bodies) and to obey the same force law. This differs from the first example in that existential claims about the same force presumably involve more than just the fact that it can be mathematically represented as an inverse-square law (i.e., that the same formal relations hold for a variety of different contexts and phenomena). Lastly, we have Maxwellian electrodynamics, in which light and electromagnetic waves are shown not only to have the same velocity but also to be manifestations of the same process; light waves are identified with electromagnetic waves. This differs from the modern unification of the electromagnetic and weak force, in which the implication is that at high energies the forces combine, but the theory nevertheless retains two distinct coupling constants, one for each of the forces, thereby making the unity somewhat of a promissory note.

I want to argue that in true cases of unification we have a mechanism or parameter represented in the theory that fulfills the role of a necessary condition required for seeing the connection among phenomena. In electrodynamics that role was played by electric displacement or the displacement current, which provided the foundation for a field-theoretic account of the phenomena. In the electroweak theory the mixing of the weak and electromagnetic fields is represented by a parameter known as the Weinberg angle. No such mechanism presents itself simply by showing that two kinds of forces can be described using the same formal relation. For example, in Maxwell's first paper on electromagnetism he utilized a formal analogy between the equations of heat flow and action at a distance, yet no physical conclusions followed from that similarity.

I also want to claim, contra Glymour, that any identification of unification and explanation that might prove possible ought to involve more than the application of a common set of principles to diverse circumstances. The reasons *why* these principles are applicable must emerge at some level within the theory if it is to be truly explanatory. My reasons for holding such a view have to do with a belief that general principles fail to be explanatory in any substantive sense. They enable us to classify and systematize phenomena and may be thought of as the starting point for scientific explanation, but they do not provide details about *how* particular processes take place over and above a descriptive account of the relations among various quantities. Take, for instance, the different ways in which classical analytical mechanics can be formulated – the Newtonian, Lagrangian and Hamiltonian approaches. Each provides a general method for handling particular aspects of the same physical problem or different kinds of problems. However, the decision to employ any one of them depends not only on the nature of the object under investigation but also on the kind of prior information we possess. If we are unsure about the forces acting on a particular system, the Newtonian method with tell us nothing about them; we will simply be unable to apply the parallelogram rule. The Hamiltonian and Lagrangian formulations will tell us something about the evolution of the system – they will allow us to characterize stable states as those for which potential energy is at a minimum – but will tell us nothing about the specific mechanisms involved in the processes that interest us. One might want to object that Newtonian mechanics explains a startling amount about the motions of falling bodies, the tides and planetary motions by showing how each is an instance of the law of universal gravitation. The explanatory relation in this case amounts to an accurate calculation of these motions based on the relations specified by the inverse-square law. But here again there is nothing specific in the theory about how or why the mechanism operates – something that was, at the time the *Principia* was published, clearly a legitimate topic for explanation. By contrast, general relativity does provide an explanatory framework for understanding gravitation. My point, then, is not just that the division between explanation and unification is not uncommon in unified theories, but on the basis of the unifying process we have no principled reason to expect it to be otherwise.

Most modern physical theories seek to unify phenomena by displaying a kind of interconnectedness, rather than a traditional reduction of the many to the one. Two distinct but related conditions are required for this interconnectedness to qualify as representing a unification. First, the mathematical structure of the theory must be general enough to embody many different kinds of phenomena and yet specific enough to represent the way in which the phenomena are combined. The second, related condition refers to the "rigidity" as opposed to the "flexibility" of a theory.[14] In the latter case the theoretical structure does little to resist the multiplication of free parameters in order to account for distinct phenomena. Rigidity, on the other hand, not only minimizes the number of free parameters in the theory's domain but also rules out the addition of supplementary theoretical structure as a way of extending the theory's evidential base. These requirements are definitive of the unifying process, but as such they have very little to say about the nature of scientific explanation.

My discussion of unification in the subsequent chapters is motivated not only by what I see as errors and omissions in current philosophical analyses of the subject but also by historical investigation of what exactly was involved in paradigm cases of unification in both the physical and biological sciences. I want to stress at the outset that my emphasis is on the process of theory unification, something I want to distinguish from a metaphysical or even methodological thesis about the "unity of science" or a "unity of nature". What I want to show is that the methods involved in unifying theories need not commit one to a metaphysics of unity, of the kind that, say, Kepler advocated. As we saw earlier, Kepler's mathematical physics was rooted in the corresponding belief that nature was harmonious; hence there was a kind of one-to-one correspondence between the mathematical simplicity of physical laws and the mathematical simplicity of nature. Although some might claim that the motivation for theory unification embodies a belief in something like Keplerian metaphysics, I want to argue that there are good reasons, despite the presence of unified theories, for thinking such a belief to be mistaken. It is perfectly commonplace to have a high-level structural unity within a theoretical domain in the presence of a disunity at the level of explanatory models and phenomena. In addition to the electroweak case, population genetics, which is discussed in Chapter 7, is a case in point.

The purpose of this overview has not been to set out particular accounts of unification as models for the cases I intend to discuss. My intention has rather been to present a brief sampling of some ways in which unity and unification have been characterized throughout the history of science and philosophy and to give some sense of the diversity present in accounts of unity. I have also attempted to lay some groundwork for my argument that unity and explanatory power are different and frequently conflicting goals. Undoubtedly, strands of each of the views I have discussed can be found in the examples I shall present, something that serves to illustrate my point, namely, that although unified theories themselves may share structural similarities, no hard and fast conclusions can be drawn from that about nature itself. This is partly a consequence of the methods involved in theory unification, but it is also due to the fact that unity in science and nature can take on many disparate and contradictory interpretations and forms.

2

Unification, Realism and Inference

The question that occupies most of this chapter is whether or not the first word in the title – unification – bears any relation to the other two, and if so, how that relation ought to be construed. As mentioned in the introductory remarks, a common approach to fleshing out the notion of unification is to link it to explanation. A unified theory is thought to be one that can explain phenomena from different domains by showing either that the phenomena are essentially the same (e.g., light waves are simply electromagnetic waves) or that diverse phenomena obey the same laws, thereby suggesting some link between them. This explanatory power supposedly provides good evidence that the theory is true; hence, the best explanation, which typically will be the one that reveals some unity among the phenomena, should be seen as more likely to be true than its competitors. Of course, not all "best explanations" will perform a unifying function. There may be only one explanation of a particular phenomenon, and hence, by default, it will have to be considered the best. So embedded in the debate are two issues, one linking unity to explanatory power, and the other linking the concept of "best explanation" to increased likelihood of truth. This practice of drawing inferences to truth on the basis of explanatory power has been dubbed "inference to the best explanation" (IBE) and has been advocated by, among others, Harman (1965) and Thagard (1978).

More recently, however, there have been forceful criticisms by van Fraassen (1980), Cartwright (1983) and Friedman (1983) of the link between IBE and truth and its use as a methodological rule that forms the basis for inference. The complaints are varied. Some, particularly van Fraassen, emphasize the fact that explanation has to do with providing answers to "why" questions or organizing and systematizing our knowledge – pragmatic features that do not provide evidence for the literal truth of the background theory used in the explanation. Cartwright has argued that truth and explanation are, in fact, inversely related: Explanatory power requires broad general laws that do not accurately describe physical processes. But even for those who disagree about the pragmatic status of explanation or its relation to truth, the best available explanation may not be the one that we would want to accept, even provisionally. Friedman opposes IBE on the ground that it provides no guidance on the issue of whether we should construe theoretical structure literally or instrumentally. It simply fails to explain why theoretical structure should *ever* be taken literally. For example, consider two attitudes one might have toward the molecular model of a gas: Either one can be a realist and claim that gases really are just configurations of molecules, and the former can be reduced to or identified with the latter, or one can simply believe that the function of the kinetic theory is to supply a mathematical model for the observable behaviour of gases by associating gases or their properties with mathematical aspects of the model. In this case there is a mapping or correlation of the two domains, but not a literal identification; we have a representation, but not a reduction. We can think of the phenomenological and theoretical domains as being two structures B and A. The realist sees the relation between these two as that of model to sub-model; B is a sub-model of A, and hence the objects in B are identified with their counterparts in A. The anti-realist, however, claims only that B is embeddable into A; there is a mapping from one domain to the other, but no literal identification is made.

The important question, of course, is when to adopt one attitude rather than another. Part of Friedman's objection to IBE is that it provides no guidance on this issue. Regardless of whether we interpret theoretical structure as a mere representation of observable phenomena or as a literal reduction, we enjoy the same consequences vis-à-vis the observable realm. That is, we get the same explanations of the observable phenomena, the only difference being that the anti-realist says that the phenomena behave "as if" they were composed of molecules, rather than actually believing that to be so. In addition, we may have only one explanation of a particular phenomenon, one that might not be acceptable for a variety of reasons; nevertheless, if we apply the rule of IBE we are forced to accept it. Friedman's solution to this problem consists not in giving up this method of inference but rather in restricting its applicability. He argues that theoretical inference can be sanctioned when accompanied by unification, thereby linking unity, explanation and truth. Inference to the "unified explanation" is touted as superior because we get an accompanying increase in the confirmation value of the phenomena to be explained and greater confirmation than would accrue to the previously unconjoined (or non-unified) hypotheses. For instance, if we conjoin the atomic theory of molecular structure and the identification of chemical elements with different kinds of atoms, we can explain chemical bonding. This imparts more confirmation to the assumption that gases are simply molecular systems, a hypothesis that is also confirmed by the gas laws themselves.

Friedman provides persuasive arguments to suggest why one ought to be a realist about certain bits of theoretical structure that figure in the process of unification. Realism allows a literal interpretation of the relevant structure, which in turn affords our theories their unifying power and subsequently their confirmation. In other words, we simply cannot conjoin or unify hypotheses that we do not interpret literally, and, on his view, a literal interpretation requires realism. Without this ability to unify, there is no basis for increased confirmation and hence no basis for belief. Any theoretical structure not participating in unification can be treated as purely representational, without any adverse consequences for the theory in question.

An important part of Friedman's programme is his characterization of the relationship between observational and theoretical structures as that of sub-model to model. For example, we can literally reduce (using an identity map in the model-theoretic characterization) the observational properties of gases to their molecular configurations, with the observable structure of kinetic theory construed as a sub-model of the larger theoretical structure (Friedman 1983, p. 240). Once the reduction is complete, we can then conjoin this theoretical structure with others to form a unified theory. Consequently, much of Friedman's discussion of unification focuses on the role of conjunctive inference and reduction in facilitating the process of unification.[1]

Before moving on to the details of Friedman's proposal, let me briefly mention what seem to be some obvious difficulties with the account. On the face of it, Friedman's argument seems problematic. He claims that we need realism for unification (1983, pp. 11, 244–5), but only if we have unification can we be realists (1983, pp. 7, 259–60). In looking at the details of his argument, particularly the association between unification and conjunction, one needs to distinguish between the logical and methodological aspects of conjunctive inference. Because any account of theory unification must respect, to some extent, the methodological practices of science, it is important to see whether or not the logical model of conjunctive inference as described by Friedman has any bearing on how theories are actually unified. I want to argue that neither a logical account nor a methodological account of *conjunction* is appropriate as a way of explicating unification.

A further criticism centres on Friedman's use of the model/sub-model approach to represent the relation between observational structure and theoretical structure. If the observable structure of our theories is to be construed as a sub-model of the larger theoretical structure (model) in the way that Friedman suggests, then it isn't immediately clear how we are to account for changes in the relationship between the observable and theoretical structures that result from theory evolution. That is, how do we relax the requirements on the identity relation to accommodate changing views on the nature of theoretical structure? Moreover, it is frequently the case that a particular empirical law (e.g., Boyle-Charles) and its approximations (the van der Waals law) require different theoretical models, depending on the context of application. My point here is simply that the ideal-gas law presupposes a different model of the molecular system than does the van der Waals law, and if we utilize the identity map in facilitating theoretical reductions, then we are committed to a unique interpretation of how the molecular system is constituted. That, in turn, makes it difficult to envision how different and sometimes incompatible aspects of theoretical structure are to be accounted for. Friedman claims that his approach accounts well for the way theories get confirmed over time, but the historical record shows that as the kinetic theory evolved, the molecular models underwent drastic changes, with the behaviour of gases often being explained by a variety of different and seemingly incompatible models of the phenomena. In light of these kinds of considerations it is difficult to see how the model/sub-model approach is an effective way to characterize scientific theories; the logical constraints on the model/sub-model relationship (the identity condition) are too tight to allow for the kind of looseness of fit that exists between the theoretical and observable structures of theories. In contrast, not only does the embedding account that Friedman attributes to the anti-realist provide a more accurate characterization of scientific practice, but also it need not fall prey to the criticisms he advances against it.

Finally, Friedman emphasizes the connection between his views on unification and those of Whewell on the consilience of inductions. However, a traditional Whewellian view incorporates a great deal of conceptual change in the evolution from one theoretical picture to its successor, something that cannot be accommodated on the simple "logical" picture of theory conjunction that Friedman puts forth.[2]

In the end, however, regardless of whether or not Friedman's model of theory unification is viable, there are independent reasons for rejecting unification as a justification for scientific realism or a basis for inference. Of course, this still leaves open the question of whether or not unification should be equated with explanatory power. That issue will be taken up in some of the remaining chapters.

2.1. The Friedman Model

A typical scientific explanation is characterized in the following manner: We postulate a theoretical structure $A = \langle A, R_1, \ldots, R_n \rangle$ (where A is the domain of individuals, and $R_1, \ldots, R_n$ are physical relations defined on A) possessing certain mathematical properties. We also have an observational substructure $B = \langle B, R'_1, \ldots, R'_m \rangle$ ($m \leq n$), and A functions as an explanation or reduction of the properties of B. Using the kinetic theory, we can explain the observable properties of gases characterized by B by embedding them in A, where A is literally construed as the world of molecular theory. This enables us to account for the behaviour of gases by identifying them with large configurations of molecules that interact according to the laws of Newtonian mechanics. Because of the properties and relations provided by the theoretical structure, we can derive laws that govern the behaviour of observable objects. By contrast, if we remained strictly on the phenomenological level we would not be able to accurately formulate a law like the van der Waals gas law because we would be unable to appeal to the account of intermolecular forces provided by the higher-level theoretical structure.

Friedman sees the correct relationship between A and B as that of model to sub-model, where $B \subseteq A$ and $R'_i = R_{i/B}$ ($i \leq m$). This characterization affords us a literal identification of the elements in B and A, which in turn results in the larger structure A "inducing" theoretical properties and relations on objects in B, properties necessary for stating accurate laws about observable objects (Friedman 1983,

p. 240). Contrast this with what Friedman terms the representational account. On that view we do not interpret $\mathcal{A}$ literally (as the molecular world); rather, it is construed as a mathematical representation. Instead of asserting that $\mathcal{B}$ is a sub-model of $\mathcal{A}$, we claim only that $\mathcal{A}$ is embeddable into $\mathcal{B}$; there exists a one-one map $\phi : B \to A$ such that $\phi(R_i') = R_{i/\phi(B)}$ $(i \leq m)$. $\mathcal{A}$ does not "induce" the necessary theoretical properties on objects in $\mathcal{B}$ unless, of course, those properties are definable from the observational properties $R_1, \ldots, R_m$. Consequently, we could have two different embeddings ϕ and ψ of $\mathcal{B}$ into $\mathcal{A}$ such that for some property R_j $(j > m)$ and some $b \in B$, $R_j[\phi(b)]$ & $-R_j[\psi(b)]$. This difficulty is avoided on the sub-model interpretation because of the uniqueness of the mapping (the identity map).

As Friedman points out, the representationalist account does not prevent us from generating accurate laws; we simply do so by adding new primitive properties and relations to $\mathcal{B}$ instead of deriving them directly from higher-level structure. However, on this account we provide explanations only in response to particular observable events. There exists no background structure that can be appealed to in attempting to furnish a unified account of various observable phenomena. As a result, the representationalist account provides explanations that are less powerful, and hence it proves unhelpful when confirmation of laws is at issue. The literal construal is preferred because it yields greater unifying power and increased confirmation; for example, we can *conjoin* molecular theory with atomic theory to explain chemical bonding, atomic energy and many other phenomena. Consequently, the molecular hypothesis will pick up confirmation in all the areas in which it is applied. The theoretical description then receives confirmation from indirect evidence (chemical, thermal and electrical phenomena) that it "transfers" to the phenomenological description. Without this transfer of confirmation the phenomenological description receives confirmation only from the behaviour of gases. So in cases where the confirmation of the theoretical description exceeds the prior probability of the phenomenological description, the latter receives the appropriate boost in confirmation as well. Hence the phenomenological description is better confirmed in the context of a total theory that includes theoretical description than in the context of a theory that excludes such description. The literal interpretation can thereby be seen as better confirmed, more plausible and less *ad hoc* (Friedman 1983, p. 241).

2.2. The Importance of Conjunction

Friedman claims two virtues for his reductivist programme. First, there is a type of theoretical inference (specifically, conjunction) that is valid on the hypothesis of a genuine reduction, but not in the case of a representation. This is because the structural relations that hold between the model and the sub-model guarantee the identity of the observational and theoretical structures. Because Friedman sees $\mathcal{A}$ as a "real particular world in its own right" (1983, p. 246), it follows that the identity between $\mathcal{B}$ and $\mathcal{A}$ is also one that should be understood as literally true. For example, where Δ_1 and Δ_2 are classes of models, a reduction facilitates the inference

$$
\frac{\langle B, R_1 \rangle \subseteq \mathcal{A} \quad \text{and} \quad \mathcal{A} \in \Delta_1}{\langle B, R_1, R_2 \rangle \subseteq \mathcal{A} \quad \text{and} \quad \mathcal{A} \in \Delta_2}
\\
\langle B, R_1, R_2 \rangle \subseteq \mathcal{A} \quad \text{and} \quad \mathcal{A} \in \Delta_1 \cap \Delta_2
$$

whereas the non-literal interpretation restricts the inference to the following form:

$$
\frac{\exists \mathcal{A} \exists \phi \colon \langle B, R_1 \rangle \to \mathcal{A} \quad \text{and} \quad \mathcal{A} \in \Delta_1}{\exists \mathcal{A}' \exists \psi \colon \langle B, R_2 \rangle \to \mathcal{A}' \quad \text{and} \quad \mathcal{A}' \in \Delta_2}
\\
\exists \mathcal{A}'' \exists \chi \colon \langle B, R_1, R_2 \rangle \to \mathcal{A}'' \quad \text{and} \quad \mathcal{A}'' \in \Delta_1 \cap \Delta_2
$$

This latter inference is invalid, because $\mathcal{A}$ and $\mathcal{A}'$ are different models, but even if they were the same, we would require some guarantee that the mappings ϕ and ψ had the mapping χ in common. This, of course, is not needed on the reductionist account, because in all cases the mapping (the identity map) is the same. Consequently, we are able to obtain a single joint reduction that is already entailed by our original hypothesis, while on the representationalist account we require the addition of some new piece of theoretical structure.

The second virtue also concerns the utility of the conjunctive inference rule. According to Friedman, our theories evolve by conjunction. Certain assumptions about molecular structure play a role in the explanation of the gas laws, and these, together with further assumptions, figure in the explanation of chemical combination. As a result, the theoretical assumptions receive confirmation at two different times. These advantages also extend to the case of observational predictions. Suppose we have two reductions $\mathcal{A}$ and $\mathcal{B}$, each of which receives an individual boost in confirmation at time t_1 and t_2, respectively. If their conjunction implies a prediction P at t_3 that does not follow from either conjunct individually, then both conjuncts receive repeated boosts in confirmation at t_3 if the prediction is borne out. On the representational schema $[\exists \phi(A), \exists \psi(B)]$ we cannot derive the same observational prediction at t_3 by a simple conjunction; we need a new joint representation $\exists \chi(A \& B)$. The disadvantage of this approach is that the joint representation is formulated as a response to a new observational situation, rather than as a result of the theory's evolution over time by conjunction of hypotheses. Consequently, there is no common unified structure whose parts could participate in the increased confirmation.

Friedman is certainly right about the shortcomings of the representationalist/ instrumentalist programmes in accounting for the role of conjunctive inference. The original version of the conjunction objection advanced by Putnam (1975) was roughly the following: When a scientist accepts a theory, he believes it to be true; it is only by having such a belief that he is able to perform the appropriate conjunctions. Thus, because theory conjunction is a desirable virtue, our epistemic attitudes must be such as to allow for this practice. Friedman changes the tone somewhat by taking the argument one step further, claiming that the product of these conjunctions, the part of the theoretical structure that unifies the other parts, is what is to be believed or interpreted literally. However, because both he and Putnam claim that our theories evolve by conjunction, it appears that we must have some *prior* belief in the truth of our hypotheses in order to achieve the desired outcome. Although Friedman cites unification as a justification for the literal interpretation of theoretical structure, it is interesting to note that on his account we cannot achieve a unification unless we *first* adopt a reductivist approach that construes the theoretical structure as literally true. In other words, in order to have a unified theoretical structure, we must be able to conjoin our theories, which in turn requires the belief that they are true. But this was the *same* belief for which unification was thought to provide a justifying condition. Hence it appears as though we cannot simply limit belief to the unifying part of the theory; we need a stronger form of realism to motivate this model of unification.

One of the difficulties in Friedman's discussion of conjunction is his failure to distinguish between semantic realism and epistemological realism. This is especially important, because the semantic realist can quite consistently provide a literal interpretation of theoretical structure while withholding belief that the interpretation is in fact *true* (e.g., that what the kinetic theory says about molecules is a true description of their constitution).[3] This version of realism denies the tenets of classical instrumentalism by interpreting theoretical structure literally, thereby enabling us to appeal to theoretical structure for derivations and entailments of phenomenological laws without relying solely on phenomenological properties as the basis for theoretical explanation. But this need not entail belief that the structure is real. It requires only that it be interpreted literally – without the "as if" clause. It may be real, but there is no commitment either way. But this latter approach does not conform to the demands of the original conjunction objection, namely, that belief in the truth of our theories is the only way to make sense of conjunction. Because we do not claim truth for our hypotheses, we cannot motivate conjunction. In other words, conjunction seems to require epistemological realism as well. But in light of some recent literature is seems fairly obvious that this strong form of realism is no longer tenable; it is simply too strong a requirement to claim truth for theoretical hypotheses. Hence, we should perhaps interpret Friedman's conjunction model as distinct from Putnam's model, advocating the semantic version rather than the epistemological version of realism. But what are the consequences of this interpretation? Perhaps most important, it would render Friedman's position no different from that of van Fraassen (1980), one that he rejects (Friedman 1983, p. 220) because it fails to sanction the kind of realism about theoretical structure that Friedman needs for conjunction and unification. As he points out (1983, p. 246), it is this kind of epistemological anti-realism that has the undesirable consequence of allowing for different embeddings of observational structure into theoretical structure.

Independent of the logical issues about truth and inference is the question whether or not the actual practice of science and the evolution of theories can be modelled on the approach Friedman describes. It is important here to distinguish between the "in principle" use of conjunctive inference (the logical issue) and its legitimation in specific instances (whether or not science actually proceeds in this way). No one objects to the use of conjunction as a logical rule that guarantees truth, *if* we begin with true conjuncts. But making this assumption in the case of scientific theories is simply to ignore the very problem of realism. Moreover, when we conjoin theories, we rarely, if ever, do so strictly on the basis of logical principles. A complicated process of testing and manipulation is involved in order to ensure a relatively successful outcome. The criticism levied against the anti-realist by Putnam and Boyd emphasized that by denying knowledge of the truth of our theories, we have no guarantee that the conjunction will be successful and no way to make sense of the practice of theory conjunction. But isn't that exactly the point? We do not and should not have any guarantee that the practice will be successful. At the level of theory construction, unification is nothing like the kind of straightforward process that this account suggests. Moreover, if we consider what is involved in theory conjunction, realism presents no methodological advantage. Scientists initially bring theories together with an eye to further testing and successful prediction; only if the conjunction survives empirical tests and yields accurate predictions will the theory be believed or accepted. Hence, our expectations for theory conjunction are validated on the basis of predictive success; but to equate that predictive success with truth would yield the kind of instrumentalism that realists typically object to because it cannot account for the theory conjunction.

A realist might want to maintain that successful conjunction provides evidence for or is an indicator of truth; and indeed Friedman himself seems to suggest that in claiming that through conjunction, theories and pieces of theoretical structure pick up boosts in confirmation (1983, p. 254). Although no "confirmation theory" is provided by Friedman, it is reasonable to assume that, like most realists, he equates increased confirmation with increased likelihood that the theory is true. If one interprets this structure as a literally true description of reality, then difficulties arise in cases where there is more than one way of identifying observational structure with theoretical structure. Approximate truth fares no better, because logical laws like conjunction and transitivity cannot always be successfully applied to terms and hypotheses that are true only in a limited domain. Consequently there is no reason to assume that theoretical conjunctions involving these terms will be truth-preserving. Similarly, any theory that defines truth epistemically or instrumentally will also prevent application of the conjunction rule.

But is this all that can be said about the issue of conjunction? Demopoulos (1982) claims that no realist would deny that correction often occurs prior to conjunction – a point that leaves the anti-realist with the problem of accounting for the conjunction of corrected theories.[4] It seems, however, that because the issue of correction has been introduced, the logical and methodological aspects of Putnam's original conjunction objection have changed. The point of the conjunction objection, in its early formulation at least, was that the truth predicate gave realist epistemology a distinct advantage over its rivals who denied the truth of scientific theories. If one could not expect true predictions, then what could possibly be the purpose of conjoining theories?[5] But if the motive for correction is to facilitate theory conjunction, which presumably it is, then the truth predicate that was initially applied to our theories has little, if anything, to do with the methodological process. The conjunction of corrected theories then becomes an empirical process of bringing together two theories that have been *designed specifically* for that purpose. If they have previously been corrected to ensure, as it were, successful prediction, then there is no reason for the realist to claim any kind of epistemic or methodological superiority. The anti-realist simply explains the practice of conjunction as one that is crucial in the search for theories that are equipped to explain and predict a variety of phenomena. The issue (and practice) becomes a methodological one that involves trial and error, rather than simply a logical operation encompassing semantical and epistemological considerations.

It seems, then, that if Friedman's account of theory conjunction shares the presuppositions of the original Putnam-Boyd formulation, it is fraught with many of the difficulties that led many realists to soften their line on whether or not certain theories could or should be believed to be true in the sense required by conjunction. If, on the other hand, he intends his account to involve no more than semantic realism, an argument for a literal interpretation of theoretical structure (in the manner of van Fraassen), then he cannot motivate the kind of uniqueness that the model/sub-model approach and the identity map guarantee.

Much of Friedman's discussion of conjunction depends on the viability of his model/sub-model account of the relationship between observations and theoretical structure. Ironically, it is possible to show that the kinds of reductions achieved by means of the identity map can actually *prevent* a literal interpretation of theoretical structure – a result that seems to leave the embedding approach as the more accurate representation of scientific practice.

2.3. Reduction versus Representation

2.3.1. Is Reduction a Viable Approach?

The traditional philosophical problems associated with reduction focused on the relationship between thermodynamics and statistical mechanics and dealt with the identification of concepts such as temperature, mean kinetic energy and entropy. However, the idealized nature of the assumptions of statistical mechanics exposes a more serious problem than those linguistic debates would indicate. Consider the following case, in which we have a geometric representation of a mechanical system. From a physical point of view, we describe the mechanical system G with s degrees of freedom by values of the Hamiltonian variables: $q_1, q_2, \ldots, q_s; p_1, p_2, \ldots, p_s$. The equations of motion assume the following form:

$$dq_i/dt = \partial H/\partial p_i, \qquad dp_i/dt = -\partial H/\partial q_i \qquad (1 \leq i \leq s)$$

where H is the so-called Hamiltonian function of the $2s$ variables $q_1, \ldots, p_s$. The Hamiltonian function expresses the energy of a system in terms of momenta p and positional coordinates q. Now consider a Euclidean space Γ of $2s$ dimensions whose points are determined by the Cartesian coordinates $q_1, \ldots, p_s$. To each possible state of the mechanical system G there corresponds a uniquely determined point of the space Γ that can be called the image point of the given system. The whole space Γ is the phase space of the system. The dynamic coordinates of a point in Γ are simply the Hamiltonian variables of the given system G. Any function of these variables is called the phase function of the system. The most important phase function is the Hamiltonian function $H(q_1, \ldots, p_s)$, which determines the mechanical nature of the system (in virtue of the fact that it determines the equations of motion). The total energy E of the system can be represented as $E(q_1, \ldots, q_s; p_1, \ldots, p_s)$. For an isolated part of the system, this function has a constant value; hence for any constant a, the region of the phase space for the point where $E = a$ is an invariant part of phase space. Such regions can be referred to as surfaces of constant energy. Hence, $\sum_\chi$ is the surface of constant energy where $E = x$; for $x_1 < x_2$, the surface $\sum_{\chi_1}$ is situated entirely inside $\sum_{\chi_2}$. The family of surfaces of constant energy can be represented as a family of concentric hyperspheres. The structure function of the given system can be defined as the measure (volume) $\Omega(x)$ of the surface of constant energy $\sum_\chi$. This structure function determines certain features of the mechanical structure of the corresponding physical system, as well as geometrical aspects of phase space.

Assume that the total energy $E(x_1, \ldots, x_n)$ of a system can be represented as the sum of two terms E_1 and E_2, and $(x_1, \ldots, x_n)$ denotes the dynamical coordinates of a point of the space Γ (the product of the phase spaces of all the components). Each phase function and the total energy E of the given system are functions of these n variables. $E_1 = E_1(x_1, \ldots, x_k)$ depends on *some* of the dynamical coordinates, whereas $E_2 = E_2(x_{k+1}, \ldots, x_n)$ depends on the remaining coordinates. Given this characterization, we say that the set of dynamical coordinates $(x_1, \ldots, x_n)$ of a particular system is decomposed into the components $(x_1, \ldots, x_k)$ and $(x_{k+1}, \ldots, x_n)$. However, a peculiarity results when we try to interpret each component as a separate physical system contained in the given system.[6] Although each materially isolated part of the system usually determines a certain component of the system, some components or sets of coordinates do *not* correspond to any *materially isolated* part of the system. The isolated character of these components defines (in the sense given earlier by the definition of a component) pure energetical aspects of the system. For example, consider a system of one material particle, with the components of velocity and mass being u, v, w, m; if its energy E reduces to kinetic energy (because there are no intermolecular forces), we have $E = m/2(u^2 + v^2 + w^2)$. Although u is a component of the system whose energy is $mu^2/2$, it doesn't correspond to any material aspect of the system. Although it *mathematically* represents the u component of velocity, its relation to an isolated material aspect of the mechanical system is more problematic. First of all, it is not even possible to say that any particular molecule has exactly some stated velocity; instead, because of the way in which probabilities enter in the calculation of the velocity distribution function, it cannot be applied to a singular situation. In order to make a substantial claim about velocities, we must consider a number of molecules in some range of velocity (Feynman 1963, vol. 1, pp. 40–5).

Even if we consider a gas with a wide distribution of molecular velocities, it isn't clear that separation of velocity into its three components would result in anything that could be isolated in any *physical* way within the mechanical system. Because velocity is a vector quantity, we assume that it is possible to *physically* isolate each component because we can do so *mathematically*.[7]

What are the consequences of an example like this for Friedman's view? That is, if we subscribe to reduction as a theoretical goal, how do we interpret a theoretical situation of this sort? If we literally identify the energy of a particular molecule with the mathematical representation $E = m/2(u^2 + v^2 + w^2)$, then there are aspects of the latter that seem to lack any materially isolatable counterparts having physical significance. In other words, the notion of a literally true identification in this case seems too strong.[8] Nevertheless, as a mathematical representation it contains certain parameters that are crucial for modelling a statistical system. Each component is a group of dynamic coordinates, and it has a definite energy, with its own phase space (the phase space of the system Γ being the product of the phase spaces Γ_1 and Γ_2 of its two components). Moreover, each component also has its own structure function, and taken together, they determine the structure function of the given system. Indeed, the law governing the composition of the structure function is one of the most important formulas in statistical mechanics.

An additional problem is the methodological paradox that arises from decomposition of the system into components, something that results in exclusion of the possibility of any energetical interaction between particles defined as components. The irony is that statistical mechanics invariably assumes that particles of matter are in a state of intensive energy interaction, where the energy of one particle is transferred to another through the process of collisions. In fact, its methods are based precisely on the possibility of such an energy exchange. Quite simply, if the total energy of a gas is expressed as the *independent* energies of the two components (the energies of the molecules), then the assumptions of conservation and velocity distribution are violated, because each assumption requires that the particles interact. If the Hamiltonian expressing the energy of the system is a sum of functions, each of which depends only on the dynamic coordinates of a single particle (and represents the Hamiltonian of this particle), then the entire system of equations governing the motion of the system splits into component systems. Each component system describes the motion of some separate particle and is not connected to any other particle (Khinchin 1949). As a result, the energy of each particle expressed by its Hamiltonian function appears as an integral of equations of motion and remains constant.[9] From the fact that the particles are independent and the fact that the sum of the energies is constant, it follows that the individual energies must be constant as well. But because this conclusion violates conservation of energy, we must deny the claim that the total energy is the sum of n independent individual energies. In other words, the way the mathematical model describes the system violates some of the structural constraints of statistical mechanics; hence there is good reason not to interpret the mathematical representation as a literally true account of the mechanical system.

For practical purposes, the difficulty is resolved by idealizing assumptions that consider particles of matter as *approximately* isolated energetical components. Although the precise characterization of energy contains terms that depend simultaneously on the energy of several particles, and also allows for energy interaction between them, these forces of interaction manifest themselves only at very small distances. Consequently, the "mixed terms" in the energy equation (those that represent mutual potential energy of particles) will be negligible compared with the kinetic energy of the particles and therefore will be of little importance in the evaluation of averages. In a majority of cases, such as calculation of the Boyle-Charles law, we can neglect these terms and still arrive at a good *quantitative* approximation; we simply assume that the energy of the system equals the sum of component energies. However, on a *qualitative* analysis the mixed terms are extremely important, because they provide the basis for an understanding of energy exchange between particles, the very core of statistical mechanics. Hence, we sacrifice explanatory power for predictive success.

These examples raise some fairly serious difficulties for the kind of literal reductivist approach outlined by Friedman. Even if we disregard the problem of identifying temperature and mean kinetic energy across theoretical boundaries, a more significant difficulty arises in the case of *identifying*, in the way suggested by the model/sub-model approach, the constituents of the system postulated by classical statistical mechanics with its individual particles. The structural presuppositions involved are radically different in each case. Although we can ignore these assumptions in some cases of quantitative prediction, that is not the important issue. As Friedman himself suggests, if we are interested in purely phenomenological laws, then there is no reason to prefer a reduction to a representation (1983, p. 241). But the motivation for Friedman's account is to achieve a literal interpretation of theoretical structure, which in turn will yield greater confirmation of hypotheses, something he sees as guaranteed by the model/sub-model approach.

If we recall the constraints involved in the relationship between a model and its sub-model, we see that they are structurally similar insofar as the interpretation of each relation, function and constant symbol in the sub-model B is the restriction of the corresponding interpretation in the model A. Equivalently, for every atomic formula ϕ and assignment (s) in B, $B \models \phi(s)$ iff $A \models \phi(s)$. Applied to our physical example, we see that a *literal* identification of the properties of individuals of the mechanical system B cannot be accomplished given the structural constraints on A, the mathematical representation of the statistical system. A literal identification of B with A would preclude the formal mathematical model of the statistical theory from accounting for specific parameters (the possibility of energy exchange between particles) that must be interpreted literally if we are to have a proper understanding of the theory's physical foundations. This difficulty can be countered on a representationalist account, where we have an embedding of the properties in B into A. We do not claim a literal *identification* of one with the other, but instead correlate, by way of an embedding map, certain features of B with features of A. Every aspect of B need not have a counterpart in any one model of the statistical theory. Instead, the theory may have several models, each suited to a particular application. In this case the relationship between corresponding elements of A and B is not uniquely specified by the identity map, and hence there can be a variety of ways that the "reduced" entities/theory can be correlated with the reducing theory or model. Given the logical properties of the model/sub-model relationship, we demand that the relations and functions specified by the identity map be preserved over time, in the way we think of inference rules as truth-preserving. But the representational approach allows the relationship between A and B to change over time, something that is *prima facie* ruled out by a literal identification of their corresponding elements. Although these problems do not necessarily deal directly with the straightforward reduction of observational structure to theoretical structure, they do deal with the idea of reducing and identifying physical concepts/properties with theoretical or mathematical representations. What the examples expose are the difficulties associated with reduction as a methodological strategy. But as we shall see later, the demand for different models to account for the same phenomena also arises in the more narrowly defined context in which we have a simple reduction of observable entities to their theoretical counterparts. This situation poses obvious problems for the model/sub-model approach and the accompanying idea that we can correlate the elements in each model by means of an identity map.

2.3.2. The Problem of Many Models

In Friedman's kinetic-theory example he claims that given an appropriate theory of molecular structure and intermolecular forces, we can explicitly define the a and b terms (those representing molecular size and intermolecular forces) and go on to derive the van der Waals law from the kinetic theory – something we cannot do if we remain at the phenomenological level. Although I am in agreement with his claims about the disadvantages of a purely phenomenological approach, even if we acknowledge the need for a literal interpretation of theoretical structure it does not follow that the model/sub-model approach can be vindicated. We shall see that in different contexts the solutions to the problems addressed by this law indicate the need for more than one molecular model. Hence there exists a "looseness of fit" between the phenomenological and theoretical structures that cannot be accommodated on the model/sub-model account.

Friedman contrasts the van der Waals law, $(p + a/V^2)(V - b) = RT$, with the Boyle-Charles law, $pV = RT$, claiming that the latter is false, whereas the former presents a more accurate account of real gases (1983, p. 239). He goes on to claim that it is the structure of the kinetic theory that supplies us the properties and relations that enable us to formulate the more accurate law. As a point of historical interest, if we look at the details of this "deduction" we can see that the van der Waals method for deriving his equation of state departed from the kinetic principles illustrated by the virial theorem, and as such his equation was unsatisfactory as a deduction from the kinetic theory.[10] It is also interesting to note that although the van der Waals theory suggested the possibility of explaining the gas–liquid transition in terms of intermolecular forces, it was not really an application of statistical mechanics (see the Appendix to this chapter). The first and simplest example of a phase transition derivable from statistical mechanics was the famous condensation of an Einstein-Bose gas at very low temperatures. And although that discovery was made in 1924, its physical significance was not appreciated until almost 10 years later (Brush 1983).

Historical points aside, the important issue here is the possibility of reconstructing this example according to the model/sub-model strategy. In order to do this, we require, minimally, that the molecular assumptions required for the corrected van der Waals law to hold not contradict those required of the Boyle-Charles law. Not only is this condition not met, but also when we move to further refinement of the gas laws we need additional assumptions that conflict with those initially postulated. Despite its experimental corroboration, when applied to cases of greater than first-order deviation from Boyle's law, the molecular model suggested by the van der Waals approach was seen to be insufficient (Tabor 1979; Jeans 1917, esp. ch. 3). Basically, the model overlooked the fact that when cohesive forces exist between the molecules, some molecules never reach the boundary (the wall of the container). As a result, van der Waals assumed that those molecules exerted a negative pressure, an assumption that implied negative values for p. Because an examination of physical conditions showed that the true value for p had to be positive, an alternative formulation and molecular model were proposed by Dieterici: $p(V - b) = RT \exp(-a/RTV)$ (Tabor 1979; Jeans 1917). That model assumed a constant temperature of the gas molecules, so that the total energy distribution applied to molecules striking the wall as well as those that did not. Although both equations imply the existence of what is termed a critical point, a point where the liquid, gaseous and vapour states meet,[11] they make different predictions as to the existence of this point; with the Dieterici equation appearing to be more accurate for heavier and more complex gases. Generally speaking, however, neither one comes particularly close to actual observations of critical data.[12] The reason for the discrepancy is that both equations are true only when deviations from Boyle's law are small, with the critical point representing a rather large deviation.[13]

Various attempts have been made to improve the van der Waals equation by the introduction of more readily adjustable constants to supplement a and b, constants referring to molecular size and forces that can be chosen to make the equation's results agree more closely with experiment. One approach introduced a term a' to replace a. Because a' specified that a vary inversely as the temperature for some gases, it provided a better fit with the observations than did the original van der Waals equation.

The overall difficulty seems to be one of specifying a molecular model and an equation of state that can accurately and literally describe the behaviour of gases. We use different representations for different purposes: the billiard-ball model is used for deriving the perfect-gas law, the weakly attracting rigid-sphere model is used for the van der Waals equation and a model representing molecules as point centres of inverse-power repulsion is used for facilitating transport equations. What the examples illustrate is that in order to achieve reasonably successful results we must vary the properties of the model in a way that precludes the kind of literal account that Friedman prescribes (an account that assumes that our model is a literally true description of reality). Instead, an explanation of the behaviour of real gases (something the van der Waals law is designed to explain) requires many different laws and incompatible models. And there are other difficulties with the accuracy or so-called truth of the van der Waals law. In general, the equation tends to smooth out the differences between individual substances and predicts that they will behave more uniformly than they do.[14] In fact, very accurate experiments have shown quantitative discrepancies from the results predicted by the van der Waals law. In calculations of the difference in density between the liquid and gaseous phases, the equation predicts that the difference should go to zero as the square root of the difference between the temperature and T_c. In reality, that difference varies nearly as the cube root, a result that suggests differences in the microstructure of fluids. In what sense, then, can we link the van der Waals gas law with a molecular model that truly describes or can be identified with the behaviour of gases at the phenomenological level?

If the relationship between the behaviour of gases and their molecular model is one of sub-model to model, then the same relations and properties that hold in the latter must hold in the former (with the sub-model being a restriction of the relations in the model). So if the van der Waals equation requires a specific molecular model to establish its results, and the Dieterici equation requires a different model, it seems that we are unable to claim that either provides a literally true account of molecular structure. The so-called derivation of the van der Waals law can be achieved using a particular model that we know to be inapplicable in other contexts. Hence it appears that the uniqueness of the mapping in the model/ sub-model account is actually a drawback rather than an advantage.

The fact that the observable behaviour of gases requires more than one model for its explanation again seems to favour the embedding approach over the sub-model account as a way of understanding physical theory.[15] The advantage here is that one is not committed to the literal truth of the theoretical claims, some-thing that allows for an evolution in our views about the nature of physical systems without falling into logical and semantic difficulties over interpretations of truth. Because the embedding account does not require commitment to a strictly defined and unique identity relation holding between observable properties of physical systems and their molecular configurations, we can accept many different possible interpretations of theoretical structure at the same time. As pointed out earlier, this does not rule out a literal interpretation of the molecular structure postulated by the kinetic theory; we need not become instrumentalists in the way Friedman suggests. The sceptical problem regarding our ability to assign a particular truth value to the theory or model in question is an epistemological issue, not a semanti-cal issue. Whether or not we have sufficient evidence or justification to claim that our theory is true is a separate concern, and an answer is not required for a literal interpretation of the theory's assumptions.[16]

There is, however, another line of defence that the realist could use to vindi-cate the position. One could claim in this case that the van der Waals example shows that a single *general* molecular theory that does *not* incorporate any *specific* assumptions about when a fluid is gaseous and when it is liquid could be used to explain the transition from one state to the other. The idea of continuity be-tween the gaseous and liquid states of matter supplies a kind of ontological unity that forms the theoretical basis for the van der Waals equation. One could then interpret this as suggesting that there is one overarching model of molecular the-ory, the details of which change over time, allowing for corrections and revisions to the same basic structure. If an account like that could be motivated, then one could claim that despite the changes in some properties, the core of the model would remain unchanged. Such a view clearly would go some way toward vindi-cating Friedman's model/sub-model interpretation of theories, and it would solve the problem of incorporating the substantial changes that took place within the framework of the kinetic theory over a period of time. For instance, in addition to the fact that the van der Waals law and the properties of real gases require a number of different models for their explanation, several changes in the structural presuppositions of the theory were necessary to account for the problem of specific heats at low temperatures. Because there occurred a "falling off" of specific heat for diatomic gases such as hydrogen, a phenomenon that could not be attributed to the fact that hydrogen ceases to be a perfect gas at low temperature, the kinetic theory needed to be modified in the domain of idealized perfect gases. The the-ory was subsequently improved by reinterpreting it in terms of the mechanics of relativity. The foundations were left unchanged (i.e., the Gibbs theorems on the conservation of extension and density in phase, and the ergodic hypothesis), but the law of equipartition was rejected and replaced by a different law of partition. It was soon noticed that those refinements were not applicable when the motions of molecules were slow, a situation that arises at low temperatures when relativistic theory merges into the classic theory. The required modification was furnished by the quantum theory and the statistics of Fermi and Bose.

The difficulty with adopting the kind of restrictive yet generalized realism described above is that it requires a separation of the entities like molecular structure and the properties that define that structure, properties that supposedly give rise to many of the empirical phenomena we are concerned to explain. The idea is that there exists a molecular structure that forms the basic core of the model, but we have no stable, realistic account of the details of the structural properties, because our account of them changes with shifts in theoretical knowledge. However, it is exactly these properties that figure importantly in the derivations of phenomenological laws from the higher-level theoretical structure. Postulating the existence of a molecular structure defined by a model that is devoid of specific properties and relations allows us to maintain our structure over time, but in return provides none of the advantages the model was designed to create. On the other hand, a molecular model endowed with specific features cannot be interpreted as a literally *true* description of theoretical structure, because such an interpretation would provide no mechanism for changes in the model over time and no account of the nature of incompatible models – situations that are necessary for a realization of the various contexts and possibilities envisioned by the theory. Hence, neither option is a possibility. Because the embedding approach allows for a variety of models of the phenomena, it seems closer to actual scientific practice. Not only is Friedman's model/sub-model account too restrictive in its requirement that the model be literally true, but also the relationship between the phenomenological structure and theoretical structure cannot be accurately depicted by the kind of stringent logical requirements (furnished by the identity map) that hold between models and their sub-models.[17]

The embedding approach is a significant feature of van Fraassen's semantic view of theories. The importance of the semantic view for our discussion rests not only with the details discussed earlier but also with the way models function as part of scientific practice. According to the semantic view, a theory is simply a family of models that represent actual phenomena in a more or less idealized way; so when we talk of embedding we refer to the process of mapping the phenomena into one of the many models that might be used to account for them. For example, there are many models that can describe a quantum-mechanical particle in a potential well, and we can model the pendulum as an undamped or a damped harmonic oscillator, depending on how "realistic" we need our model to be. The laws of Newtonian mechanics tell us how to add corrections to the model of the undamped pendulum in order to make it more like the physical apparatus, and in that sense the resulting model can be seen as a model of the theory, an application of the theory's laws in a specific context as well as a model of the physical pendulum. We don't simply apply the theory directly to nature; instead, we construct a model that resembles some aspects of a natural phenomenon to see how the laws apply to an object so defined. The damped oscillator gives predictions that more accurately approximate the phenomena, but sometimes that level of accuracy is not needed to solve the problem at hand. The important point here is that the semantic view seems better equipped to handle the diversity of models in scientific practice and their role of providing approximations to physical systems. In other words, scientific models are never exact replicas of either the theories or the phenomena. They can approximate either the theories or the phenomena or both. In that sense it would be misleading to claim, as the model/sub-model account does, that there is a literal identification between the elements of a model (in that case understood as the observable phenomena) and the theoretical structure. Neither the natural sciences nor the social sciences view models in that way. Not only are there many ways the phenomena can be modelled, but also theoretical structure, by its very nature, will always present a more abstract picture than the phenomena themselves.[18]

Where does that leave us with respect to achieving or defining unity? I shall have more to say about this in later chapters, particularly on the connection between explanation and unification, but for now let me briefly recap the basic strategy of Friedman's argument. On his view, we achieve theoretical unity through a process of reduction and conjunction. The observational structure of a theory is reduced to its theoretical counterpart by identifying the two in the appropriate sort of way. This process can be represented in logical terms by means of the identity map that relates a sub-model to its model. Once the reduction and identification are complete, the theoretical structure can then be conjoined to other structures to achieve a unified theoretical account of diverse phenomena. What I have tried to show is that neither the theories that incorporate observational and theoretical structure nor the phenomena themselves are literally reducible in the way the model/sub-model approach suggests. The use of the reductivist strategy as well as the use of conjunction as important features in unification become either simply inapplicable or at best questionable. The apparatus of formal model theory is simply too rigid to capture the rather messy relations that are part of the modelling of scientific phenomena.

If we look at the pattern of unification suggested by William Whewell under the title "consilience of inductions", an approach that Friedman likens to his own account (1983, p. 242, n. 14), we quickly see that neither conjunction nor reduction plays a role in the unifying process. As we saw in Chapter 1, we find instead a rather complex process of reinterpretation of basic aspects of theoretical laws and structures, a reinterpretation that extends far beyond the product of the simple conjunction of existing theories. Although Whewell's account of the process squares better with scientific practice, he also sees unity as a "stamp of truth", something that, I want to claim, need not be the case.

2.4. Consilience and Unification

According to Whewell, a consilience of inductions is said to occur when a hypothesis or theory is capable of explaining two or more classes of known facts, when it can predict cases that are different from those the hypothesis was designed to explain/predict, or when it can predict/explain unexpected phenomena.[19] In each case a consilience of inductions results in the unification or simplification of our theories or hypotheses by reducing two or more classes of phenomena that were thought to be distinct to one general kind, or by showing their behaviours to be describable by one theory. In addition, this unification results in a reduction in the amount of theoretical structure required to account for the phenomena. Reduction is an important element, but not the kind of reduction characteristic of the model/sub-model approach. Although the deductive-entailment content of consilient theories is very high, often that virtue is achieved only through a process of *reinterpretation* of the laws and key terms:

When we say that the more general proposition includes the several more particular ones ... these particulars form the general truth not by being merely enumerated and added together but by being seen in a new light. (Butts 1968, pp. 169–70)

Whewell goes on to point out that in a consilience of inductions

there is always a new conception, a principle of connexion and unity, supplied by the mind, and superinduced upon the particulars. There is not merely a juxta-position of materials, by which the new proposition contains all that its component parts contained; but also a formative act exerted by the understanding, so that these materials are contained in a new shape. (Butts 1968, p. 163)

Perhaps the most frequently cited example of a consilience (both by Whewell and by contemporary philosophers of science) is the unification of Kepler's and Galileo's laws under the inverse-square law. Newton's theory of universal gravitation could explain terrestrial phenomena like the motions of the tides and celestial phenomena like the precession of the equinoxes, classes of facts that were thought to be disjoint. The corrections applied to lower-level laws such as Galileo's laws of falling bodies and Kepler's third law of planetary orbits were motivated strictly on the basis of the overarching theory, rather than as generalizations from phenomena, as was the case in their initial formulations. Alternatively, one could say that Newton's theory showed how terrestrial and celestial phenomena were the same kind of entity (i.e., gravitating bodies).

Traditionally, many philosophers of science (including Duhem, Hesse, Laudan and Butts) have argued that the relationship between Newton's theory and Kepler's laws (as well as Galileo's laws) is not one of conjunction or even entailment, because universal gravitation *contradicts* the conclusions that those individual sets of laws provide. More recently, Malcolm Forster (1988, pp. 88–91) has pointed out that Newton himself was explicit in denying that view, maintaining that Kepler's third law held exactly, because all observed deviations from the law could be ascribed to other causes (which he afterward explained as the result of other gravitating masses). Forster claims that Newton proves, in the *Principia*, that Kepler's three laws for the earth's motion around the sun or the moon's motion around the earth can be expressed as instances of the inverse-square law, with the equation for each of the respective motions representing the Keplerian component of the motion as described by the three laws; that is, the solutions for the equations are elliptic orbits satisfying the area and harmonic laws. Conversely, Kepler's area law implies that the acceleration of each body is toward the other, and the ellipticity of the paths proves that the acceleration is inversely proportional to the square of the distance between the two bodies. In that sense, Kepler's laws can be seen to entail instances of the inverse-square law, suggesting the conclusion that Newton's theory of gravitation *entails* that Kepler's laws are true. Although they do not provide a complete description of the phenomena, they do give a "description in full agreement with Newton's theory" (Forster 1988, p. 89) and play an essential part in Newton's reasoning.

It is important, however, in this context to look beyond the mathematics to the physical interpretation that each theory furnishes. As we saw in the statistical-mechanics examples discussed earlier, we often need to supplement abstract mathematical laws and models with a qualitative understanding of the theory in order to appreciate the implications for concrete physical systems. In the case of Newtonian mechanics and Kepler's third law, the latter states that the cube of the mean distance of a planet from the sun divided by the square of the period of revolution is a constant for all planets (a^3/T^2 = constant). The Newtonian version of the law states that $a^3/T^2 = m + m'$, where m is the mass of the sun, and m' is the mass of the planet in question. By ignoring m' on the ground that it is much smaller than m (at least for our solar system), we can assume that the two laws are roughly the same. In what sense is the Keplerian formulation true? It is true only if we ignore the fundamental qualitative aspects of Newton's theory that serve to differentiate it from Kepler's account of celestial mechanics. If we consistently ignore m', it becomes impossible to apply Newton's theory, because there is no gravitational force on a body with zero rest mass. On the other hand, if we make the simplifying assumption that m' is the same for all planets, we can then apply Newton's theory to get Kepler's laws within all observational accuracy. But here again the point is that the masses of the planets *are* different, and it is one of the benefits of Newtonian mechanics that we are able to calculate planetary orbits based on that information.

A similar situation holds for the case of Kepler's first and second laws. Newton himself remarks (*Principia*, bk. 1, prop. lxv, theorem xxv) that in cases where more than two bodies interact, Kepler's first two laws will be approximately valid at best, and even then only in very special cases. Although it is true that in some instances we can arrive at the same numerical values using both Kepler's law and Newton's law, the relationship between them, within the context of an overall mechanical theory, is not one of entailment. Nor is the latter the result of a conjunction of the former plus Galileo's laws.[20] From neither of these two groups of laws taken separately is there any indication of how they can be conjoined to produce a theory like Newtonian mechanics.

What about other instances of unification, such as relativity theory and Maxwell's electromagnetism? Is it reasonable to assume that there has been significant conjunction in each case? In the case of electromagnetic theory, there was a relatively straightforward identification of light waves and electromagnetic waves that resulted from Maxwell's introduction of the displacement current, a phenomenon for which there was no experimental/physical justification. That led to modification and reinterpretation of the physics behind Ampere's law, which described the relationship between an electric current and the corresponding magnetic field, a law that had been central to the earlier theories of electromagnetism. In addition to those changes, there were laws of physical optics that could not be accounted for by the new field theory, most notably reflection and refraction (dispersion). The sense in which optics and electromagnetism were simply conjoined and corrected is remote at best. The same is true of special relativity. Einstein left the mathematical form of Maxwell's equations virtually intact in showing them to be Lorentz-covariant. There were, however, several changes made to Newtonian kinematics and dynamics. And although for cases of low velocity Newtonian mechanics gives completely accurate quantitative results that are indistinguishable from those of relativity, the unification of Newtonian mechanics and electrodynamics involved substantial reinterpretation of classical physical magnitudes, including the nature of space and time.

Those instances, as well as the unification achieved in general relativity and our current attempts at bringing together quantum mechanics and relativity, exemplify a much more complicated process than simple conjunction and correction. In each case there is what Whewell calls the "introduction of a new mental element" (Butts 1968, p. 170). Whewell is careful to point out that the inductive truth is never merely the "sum of the facts"; instead, it depends on the "*suggestion* of a conception not before apparent" (Butts 1968, p. 170). The generality of the new law or theory is constituted by this new conception. This issue speaks not only to the unification of specific theories but also to the wider issue of convergence and unity in physics as a whole. The presence of limiting cases (e.g., Newtonian mechanics as a limiting case of relativity for low velocities) does not, in and of itself, suggest unity or convergence, any more than conceptual reorganization suggests incommensurability. It seems reasonable to expect that many low-level laws describing the behaviour of macroscopic objects will remain relatively unchanged in the development of scientific theories. Where significant change will occur is in our understanding of fundamental aspects of nature. This is the point at which our concepts and theories are becoming drastically altered, often with a complete reorientation as to what counts as an explanation of a natural process. The case for unity and convergence across the history of physics is a difficult one to make, most likely because it simply can't be made in anything but a trivial way.

Whewell's own work indicates a lack of unity across the sciences, but argues strongly for unity within each one. However, the *kinds of arguments* he offers for that unity have limited, if any, application to contemporary practice. Contemporary realist approaches take after Whewell in attributing a higher degree of confirmation to the unifying theory on the assumption that we can explain and possibly predict a variety of phenomena. The criterion of diversity emphasized by Whewell is the key to understanding his notion of consilience. When a theory was found to be applicable to a body of data other than that for which it was designed, the additional data were seen as providing independent evidence for the theory. So, for example, in the development of the theory of electromagnetism by Maxwell, many changes were made to existing notions of electromagnetic-wave propagation. But in addition, Maxwell found that the velocity of wave propagation coincided with the velocity of light and that the theory was able to account for various optical phenomena. According to Whewell (but not necessarily Maxwell himself, as we shall see in the next chapter), that would constitute independent evidence for the theory. A consilience is similar to the testimony of two witnesses on behalf of a hypothesis:

... and in proportion as these two witnesses are separate and independent the conviction produced by their agreement is more and more complete. When the explanation of two kinds of phenomena, distinct and not apparently connected leads to the same cause such a coincidence does give a reality to the cause, which it has not while it merely accounts for those appearances which suggested the supposition. This coincidence of propositions is ... one of the most decisive characteristics of a true theory ... a consilience of inductions. (Whewell 1847, II:285)

When two different classes of facts lead to the same hypothesis, we can assume that we have discovered a *vera causa*.

We must, however, be cautious when characterizing the so-called independent evidence that is cited on behalf of a consilient theory. Although the evidence may be drawn from a variety of different and supposedly independent domains, there may nevertheless be no independent evidence for the theory itself or its particular unifying structure other than its ability to present a unified account of disparate phenomena. Again, Maxwell's electromagnetic theory was highly successful in unifying electromagnetic and optical phenomena. However, the mechanical model of the aether and the displacement current that initially facilitated that unification could not be justified on experimental or independent grounds. It was the failure of the aether model that led Maxwell to reformulate the theory using the abstract dynamics of Lagrange, with electrical and mechanical concepts occupying merely illustrative roles. As mentioned earlier, a crucial component in Maxwell's aether model and in the unification of electricity and magnetism was a phenomenon known as the displacement current. It was introduced to augment the usual conduction current and to create a field-theoretic explanation of propagation, which further enabled him to calculate the velocity of electromagnetic waves travelling through the hypothetical medium. When the early aether models were abandoned, electric displacement remained as a designated quantity, yet he remained agnostic about any qualitative account that might be given of either its nature or operation. Although the theory unified a great many phenomena, there was no evidence for the existence of electromagnetic waves themselves, nor any explanation of how they could be propagated.

Many of Maxwell's contemporaries, including Kelvin, thought of displacement as little more than an *ad hoc* postulation. And as the history of the period reveals, Maxwell's theory was not well received, despite the rather remarkable unification it achieved; nor was there increased support for the displacement current as a crucial theoretical structure that facilitated the unification. Maxwell's final account of electromagnetism, presented in his *Treatise on Electricity and Magnetism*, was a theory that achieved its unifying power from the Lagrangian formalism; a slightly revised version was fully embraced by the scientific community some 13 years later, after Hertz's famous experiments on electromagnetic waves. The vindication of the theory came not as a result of its ability to explain or unify a variety of independent phenomena through the postulation of theoretical structure, but from independent experimental tests showing the existence of electromagnetic waves and the use of the field equations to describe their propagation. So although these examples seem to follow the *process* of unification described by Whewell, there is little historical evidence to suggest that the epistemological aspects of his views were compelling or that people saw unity as a stamp of truth or as a reason to accept a particular theory.

On Friedman's account of unification he repeatedly emphasizes the importance of the confirmation that results from the ability to predict and explain a variety of phenomena. To that extent his views are similar to Whewell's and to traditional philosophical accounts of confirmation. However, according to Friedman this confirmation results from the fact that our theories evolve by conjunction. Although he claims that it is unifying power that serves as the criterion for a realist interpretation of theoretical structure, his case involves a more complicated methodology. The persistence or stability of particular structures through time enables our theories to evolve through a conjunctive process, thereby increasing their confirmation value (Friedman 1983, p. 245).

This approach is, however, at odds with most cases of consilience/unification, in which significant changes have been made to the laws and terms of the hypotheses involved, with the addition of new structures and entities. Although simple conjunction of hypotheses does not occur in cases of consilience, there *is* a reduction of entities, structures or hypotheses under the umbrella of one theory. This reduction obviates many of the traditional problems associated with Friedman's model/sub-model account because it involves the unification of two groups of phenomena by means of a mechanism that involves theoretical, mathematical and semantical changes. No account of unification can be complete without recognition of the need for and the implications of this conceptual reshuffling. Consequently, much of the discussion of conjunction, together with the model/sub-model account of theoretical and observational structure, cannot be used to motivate a philosophical account of how unification, confirmation and realism are connected.

2.5. Unification as an Evidential or Epistemic Virtue

Earlier I claimed that as a matter of historical fact unity seemed to provide little in the way of empirical support for theories. The cases detailed in the following chapters will bear out that conclusion in what I hope is an unambiguous way. But my analysis here would be incomplete without some discussion of the philosophical status of unification itself as a justification for realism. Regardless of whether or not one adheres to the conjunction model of unification, there seem to be independent reasons for thinking that unification can sustain or motivate only a contextually based form of realism. Let me briefly explain why I take this to be so.

In discussing consilient theories it is important to note that a theory becomes consilient when it shows that phenomena originally thought to be of different kinds are in fact the same kind. This occurs only in relation to some other theory or set of particular beliefs and background knowledge about the phenomena, conditions that usually take the form of the currently accepted theory. For example, Newton's theory was consilient at the time of its emergence because celestial and terrestrial phenomena were regarded as distinct types. Had universal gravitation been proposed within the context of a Cartesian system, it would not have been considered consilient, because Descartes regarded both kinds of phenomena as due to the actions of similar types of vortices.[21] Friedman himself emphasizes that unifying power is a relative notion (1983, p. 249). He points out that absolute rest had no unifying power in the context of Newtonian gravitation theory. However, in the context of classical electrodynamics, absolute rest did have unifying power and therefore should be interpreted literally; but because we now no longer subscribe to classical electrodynamics, we can assign absolute rest a purely representative status.

This issue of historical relativism is perhaps the most important difficulty for a realist account of confirmation like Friedman's (one that equates consilient or unified theories with reasons for true belief or ontological commitment). Because the question of whether or not a particular piece of theoretical structure plays a unifying role becomes relativized to a specific context, it is difficult to see how this kind of virtue could be taken as evidence for truth or realism (unless, of course, both are construed in a purely local context). Specific ontological claims are legitimated on the basis of unifying power, but because these entities/structures perform a unifying role in some contexts and not in others, our beliefs become dictated solely on the basis of the historical contingencies involved in the unifying process. However, to talk of ontology and truth from within the confines of a particular theory is to collapse talk of truth and ontology into talk about the theory.[22] A strategy of that kind simply rejects what is right about realism, namely, the search for theory-neutral facts that can act as arbiters in theory choice and remain relatively stable in the face of theoretical restructuring. This notion of independence or neutrality seems to be sacrificed in an account that uses unification (in the context-dependent way Friedman describes) to motivate realism. Indeed, it is difficult to see how any distinction between the real and the representational at this local level could be extended to a global epistemology of science. Because a realism based on criteria like explanatory unification provides no ontological stability over time, it would seem that as a justification for *belief*, even at the local level of scientific practice, it proves unsatisfactory for realists and anti-realists alike. To the extent that theory confirmation depends on beliefs about the relative merit of the new theory as compared with its predecessor, it may always involve a certain amount of historical localization; but that is a different issue from the kind of variability that results from isolating a methodological process like unification as the criterion for a realistic interpretation of theoretical structure/entities. In the latter context, our commitments and beliefs become doubly abstracted; they are not simply relativized to a particular epistemic community, but to the domain of a unifying theory.

As mentioned earlier, one of Friedman's justifications for a literal construal of theoretical structure is that it allows for a persistence of that structure over time. But given the historical variability present in his own account, it is unlikely that such persistence could be guaranteed independently of a particular context.[23] Thus far I have tried to show some practical and philosophical difficulties with Friedman's particular account of unification. Not only is conjunction and the model/submodel approach ill-equipped to handle the ways in which theories and models function in practical contexts, but also the general strategy seems unable to support the kind of realism it was designed to defend. And once we acknowledge the relation between unification and particular theoretical contexts, it becomes difficult to provide an argument that links unity, truth and realism.

I have not yet said much about explanation, except to criticize Friedman's "inference to the unifying explanation". I have, however, suggested that unification is not the kind of criterion on which to base arguments for realism, and I have also hinted that it may not be important for theory acceptance either. In order to substantiate that argument with empirical evidence, I want to examine some specific and paradigmatic cases of theory unification in both the physical and biological sciences. One of my claims is that unification typically was not considered to be a crucial methodological factor in either the development or confirmation of the physical theories. And even in cases where it was a motivating factor, such as the unification of Darwinian evolution and Mendelian genetics, the kind of unity that was produced could not be identified with the theory's ability to explain specific phenomena. This is not to say that there is no evidence for unification in science or that unity is largely a myth, as some contemporary writers on disunity suggest (e.g., Dupré 1993, 1996). Rather, I want to demonstrate that the ways in which theory unification takes place and the role it plays in scientific contexts have little to do with how it has been characterized in traditional philosophical debates.[24] Once we have a clearer understanding of the unifying process, we can begin to see where its importance lies, what its connection is, if any, to explanation and the way unity functions in particular domains as well as in the broader context of scientific inquiry.

Appendix

Derivation of the van der Waals Law: Historical Details

The van der Waals law was originally formulated as a response to the idealizing assumptions of the Boyle-Charles law, which maintained that the sizes of molecules and the forces between them were negligible.[25] Although the molecules of real gases are assumed to have finite sizes and exert intermolecular forces at ordinary temperatures and pressures, real gases behave very much like ideal gases and thus in some situations obey the Boyle-Charles law. However, at sufficiently high and low temperatures, real gases can become liquified, thereby invalidating the application of ideal-gas laws to real gases. The foundations of the kinetic theory disregard the volumes and mutual attractions of the molecules, yet it is these attractions that account for such phenomena as cohesion, surface tension, the existence of a critical point and the phase transition (condensation).

It was Rudolf Clausius who initially suggested that the intermolecular forces that account for cohesion of the liquid phase must act throughout the range of temperatures and pressures. Their effects should be appreciable even in the gaseous phase, when molecules closely approach one another in collisions. In other words, because the nature of the substance was defined by its molecular model, the properties of the model should be present under all temperatures and pressures.

Van der Waals came upon this idea of continuity of the liquid phase and gaseous phase as a result of the work of Clausius on the virial theorem, which was an attempt to reduce the second law of thermodynamics to a purely mechanical form.[26] The theorem states that for a system of material points in which the coordinates and velocities of all the particles are bounded, the average (taken over long times) of the total kinetic energy is equal to the average of the virial:

$$\left\langle \sum \left(\frac{1}{2} m_i \mathbf{u}_i^2 \right) \right\rangle = \left\langle -\frac{1}{2} \sum \mathbf{F}_i \cdot \mathbf{r}_i \right\rangle \qquad (A2.1)$$

The brackets denote time averages, and the quantity on the right-hand side is what Clausius defined as the virial. The $\mathbf{r}_i$ denotes the coordinates of the ith particle, whose mass is m_i and whose velocity is $\mathbf{u}_i$, and $\mathbf{F}_i$ is the resultant force acting on it. The object was to trace the actual motions of the molecules that constituted the heat and to show that the effective force of heat was proportional to absolute temperature. The problem remained a purely mechanical one, with no appeal to probabilistic arguments to investigate the motions. Quite simply, the theorem states that for a system of material points in which the coordinates and velocities of all the particles are bounded, the mean kinetic energy of the system is equal to its virial. If the forces on the particles confined in a volume V can be divided into a uniform external pressure (from the container walls) and the central forces $\phi(\mathbf{r}_{ij})$ acting between particles, the theorem will take the following form:

$$\left\langle \sum \left(\frac{1}{2} m_i \mathbf{u}_i^2 \right) \right\rangle = \frac{3}{2} P V + \left\langle \frac{1}{2} \sum_{ij(i>j)} \mathbf{r}_{ij} \phi(\mathbf{r}_{ij}) \right\rangle \qquad (A2.2)$$

where $\mathbf{r}_{ij} = |\mathbf{r}_i - \mathbf{r}_j|$. Using the theorem, Clausius was able to derive a form for the second law only in the case of reversible processes; van der Waals, on the other hand, saw in it implications for the properties of matter.[27]

Although the virial theorem incorporated the possibilities of both the static and kinetic molecular theories, van der Waals did not calculate the molecular pressure P' from the average virial of intermolecular forces. Instead, he used a distinctly different approach to discuss the effects of the extended molecular volume and intermolecular attraction. Using a series of assumptions about the mean free path,[28] he allowed for the fact that the molecules were of finite size.[29] In calculating the value of P, van der Waals argued that the effective force on a unit area of surface, arising from attractive forces between molecules, was the result of a thin layer of molecules below the surface. That followed from continuity considerations, which implied that the attractive forces acted over only a very short range.[30] The final step toward the completed equation of state involved replacing the average kinetic energy of the fluid by the expression proportional to the absolute temperature for one mole of ideal gas:

$$\left\langle \sum \frac{1}{2} m\, \mathrm{u}^2 \right\rangle = \frac{3}{2} RT \tag{A2.3}$$

an assumption that could be argued for only on the basis of plausibility considerations. The equation in its final form was

$$(P + a/V^2)(V - b) = RT \tag{A2.4}$$

The volume-correction term was only an approximation and was not valid at high compressions, a difficulty that van der Waals was unaware of. In addition, the pressure calculations were not completely satisfactory. The correction a arises from forces that the molecules exert on one another when reasonably near to one another. The correction b arises from forces that the molecules exert on one another when their centres are some distance apart. However, we cannot suppose that the forces acting on natural molecules can be divided up into two distinct types; they must change continuously with the distance. As a result, the a and b of the van der Waals equation ought to be different contributions from a more general correction, and so ought to be additive. The equation itself allows for no such correction. However, once a and b were determined experimentally, the isotherms (T) could be predicted for all P, V. The values arrived at using the van der Waals equation were confirmed experimentally in tests carried out by Andrews[31] on isotherms for carbon dioxide.

3

Maxwell's Unification of Electromagnetism and Optics

Maxwell's electrodynamics undoubtedly represents one of the most successful unifications in the history of science. Its demonstration that optical and electromagnetic waves travel with the same velocity and that both phenomena obey the same laws is paradigmatic of what we call the unifying power of theory. Despite enjoying some early success in Britain, there was a 15-year gap after publication of the *Treatise on Electricity and Magnetism* (Maxwell 1873) before Maxwell's theory was accepted on the Continent, fully supplanting action-at-a-distance accounts as the received view of electrodynamics. But even among Maxwell's British contemporaries there was by no means open enthusiasm. Sir William Thomson (Lord Kelvin) was highly critical of Maxwell's theory, despite being himself a proponent of field-theoretic views of electromagnetism.

The history of the development and acceptance of Maxwell's electrodynamics is interesting from the point of view of theory unification for several reasons. First, the rather striking unification of electromagnetism and optics seems to have provided little reason to embrace the theory; even its advocates (including Maxwell himself) did not mention unifying power as playing an evidential role. To that extent the case provides a counterexample to the popular philosophical argument that unification functions to increase the likelihood that the theory is true or that it even functions as a criterion for theory choice among competing rivals. Second, the theory's development took place in several stages, the first of which depended on a mechanical aether model that was given up in later formularions. Although it was that model that initially facilitated the deduction of the wave equation and hence the identification of electromagnetic and optical waves, an analysis of Maxwell's epistemological views reveals that at no point did he see the unifying power of the aether model as evidence for an ontological commitment to an electromagnetic aether. In fact, it was his desire to rid the theory of such hypothetical elements that prompted him to later jettison the aether model in favour of the abstract dynamics of Lagrange. The most important feature of the model was its incorporation of a phenomenon known as the "displacement current", which was responsible for the transmission of electric waves through space, thereby producing the effect of having a closed circuit between two conductors. The aether model explained how the displacement of electricity took place, but it was this notion of electric displacement that was the key to producing a field-theoretic account of electromagnetism.

Without the displacement current there would be no propagation of electromagnetic impulses across space. But, as with the aether itself, there was no experimental evidence for a displacement current, and indeed many critics, including Thomson, saw it as the most undesirable feature of the theory.

Here again, what we see is a methodology at odds with that portrayed in philosophical writing on unification and theory acceptance. Not only was unifying power not seen as evidence for the theory itself, but the specific unifying parameter, electric displacement current, was not given a realistic interpretation. The aether model was seen by Maxwell as purely fictitious, a way of illustrating how the phenomena could possibly be constructed. In later versions of the theory published in 1865 and 1873 Maxwell wanted to establish a more secure foundation, relying only on what he took to be firmly established empirical facts, together with the abstract mathematical structure provided by Lagrangian mechanics. That structure, unlike the mechanical model, provided no explanatory account of how electromagnetic waves were propagated through space nor any understanding of the nature of electric charge. The new unified theory based on that abstract dynamics entailed no ontological commitment to the existence of forces or structures that could be seen as the source of electromagnetic phenomena. Instead, "energy" functioned as the fundamental (ontological) basis for the theory. The displacement current was retained as a basic feature of the theory (one of the equations), but no mechanical hypothesis was put forward regarding its nature.

That change in Maxwell's methodology and the change in theoretical structure that accompanied it provide some important insights about how theory unification takes place. Because one of my goals is to highlight the role that mathematical structures play in the unifying process, Maxwell's electrodynamics is especially important. But in addition to illustrating this point, it also calls attention to three very important facts: (1) that theory unification can take many forms, (2) that the unifying power of a theory may have little to do with its explanatory power and (3) that a unified theory may not necessarily correspond to or imply an ontological unity in nature. The Maxwellian case is an example of what I want to call a reductive unification, where two seemingly disparate phenomena are reduced to one and the same kind (i.e., light waves are seen to be electromagnetic waves). But even this type of unity was expressed differently at different stages in the theory's development. A unity of this form undoubtedly has ontological implications, but these, too, are by no means straightforward. There was no evidence for the existence of electromagnetic waves, but even after their experimental discovery, no theoretical account or explanation of wave propagation was given – a situation that pointed to a significant lack of understanding regarding the nature of these waves.

The issue of explanation raises interesting problems within the context of a unifying theory. Like unity, explanatory power can take different forms. Maxwell's theory was explanatory in the sense of being able to account for optical and electromagnetic phenomena using the same laws, but not in the sense of imparting an understanding of how or why field-theoretic processes took place. But the significant point that emerges in the historical analysis of this case, as well as the others I shall discuss, is that the mechanism facilitating the *unification* may not be what provides an *explanation* of how the relevant physical processes take place. In that sense, unity and explanation may actually be at odds with each other.

To see how this is so in the context of Maxwell's theory, we need only look at the way the Lagrangian formalism functioned in allowing Maxwell to provide a dynamical theory without any explanation of the physical causes that underlay the phenomena. The generality of the Lagrangian approach makes it applicable in a variety of contexts, and it is ultimately this feature that makes it especially suited to unifying different domains. But this generality has a drawback: By not providing an account of the way physical processes take place, the unifying power is achieved at the expense of explanatory power.

Despite the importance of mathematical structures for unification, I want to claim that the mark of a truly unified theory is the presence of a specific mechanism or theoretical quantity/parameter that is not present in a simple conjunction, a parameter that represents the theory's ability to reduce, identify or synthesize two or more processes within the confines of a single theoretical framework. In Maxwell's theory the displacement current plays just such a role. It figures prominently as a fundamental quantity in the field equations, and without it there could be no notion of a quantity of electricity crossing a boundary and hence no field-theoretic basis for electromagnetism. Because displacement was not given any *ontological* primacy, energy conservation functioned as the substantial embodiment of an otherwise abstract theory. "Field energy" was the only thing to be given a literal interpretation by Maxwell, because it was the one thing that he felt certain played a role in the production of electromagnetic phenomena.

Let me begin, then, with Maxwell's first work on electromagnetic theory and follow through to its culmination almost 30 years later in 1873 with publication of the *Treatise on Electricity and Magnetism.* The history reveals in striking terms just how the theory became unified, how its unity had little impact in furthering its acceptance and what this particular case of theory unification reveals to us about the philosophical problems surrounding unity in nature.

3.1. Development of Electromagnetic Theory: The Early Stages

3.1.1. *Maxwell circa 1856: "On Faraday's Lines of Force"*

The goal of Maxwell's 1856 paper "On Faraday's Lines of Force" was to present Faraday's account of the electromagnetic theory in mathematically precise yet visualizable form.[1] Having been strongly influenced by William Thomson's use of mathematical and physical analogies, particularly his use of molecular vortices as a representation of the field, Maxwell employed a similar method as a means of illustrating Faraday's conception of lines of force.[2] He felt that Faraday's "lines" could be used as a purely geometrical representation of the structure of the field, with the directions of the forces acting in the field represented by lines of force filling space. In order to employ these lines as a quantitative expression of the forces themselves, Maxwell conceived of them as tubes carrying incompressible fluid made up of "a collection of imaginary properties" (Maxwell 1965, vol. 1, p. 157). That geometrical model defined the motion of the fluid by dividing the space it occupied into tubes, with forces being represented by the fluid's motion. Using the formal equivalence between the equations for heat flow and action at a distance, Maxwell substituted the flow of the ideal fluid for the distant action. Although the pressure in the tubes containing the fluid varied inversely as the distance from the source, the crucial feature of the model was that the action was *not* at a distance; the energy of the system was in the tubes of force.

The origin of that approach was Thomson's representation of the analogy between heat flow and electrostatic action (Thomson 1872). In keeping with Thomson's mitigated scepticism about the status of those models, Maxwell remained cautious, warning that "the two subjects will assume very different aspects" if their resemblance is pushed too far (Maxwell 1965, vol. 1, p. 157). Nevertheless, the mathematical resemblance of the *laws* was something that could be considered useful in the development of further mathematical ideas (1965, vol. 1, p. 157). The importance of Maxwell's analogy was that it showed that one could look at electrical action and magnetic action from two different yet mathematically equivalent points of view.[3] But like his colleagues Faraday and Thomson, Maxwell emphasized the gap that existed between physical theory and mathematical representation.

Maxwell went on to provide an illustration of the phenomena of electrostatics, current electricity and magnetism by drawing analogies between them and the motions of the incompressible fluid. At no time, however, did he intend the account to be anything more than an analogy to be used for heuristic purposes. In fact, the incompressible fluid was not even considered a *hypothetical* entity – it was purely fictional. Consequently, the geometrical model was not put forth as a physical hypothesis; instead, the emphasis was on mathematical rather than physical similarity, a resemblance between mathematical relations rather than phenomena or things related.[4] The advantage of the method of analogy over a purely analytical formalism was the visual representation provided by the lines, surfaces and tubes. Although Maxwell referred to his method as one that presented physical analogies, he did so because he considered it a method of obtaining physical *ideas* without adopting a physical *theory*.

By a physical analogy I mean that partial similarity between the laws of one science and those of another which makes each of them illustrate the other. Thus all the mathematical sciences are founded on relations between physical laws and laws of numbers, so that the aim of exact science is to reduce the problem of nature to the determination of quantities by operations with numbers. (Maxwell 1965, vol. 1, p. 156)

The kind of emphasis Maxwell placed on analogy was indicative of his commitment to a strong distinction between a mathematical theory and a physical theory. In an early draft of the 1856 paper,[5] Maxwell remarked that he had assumed a purely imaginary fluid because "while the mathematical laws of heat conduction derived from the idea of heat as a substance are admitted to be true, the theory of heat has been so modified that we can no longer apply to it the idea of a substance".[6] That the purpose of the analogies was purely heuristic is evident from other, more specific remarks:

By referring everything to the purely geometrical idea of the motion of an imaginary fluid, I hope to attain generality and precision, and to avoid the dangers arising from a premature theory professing to explain the cause of the phenomena. If the results of mere speculation which I have collected are found to be of any use to experimental philosophers in *arranging* and *interpreting* their results, they will have served their purpose, and a mature theory, in which physical facts will be physically explained, will be formed by those who, by interrogating nature herself, can obtain the only true solution of the questions which the mathematical theory suggests. (Maxwell 1965, vol. 1, p. 159; italics added)[7]

In essence, then, the analogy served as a means for applying specific techniques to a variety of phenomena, while at the same time carving out a middle ground between mere mathematical abstraction and full-blown commitment to a physical hypothesis. According to Maxwell, the danger of adopting a particular hypothesis lies in seeing the phenomena only through a medium; we become "liable to that blindness to facts and rashness in assumptions which a partial explanation encourages" (Maxwell 1965, vol. 1, pp. 155–6). The method of analogy provides a means of investigation that facilitates a clear physical conception of the phenomena without being carried beyond the truth because of a prior commitment to a particular hypothesis (1965, vol. 1, pp. 155–6). Instead of providing a literal interpretation of possible states of affairs, analogies furnish guidelines for constructing and developing theories, as well as suggesting possible approaches for experimentation. They supply the mechanisms for what has traditionally been termed the "context of discovery". In contrast to the method of physical analogy, a *purely* mathematical approach causes us to "lose sight of the phenomena themselves", and its lack of heuristic power prevents possible extensions of the "views on the connections of the subject" (1965, vol. 1, pp. 155–6).

But this is not to downplay the importance of mathematization of the phenomena and the benefits it has for expressing relations among them. The second part of "On Faraday's Lines" consisted of a mathematical formulation of Faraday's notion of the "electrotonic state". That concept was used to represent the electrical tension of matter that was associated with Faraday's early theory of particulate polarization that eventually gave way to the idea of lines of force. Maxwell introduced a mathematical expression (later called vector potential) that enabled one to conceive of the electrotonic state as a quantity determinate in magnitude and direction (1965, vol. 1, p. 205). Again Maxwell was quick to point out that his representation involved no physical theory; it was what he called an artificial notation.

However, he used the electrotonic state because it reduced the attraction of currents and electrified bodies to one principle without introducing any new assumptions.[8] Although the nature of the electrotonic state was at various times represented by a variety of physical entities, its essence was its mathematical form. As Maxwell pointed out,

the idea of the electrotonic state has not yet presented itself to my mind in such a form that its nature and properties can be clearly explained without reference to mere symbols. . . . By a careful study of the laws of elastic solids and the motions of viscous fluids I hope to discover a method of forming a mathematical conception of this electrotonic state adapted to general reasoning. (Maxwell 1965, vol. 1, p. 188)

In keeping with his cautious attitude toward physical hypotheses, Maxwell saw the emphasis on mathematical formulations as providing a distinct advantage over accounts that presupposed the *truth* of specific theoretical assumptions. For example, in the development of Weber's action-at-a-distance account of electromagnetism it was assumed that the attraction or repulsion of moving electrical particles was dependent on their velocities. However, as Maxwell pointed out, "if the forces in nature are to be reduced to forces acting between particles, the principle of conservation of force requires that these forces should be in the lines joining the particles and functions of the distance only" (1965, vol. 1, p. 208). That suggested that Ampere's theory of two separate electrical fluids, upon which Weber's account had been based, needed to be abandoned.[9] The mathematical account provided by Maxwell suggested an alternative conception insofar as it provided the formal apparatus necessary for the beginnings of a different and more promising approach.

Unfortunately, there was reluctance to pursue Maxwell's programme, for several reasons. Not only was there no physical/mechanical basis for the flow analogy, but also it failed to provide an image of the coexistence and interactions of electrical fields, magnetic fields, and electric currents. The flow lines of the incompressible fluid were taken to correspond to electric or magnetic lines of force, or lines of electric current, depending on the context of the problem. As a result, the analogy provided a fragmented understanding of the three electromagnetic phenomena by considering each one in isolation from the others (Seigel 1985). Thomson's theory of molecular vortices offered a possible solution to this problem by promising to provide the kind of comprehensive explanatory power that the flow analogy lacked. And unlike the imaginary fluid of the flow analogy, which was simply an illustrative device, the vortex account of the connections and interactions of the various phenomena could at least be considered a genuine physical *possibility*. Although the explanatory power of the vortex hypothesis was an important consideration, it was not seen as providing final evidence for the claim that electromagnetic phenomena were actually caused by vortical motions in the medium. As Maxwell remarked,

if, by the molecular-vortex hypothesis we can connect the phenomena of magnetic attraction with electromagnetic phenomena and with those of induced currents, we shall have found a theory which, if not true, can only be proved to be erroneous by experiments which will greatly enlarge our knowledge of this part of physics. (Maxwell 1965, vol. 1, p. 452)

So, using Thomson's molecular-vortex model, Maxwell went on to develop an account that took him beyond the flow analogy to a mechanical model of electromagnetic phenomena.

3.1.2. "On Physical Lines of Force"

Faraday's distinction between the geometrical treatment of lines of force (a descriptive account of their distribution in space) and a physical treatment (dealing with their dynamical tendencies) was carried over quite distinctively into Maxwell's work. In his first paper, "On Faraday's Lines of Force", there was no mechanism for understanding the forces of attraction and repulsion between magnetic poles. In contrast, the 1861–62 paper, "On Physical Lines of Force", was an attempt to "examine" the phenomena from a mechanical point of view and to determine what tensions in, or motions of, a medium were capable of producing the observed mechanical phenomena (Maxwell 1965, vol. 1, p. 467). One needed to ascertain the physical behaviour of the magnetic lines in order to account for magnetic forces.[10]

In Thomson's 1856 paper, Faraday's discovery of the rotation of the plane of polarization of light by magnets (the Faraday effect) was explained by a theory that construed magnetism as the rotation of molecular vortices in a fluid aether. That effect led Maxwell to the view that in a magnetic field the medium (or aether) was in rotation around the lines of force, the rotation being performed by molecular vortices whose axes were parallel to the lines. The fact that magnetic poles attracted or repelled each other led Maxwell (and Faraday) to the view that the lines of force extending from these poles represented a state of tension in the medium. The concept of polarity was then modified to include the vortices, with the polarity of the lines of force represented by the polarity of the vortices that constituted them. Maxwell went on to derive a general expression for the components of the resultant force on a unit volume of the electromagnetic medium subject to the vortex motion:

$$X = \alpha m + \frac{1}{8\pi}\mu\frac{d}{dx}(v^2) - \mu\beta r + \mu\gamma q - \frac{dp_1}{dx}$$

$$Y = \beta m + \frac{1}{8\pi}\mu\frac{d}{dy}(v^2) - \mu\gamma p + \mu\alpha r - \frac{dp_1}{dy} \qquad (3.1)$$

$$Z = \gamma m + \frac{1}{8\pi}\mu\frac{d}{dz}(v^2) - \mu\alpha q + \mu\beta p - \frac{dp_1}{dz}$$

The first term represents the force acting on magnetic poles, the second represents the inductive action on bodies capable of magnetism, the third and fourth represent the force acting on electric currents and the fifth represents the effects of simple pressure. X, Y and Z are simply the forces acting in the x, y and z directions; μ is a measure of the density of the fluid medium, and p_1 is the pressure at the circumference of the vortex; α, β and γ are defined as vl, vm and vn, respectively, where v is the velocity at the circumference of a vortex, and l, m and n are the direction cosines of the axes of the vortices with respect to x, y and z. Defined electromagnetically, μ is the magnetic inductive capacity of the medium, and α, β and γ are the components of magnetic force, with $\mu\alpha$, $\mu\beta$ and $\mu\gamma$ representing the quantity of magnetic induction; p, q and r represent the quantities of electric current per unit area flowing in the x, y and z directions. The lines of force indicate the direction of minimum pressure at every point in the medium, with the vortices considered as the mechanical cause of the differences in pressure in different directions. The velocity at the circumference of each vortex is proportional to the intensity of the magnetic force, and the density of the substance of the vortex is proportional to the capacity of the medium for magnetic induction (Maxwell 1965, vol. 1, p. 467).

Part 2 of "On Physical Lines" was an attempt to specify the forces that caused the medium to move in the way described by the model and to account for the occurrence of electric currents. In order to do that, Maxwell needed to provide an explanation of the transmission of rotation in the same direction from vortex to vortex. Because of the difficulties involved in specifying a mechanical model of vortex rotation, Maxwell was forced to introduce a layer of spherical particles residing on the surfaces of the vortices, thereby separating them from one another. The problem was roughly the following: Because neighbouring lines of force in a magnetic field point in approximately the same direction, adjacent vortices must rotate in the same way. Hence, the contiguous portions of consecutive vortices must be moving in opposite directions. By introducing a layer of particles that divided the medium into cells, one could account for the motion of the vortices by means of the rotation of the material cells. The particles separating the cells could then be seen to rotate in the direction opposite to that of the vortices they separated. Each particle revolved on its own axis and in the direction opposite to that of the neighbouring vortices. Because the particles were placed *between* contiguous vortices, the rotation of each vortex would cause the neighbouring vortices to revolve in the same direction.

These particles, known as idle wheels, rolled without slipping on the surfaces of the vortices. If adjacent vortices were not revolving at the same rate, the idle-wheel particles would acquire a translatory motion. Electric current was represented as the transference of rotary velocity of the moveable particles. In other words, the motions of the particles transmitted from vortex to vortex through the medium constituted a current. The tangential action (resulting from the motion) of the particles on the cells caused the vortices to rotate and so accounted for magnetic fields due to currents as well as electromotive force.

Maxwell next deduced a set of equations from which he could determine the relationship between the alterations of motion of the vortices ($d\alpha/dt$, etc.) and the force exerted on the layers of particles between them (or, as he says, "in the language of our hypothesis, the relation between changes in the state of the magnetic field and the electromotive forces") (Maxwell 1965, vol. 1, p. 475):

$$\frac{dQ}{dz} - \frac{dR}{dy} = \mu \frac{d\alpha}{dt}$$

$$\frac{dR}{dx} - \frac{dP}{dz} = \mu \frac{d\beta}{dt} \qquad (3.2)$$

$$\frac{dP}{dy} - \frac{dQ}{dx} = \mu \frac{d\gamma}{dt}$$

P, Q and R are the components of electromotive force exerted on the particles, α, β and γ represent the circumferential velocity of a vortex and μ is the density of the vortex medium. Because Maxwell had identified μ with the quantity of magnetic induction, that set of equations related the change in the state of the magnetic field to the electromotive force that was produced. Quite simply, P, Q and R could be calculated as the rates at which F, G and H (the resolved parts of the electrotonic state) varied, resulting in $P = dF/dt$, $Q = dG/dt$ and $R = dH/dt$. If the angular velocity of each vortex remained constant, the particles separating the vortices experienced no net force pushing them in any specific direction. However, a change in form of a part of the medium produced a change in the velocity of the vortices, and when a change in the state of rotation of the vortices was transmitted through the medium, a force resulted, thereby enabling the model to account for electromagnetic induction.[11] By generalizing to a medium subject to all varieties of motion, Maxwell deduced the following equations, which are the expressions for the variations of α, β and γ:

$$P = \mu\gamma \frac{dy}{dt} - \mu\beta \frac{dz}{dt} + \frac{dF}{dt} - \frac{d\Psi}{dx}$$

$$Q = \mu\alpha \frac{dz}{dt} - \mu\gamma \frac{dx}{dt} + \frac{dG}{dt} - \frac{d\Psi}{dy} \qquad (3.3)$$

$$R = \mu\beta \frac{dx}{dt} - \mu\alpha \frac{dy}{dt} + \frac{dH}{dt} - \frac{d\Psi}{dz}$$

The first and second terms in each indicate the effect of the motion of any body in the magnetic field, the third represents changes in the electrotonic state produced by alterations in position or intensity of magnets or currents in the field and Ψ is a function of x, y, z and t. The physical interpretation of Ψ is the electric tension at each point in space.

Maxwell provided the following explanation for the production of lines of force by an electric current (Figure 3.1): AB is a current of electricity, with the large spaces representing the vortices, and the smaller circles the idle wheels. The row of vortices gh will be set in motion in a counterclockwise ($+$) direction. The layer of particles pq will be acted upon by gh, causing them to move in a clockwise ($-$)

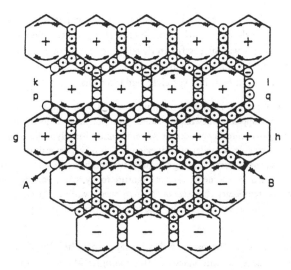

Figure 3.1. Maxwell's vortex aether model.

direction from right to left (or in the direction opposite from that of the current), thereby forming an induced electric current. If this current is checked by the electrical resistance of the medium, the rotating particles will act on the kl row of vortices, causing them also to revolve in the (+) direction. This movement continues until the vortices reach a velocity such that the motion of the particles is reduced to simple rotation, resulting in disappearance of the induced current.

Maxwell's entire model was based on a mechanical analogy with engineering. The idle wheels were those mechanisms that either rotated around fixed angles or (as was the case in certain kinds of trains) had centres whose motions were half the sum of the motions of the circumferences of the wheels between them. In Maxwell's model, the vortices played the part of the wheels. We know that the electromotive force arises from the action between the vortices and the interposed particles when the velocity of rotation is changed in any part of the field. This corresponds to the pressure on the axle of a wheel in a machine when the velocity of the driving wheel is increased or diminished. The electrotonic state is what the electromotive force would be if the currents that caused the lines of force had started instantaneously instead of gradually building up to their actual state. This corresponds to the impulse that would act on the axle of a wheel in a machine previously at rest but with the actual velocity now suddenly given to the driving wheel (Maxwell 1965, vol. 1, pp. 468–9, 478).

The basic difference between the position taken by Maxwell in "On Faraday's Lines" and that articulated in "On Physical Lines" was that in the latter the electrotonic state had come to be defined in terms of the motion of the vortices, which in turn determined the nature of the lines of force. By explaining the electromotive force in terms of the forces exerted by the vortices on the particles between them, Maxwell was able to define it as the time rate of change of the electrotonic state.

Although Maxwell was successful in developing the mathematics required for his mechanical model, he was insistent that it be considered provisional and temporary in nature. In fact, he explicitly remarked that

the conception of a particle having its motion connected with that of a vortex by perfect rolling contact may appear somewhat awkward. I do not bring it forward as a mode of connexion existing in nature, or even as that which I would willingly assent to as an electrical hypothesis. It is, however, a model of connexion which is mechanically conceivable, and easily investigated, and it serves to bring out the actual mechanical connexions between the known electromagnetic phenomena; so that I venture to say that anyone who understands the provisional and temporary character of this hypothesis, will find himself rather helped than hindered by it in his search after the true interpretation of the phenomena. (Maxwell 1965, vol. 1, p. 486)

It would seem, then, on the basis of this passage, that the idle-wheel hypothesis had much the same epistemic status as the analogies Maxwell had used in "On Faraday's Lines". Although it solved the mechanical problem raised in Part 1 of "On Physical Lines", its chief value seems to have been heuristic, serving as a way of *conceiving* the phenomena that perhaps would suggest further development of the model – something that the flow analogy could not do. It is questionable that Maxwell's reluctance to see the idle-wheel hypothesis as anything more than heuristic can be extended to the vortex hypothesis in general.[12] In the introduction to Part 2 of "On Physical Lines", Maxwell remarked that the hypothesis of molecular vortices gave a "probable" answer to the question of the mechanical cause of the differences in pressure in the medium in different directions. That answer was to be distinguished from the answers to the questions of how the vortices were set in rotation and why they were arranged according to the laws governing the lines of force:

These questions are certainly of a higher order difficulty than . . . the former; and I wish to separate the suggestions I may offer by way of a provisional answer to them from the . . . hypothesis of vortices. (Maxwell 1965, vol. 1, p. 468)

It appears, then, that even if Maxwell considered the vortices themselves as more than simply heuristic devices, he was severely constrained as to how their *properties* (specifically, rotation) were to be accounted for in a legitimate and coherent model. That difficulty was particularly evident in a remark at the end of Part 2 of "On Physical Lines":

We have now shown in what way the electromagnetic phenomena may be imitated by an imaginary system of molecular vortices. . . . [We] find here the conditions which must be fulfilled in order to give [the hypothesis] mathematical coherence, and a comparison, so far satisfactory, between its necessary results and known facts. (Maxwell 1965, vol. 1, p. 488)

At that point Maxwell had little faith in his theory as a serious alternative to Weber's action-at-a-distance programme. He had been unable to extend his model to electrostatics, and consequently it lacked the comprehensiveness of Weber's model.[13] However, that difficulty was quickly overcome when, after a period of nine months, Maxwell published Parts 3 and 4 of "On Physical Lines", which contained an account of electrostatics that proposed a new model of the aether as well as the first derivation of his theory of light.

In Part 2 of "On Physical Lines" the magnetoelectric medium was presented as a cellular structure, with each cell consisting of a molecular vortex (a rotating parcel of fluid) surrounded by a cell wall consisting of a layer of small spherical particles that functioned as idle wheels. However, in order to explain "charge" and to derive the law of attraction between charged bodies, he proposed a slightly different model of the aether. Instead of the hydrodynamic model, he resorted to an elastic solid model in which the aetherial substance formed spherical cells endowed with elasticity. The cells were separated by electric particles whose action on the cells would result in a kind of distortion. Hence, the effect of an electromotive force was to distort the cells by a change in the positions of the electric particles. That gave rise to an elastic force that set off a chain reaction. Maxwell saw the distortion of the cells as a displacement of electricity within each molecule, with the total effect over the entire medium producing a "general displacement of electricity in a given direction" (Maxwell 1965, vol. 1, p. 491). Understood literally, the notion of displacement meant that the elements of the dielectric had changed positions.

Because changes in displacement involved a motion of electricity, Maxwell argued that they should be "*treated as*" currents (1965, vol. 1, p. 491; italics added). Strictly speaking, displacement could not be equated with a current (when it attained a certain value, it remained constant), but it could be seen as the "commencement of a current" (Maxwell 1965, vol. 1, p. 491), with its *variations* constituting currents in the positive or negative direction according to whether the displacement was increasing or diminishing. Displacement also served as a model for dielectric polarization; electromotive force was responsible for distorting the cells, and its action on the dielectric produced a state of polarization.

Maxwell described his model in the following way: The electromagnetic medium is divided into cells separated by partitions formed of a stratum of particles that play the part of electricity. When the electric particles are urged in any direction, they will, by their tangential action on the elastic substance of the cells, distort each cell and call into play an equal and opposite force arising from the elasticity of the cells. When the force is removed, the cells will recover their form, and the electricity will return to its former position (Maxwell 1965, vol. 1, p. 492). The amount of displacement will depend on the nature of the body and on the electromotive force. Hence, if h is the displacement, and R the electromotive force, and E a coefficient depending on the nature of the dielectric, then displacement will satisfy the phenomenological law $R = -4\pi E^2 h$. In fact, E can be interpreted as the coefficient of rigidity, because the theory of elasticity yields an equation relating force and strain in terms of the elastic parameter. So if h is the displacement or the sphere's distortion, and if r is the value of the electric current due to displacement, then $r = db/dt$. In calculating b, we integrate over the surface of a single spherical cell (∂S) of volume V:

$$b = \int_\partial \partial S \frac{1}{2} \rho t \sin \theta / V \qquad (3.4)$$

Not only is displacement introduced in terms of a model for polarization, but also it is inserted into an experimental equation expressing the relationship between polarization and force without requiring any *theory* about the internal mechanism of dielectrics. In addition, it receives a mathematical definition that establishes it as an electric dipole moment per unit volume.[14]

As Joan Bromberg (1967, 1968) has pointed out, whenever Maxwell introduces the displacement current, he derives it from the idea of displacement simpliciter, with no attempt to justify it further. Maxwell introduced the equation for displacement as an empirical relation, claiming that it was dependent on observations of the properties of dielectrics subjected to electric forces. Given that relation, he went on to point out that it was independent of any theory about the internal mechanisms of dielectrics, and

when we find electromotive force producing electric displacement in a dielectric; and when we find the dielectric recovering from its state of electric displacement with an equal electro-motive force, we cannot help regarding the phenomena as those of an elastic body, yielding to pressure and recovering its form when the pressure is removed. (Maxwell 1965, vol. 1, p. 492)

Because the phenomenological law governing displacement expressed the relation between polarization and force, Maxwell was able to use it to calculate the aether's elasticity (the coefficient of rigidity), the crucial step that led him to identify the electromagnetic and luminiferous aethers. So, given the assumption of elasticity of the electromagnetic medium, together with the equation for displacement, Maxwell's model turned out to have a great deal of unifying power.

It is interesting to note that in Parts 1 and 2 of "On Physical Lines" there is no mention of the optical aether. However, once the electromagnetic medium was endowed with elasticity, Maxwell relied on the optical aether in support of his assumption about elasticity:

The undulatory theory of light requires us to admit this kind of elasticity in the luminiferous medium in order to account for transverse vibrations. We need not then be surprised if the magneto-electric medium possesses the same property. (1965, vol. 1, p. 489)

After a series of mathematical steps, Maxwell had the necessary tools for calculating the velocity with which transverse waves were propagated through the electromagnetic aether.[15] First of all, it was necessary to correct the equations of electric currents for the effect produced by the elasticity of the medium. By differentiating the equation of displacement with respect to t, we get

$$dR/dt = -4\pi E^2 (db/dt) \qquad (3.5)$$

which shows that when the electromotive force varies, so does electric displacement. Because this variation is equivalent to a current, it must be added to r in the equations describing the quantity of electric current. Hence, if

$$\frac{1}{4\pi}\left(\frac{d\beta}{dx} - \frac{d\alpha}{dy}\right) \tag{3.6}$$

represents the strength of an electric current parallel to z through a unit of area, and if p, q and r represent the quantities of electric current per unit of area perpendicular to the axes x, y and z, then

$$p = \frac{1}{4\pi}\left(\frac{d\gamma}{dy} - \frac{d\beta}{dz}\right)$$

$$q = \frac{1}{4\pi}\left(\frac{d\alpha}{dz} - \frac{d\gamma}{dx}\right) \tag{3.7}$$

$$r = \frac{1}{4\pi}\left(\frac{d\beta}{dx} - \frac{d\alpha}{dy}\right)$$

become

$$p = \frac{1}{4\pi}\left(\frac{d\gamma}{dy} - \frac{d\beta}{dz} - \frac{1}{E^2}\frac{dP}{dt}\right)$$

$$q = \frac{1}{4\pi}\left(\frac{d\alpha}{dz} - \frac{d\gamma}{dx} - \frac{1}{E^2}\frac{dQ}{dt}\right)$$

$$r = \frac{1}{4\pi}\left(\frac{d\beta}{dx} - \frac{d\alpha}{dy} - \frac{1}{E^2}\frac{dR}{dt}\right) \tag{3.8}$$

where α, β and γ are the components of magnetic intensity, and P, Q and R are the electromotive forces. From these equations Maxwell was able to determine e, the quantity of free electricity in a unit volume:

$$e = \frac{1}{4\pi E^2}\left(\frac{dP}{dx} + \frac{dQ}{dy} + \frac{dR}{dz}\right) \tag{3.9}$$

That, in turn, enabled him to arrive at a value for E, the dielectric constant, and then to calculate the force acting between two electrified bodies. Finally, he determined the rate of propagation of transverse vibrations through the elastic medium, on the assumption that the elasticity was due to forces acting between pairs of particles. First Maxwell used E to derive Coulomb's law:

$$F = -E^2(e_1 e_2/r^2) \tag{3.10}$$

But if one takes the same law in electrostatic units instead of electromagnetic units, it becomes clear that E is simply the ratio of electrostatic to electromagnetic units. Using the formula $V = \sqrt{m/\rho}$, where m is the coefficient of rigidity, ρ is the aethereal mass density and μ is the coefficient of magnetic induction, we have

$$E^2 = \pi m \qquad (3.11)$$

$$\mu = \pi \rho \qquad (3.12)$$

giving us

$$\pi m = V^2 \mu \qquad (3.13)$$

hence

$$E = V \sqrt{\mu} \qquad (3.14)$$

Maxwell arrived at a value for V that, much to his astonishment, agreed with the value calculated for the velocity of light ($V = 310,740,000,000$ mm/sec):[16]

The velocity of transverse undulations in our *hypothetical* [italics added] medium, calculated from the electro-magnetic experiments of Kohlrausch and Weber, agrees so exactly with the velocity of light calculated from the optical experiments of M. Fizeau that we can *scarcely avoid the inference that light consists in the transverse undulations of the same medium which is the cause of electric and magnetic phenomena.* (1965, vol. 1, p. 500)

At that point in the theory's development the equations for the electromagnetic field had roughly the following form (using modern vector notation). If we translate Maxwell's own symbols into vector form, $\mathbf{F}$ corresponds to (F, G, H) for electromagnetic momentum, $\boldsymbol{\alpha}$ to (α, β, γ) for magnetic intensity, $\mathbf{P}$ to (P, Q, R) for electromotive force, $\mathbf{f}$ to (f, g, h) for electric displacement and $\mathbf{p}$ to (p, q, r) for conduction current, with e the quantity of electricity and ψ the vector potential:

$$\mu \boldsymbol{\alpha} = \boldsymbol{\nabla} \times \mathbf{F} \qquad\qquad \mathbf{B} = \boldsymbol{\nabla} \times \mathbf{A}$$

$$\left[k_1 = \frac{1}{\mu} \right] \qquad\qquad \mu \mathbf{H} = \mathbf{B}$$

$$\mathbf{p} = \frac{1}{4\pi} \boldsymbol{\nabla} \times \boldsymbol{\alpha} + \frac{1}{4\pi \epsilon^2} \frac{d\mathbf{P}}{dt} \qquad\qquad \frac{1}{4\pi} \boldsymbol{\nabla} \times \mathbf{H} = \mathbf{J}$$

$$\sigma \mathbf{P} = \mathbf{p} \qquad\qquad\qquad \sigma \mathbf{E} = \mathbf{J}$$

$$\mathbf{P} = \left(\frac{d\mathbf{x}}{dt} \times \mu \boldsymbol{\alpha} \right) - \frac{d\mathbf{F}}{dt} - \boldsymbol{\nabla} \psi \qquad \boldsymbol{\nabla} \times \mathbf{E} = -\frac{d\mathbf{B}}{dt}$$

which yields

$$\nabla \times \mathbf{P} = -\frac{d(\mu\alpha)}{dt} \qquad\qquad \nabla \cdot \mathbf{B} = 4\pi m \quad \text{or} \quad 0$$
$$\nabla \cdot (\mu\alpha) = 4\pi m \quad \text{or} \quad 0 \qquad \nabla \cdot \mathbf{D} = 4\pi e \quad \text{or} \quad 0$$
$$\nabla \cdot \mathbf{p} = 4\pi \epsilon^2 e \quad \text{or} \quad 0 \qquad\quad \mathbf{E} = 4\pi \epsilon^2 \mathbf{D}$$

In modern notation (as presented on the right), F corresponds to the function A, $\mu\alpha$ to B (the magnetic-flux density), α to H (the magnetic-field strength), p to $\mathbf{J} + \partial\mathbf{D}/\partial t$ (where J is current density, and D is electric displacement), P to E (the electric-field strength) and f to D.

We can see, then, the importance of displacement as a fundamental quantity in the field equations. Maxwell links the equation describing it ($R = -4\pi E^2 h$) with the aether's elasticity (modelled on Hooke's law), where displacement produces a restoring force in response to the distortion of the cells of the medium, but $R = -4\pi E^2 h$ is also an electrical equation representing the flow of charge produced by electromotive force. Consequently, the dielectric constant E is both an elastic coefficient and an electric constant. Interpreting E in this way allows Maxwell to determine its value and ultimately identify it with the velocity of transverse waves travelling through an elastic medium or aether. Given the importance of displacement for producing a field-theoretic account of electromagnetism and its role in calculating the velocity of waves, one can see why it would be considered as the essential parameter or theoretical concept/quantity that functions as the unifying mechanism in identifying the optical and electromagnetic aethers. In the later stages of the theory the aether was abandoned, but displacement remained as a fundamental quantity necessary for the unity of electromagnetism and optics via the equations of the theory. But as we shall see, its status changed once it was incorporated into the Lagrangian formulation of the theory.

That the unification of optics and electromagnetism was a complete surprise to Maxwell is evident from his letter to Faraday, October 19, 1861, in which he claims that he "worked out the formulae in the country before seeing Weber's number . . . and I think we have now strong reason to believe, whether my theory is a fact or not that the luminiferous and electromagnetic medium are one".[17] That, of course, did not imply that Maxwell believed in the existence of the aether, but only that there was no reason to assume, should an aether exist, that light and electromagnetic phenomena would have different aetherial sources. Because he had set out to calculate *only* the velocity of propagation of electromagnetic waves, and given the way in which he arrived at the result, his unification of the electromagnetic and luminiferous aethers displayed all the characteristics of what Whewell would call a truly consilient theory, one that unified two diverse domains of evidence by showing how they were the results of a common mechanism or structure.

In this case it was the aether model incorporating a displacement current that facilitated that first formulation of the electromagnetic theory. The necessary calculations resulted from using electromagnetic laws together with the laws of mechanics that governed the model. What Maxwell had in fact shown was that given the specific assumptions employed in developing the mechanical details of his model, the elastic properties of the electromagnetic medium were just those required of the luminiferous aether by the wave theory of light (Bromberg 1967). Hence, what was effected was the reduction of electromagnetism and optics to the mechanics of *one aether*, rather than a reduction of optics to electromagnetism simpliciter, what Bromberg refers to as the "electromechanical" theory of light.

In that sense, the first form of Maxwell's theory displayed a reductive unity, but the more interesting question is whether or not, in the absence of the aether, the identification of electromagnetic and optical waves still constitutes a reduction of two different processes to a single natural kind. Alternatively, perhaps the theory is better characterized as a synthetic unity, in which two processes remain distinct, but are simply governed by the same laws. The issue is important not only for differentiating different kinds of unity but also for providing a basis for moving from claims about theoretical unity to ontological claims about unity in nature. Clearly, the impulse to argue for an ontological basis for reductive unity is much greater than for synthetic unity. In other words, if two processes can be reduced to one, then it is tempting to argue that the unifying or reducing theory represents a unity in nature. What I want to suggest is that the nature of the Lagrangian formalism prohibits any move from what appears to be a reductive unity at the theoretical level to an accompanying ontological unity. The reason is simply that one can achieve a level of theoretical unity using a Lagrangian approach *exactly because* one need not take account of the underlying causes that produce the phenomena. Unification becomes more difficult the more mechanisms and theoretical structures one has to accommodate.

Before going on to look at the later formulations of the theory, it is important to keep in mind that historically Maxwell's unification provided virtually no support for a realistic interpretation of the aether model, either by Maxwell himself or by his contemporaries. (Others clearly believed in an aether, but not Maxwell's aether.) What the theory lacked was some independent evidence not only for the existence of electromagnetic waves but also for displacement and other crucial components of the model employed in the derivation. As we shall see, the very structure that enabled Maxwell to achieve his spectacular unification was soon abandoned for a purely formal account devoid of commitment to any particular mechanical model. Some of the philosophical arguments we have looked at thus far (especially those of Friedman and Glymour) have suggested that unification functions as a criterion for theory acceptance and consequently underwrites a realistic construal of the entities that play roles in the unifying process. Not only does the early development of electrodynamics fail to conform to that account of realism and theory choice, but also later developments in the theory illustrate important mechanisms involved in unification that have been systematically overlooked in the philosophical literature,

specifically the significance of abstract mathematical structures in producing theoretical unification, in bringing together diverse phenomena under the constraints
of one theoretical framework.

3.2. Unification and Realism: Some Problems
for the Electromagnetic Theory

At the beginning of "On Physical Lines", Maxwell referred to his vortex aether
model as providing a mechanical account that would direct experiment and indicate features that a true theory would have to incorporate:

> My object in this paper is to clear the way for speculation in this direction, by investigating
> the mechanical results of certain strains of tension and motion in a medium, and comparing
> these with the observed phenomena of magnetism and electricity. By pointing out the me-
> chanical consequences of such hypotheses, I hope to be of some use to those who consider
> the phenomena as due to the action of a medium. (1965, vol. 1, p. 452)

One of the main difficulties was that there was no *independent* support for the vortex
hypothesis over and above its explanatory role in the model. Part 4 of "On Physical
Lines" was intended as a proof of the rotary character of magnetism, an argument
that if successful, would go some way toward supporting a realistic interpretation of
the vortices and their properties as described by the model. But it became evident in
Maxwell's later work that the general force of the argument remained independent
of any specific mechanical model of the medium. In other words, the details of the
theory of molecular vortices *were not supported* by the argument for the rotational
character of magnetism; the latter could be given independently of the specifics of
the theory of molecular vortices (Seigel 1985).

But the status of molecular vortices was not the only problem that Maxwell's
theory faced. Earlier I mentioned difficulties with the notion of electric displacement. Because displacement was criticized even by proponents of field theory like
Kelvin, we should look more closely at the nature of the problem. The idea that
the electrostatic state was a displacement of something from equilibrium was not
new. Thomson had compared electric force to the displacement in an elastic solid,
and Faraday had hypothesized that when the dielectric (which is made up of small
conductors) was subjected to an electrostatic field, there would be a displacement
of electric charge on each of the small conductors. An electric current could then
be defined as the motions of these charges when the field was varied. It was from
that account that Maxwell arrived at the idea that variations in displacement were
to be counted as currents. However, in using Faraday's idea, Maxwell completely
transformed it (Bromberg 1968). According to Faraday's notion, displacement was
applicable only to ponderable dielectrics and was introduced specifically to explain
why the inductive capacity of these dielectrics was different from that of free aether.
On Maxwell's account, displacement occurs *wherever* there is electric force, regardless of the presence of material bodies.

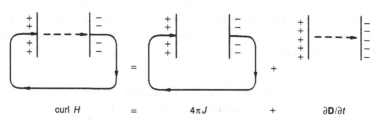

$$\text{curl } H \quad = \quad 4\pi J \quad + \quad \partial D/\partial t$$

Figure 3.2. The displacement term $\partial D/\partial t$ modified the original Ampere law.

In addition to that reinterpretation, the term specified as the displacement term was not a natural consequence of the mechanical model. As a result, Maxwell's analogy between electric displacement and elastic displacement could not adequately account for the electrostatic phenomena without additional assumptions. According to Maxwell's model, electricity was constituted by small particles, with charge linked to an accumulation of those particles.[18] The motion of idle wheels that represented electricity was governed by Ampere's law, which related electric flux and magnetic intensity (curl $H = 4\pi J$, where H is the magnetic field, and J is the electric-current density). A consequence of that law was that it failed to provide a mechanism for the accumulation of electric charge, because it applied only in the case of closed currents. As a result, a component $\partial D/\partial t$, where $D = (1/c^2)E + 4\pi P$ (the polarization vector), must be added to the current so that it is no longer circuital. The dielectric between the coatings of a condenser fulfilled that need and was seen as the origin of the process known as the displacement current[19] (Figure 3.2). The force of that current was proportional to the rate of increase of the electric force in the dielectric and therefore produced the same magnetic effects as a true current. Hence the charged current could be *regarded* as flowing in a closed circuit. The major difficulty with the account was that there were no experimental data that required the introduction of the displacement current. Because the term had the value zero for the case of steady currents flowing in closed circuits, Ampere's law, which gives the relationship between an electric current and the corresponding magnetic field, remained sufficient for dealing with the available data. However, for the case of open circuits the term took non-zero values and thus gave definite predictions for the magnetic effects of open circuits through a modification of the mathematical structure of the law.[20] But again the problem was that there were no available data against which to test the predictions provided by the alteration to Ampere's law. In that sense the introduction of displacement was a purely theoretical move devoid of any experimental justification (even though Maxwell claimed that its postulation was based on empirical considerations).[21]

However, it was not merely the *postulation* of a displacement current and its physical presuppositions that proved to be a source of difficulty.[22] The mechanical model that Maxwell used forced a dual interpretation of the law describing displacement. In his derivation of the wave equation the relation between the electromotive force and displacement ($R = -4\pi E^2 h$) functioned in two ways, as an equation of elasticity and as an electrical equation, a situation that resulted in R being interpreted as an electromotive force in the direction of displacement as well as an elastic restoring force in the opposite direction. Similarly, E was considered an electric constant as well as an elastic coefficient, and h was interpreted as a charge per unit area and a linear displacement. In virtue of its double interpretation, the equation served as a bridge between the electrical and mechanical parts of the model. The electrical part included the introduction of the displacement *current*, the calculation of a numerical value for E (a coefficient that depended on the nature of the dielectric) and the derivation of equations describing the quantity and displacement of an electric current per unit area. The mechanical part required that E represent an elastic force capable of altering the structure of the aether. It was not until the mechanical part of the model (the derivation of $R = -4\pi E^2 h$ and the calculation of the wave velocity from $V = \sqrt{m/\rho}$) was expressed in purely dynamical terms (in "A Dynamical Theory of the Electromagnetic Field" and the *Treatise on Electricity and Magnetism*), thereby eliminating the need for a bridge between the mechanical and electrical parts of the model, that these ambiguities were resolved.

We can see, then, that the substantial or material aspects of the unification suffered from fundamental theoretical difficulties. Although the internal mechanisms of the model provided a way of reducing electromagnetism and optics to waves in an aether, the assumptions that made it possible were inherently problematic over and above the obvious lack of experimental evidence. In that sense it would have been almost foolhardy to see the unification, despite its remarkable but limited success, as supporting any kind of ontological commitment. Simply formulating a model whose product was a unified account of electromagnetism and optics was not sufficient for realism about the model itself, the particular assumptions that facilitated the unification, nor its predictions about the existence of electromagnetic waves. In addition to resolving interpretive ambiguities, independent evidence for various aspects of the model was crucial for considering it theoretically viable.

3.3. The Electromagnetic Theory: Later Developments

3.3.1. The Dynamical Theory

In addition to the problems mentioned earlier, including the awkward and provisional nature of the idle-wheel hypothesis, the idea of electrical particles or substances in field theory was not looked upon favourably by Maxwell nor Faraday nor others.[23] Hence, the goal was to "clear the electromagnetic theory of light from any unwarrantable assumptions, so that we may safely determine the velocity of light by measuring the attraction between bodies kept at a given difference of potential, the value of which is known in electromagnetic measure" (Campbell and Garnett 1969, p. 340).

Maxwell intended to connect the experiments on the ratio of units with measurements of the velocity of light in a direct way that would avoid unnecessary hypothetical elements. In 1864 he wrote to Stokes, claiming that he had

materials for calculating the velocity of transmission of a magnetic disturbance through air founded on experimental evidence, without any hypothesis about the structure of the medium or any mechanical explanation of electricity or magnetism. (Larmour 1907, vol. 2, p. 26)

Maxwell's new theoretical framework in "A Dynamical Theory" (1865) was based on experimental facts and a few general dynamical principles about matter in motion, as characterized by the abstract dynamics of Lagrange. From those he was able to derive the basic wave equations of electromagnetism without any special assumptions about molecular vortices or forces between electrical particles. [24]

Although Maxwell abandoned his aether model as the mechanism for deriving his theory of light, he remained committed in some sense to the *idea* of a medium filling space. In fact, he explicitly remarked that from the Faraday effect, the polarization of dielectrics, and the phenomena of optics we are led to the *conception* of a complicated mechanism capable of a vast variety of motions (Maxwell 1965, vol. 1, p. 533). What was particularly important, however, was the role that the medium played in "A Dynamical Theory". Maxwell claimed that the theory he proposed could be called a theory of the electromagnetic field because it concerned the space in the neighbourhood of electric or magnetic bodies. Moreover, it could be called a dynamical theory because it assumed that in that space there was matter in motion that produced the observed electromagnetic phenomena (1965, vol. 1, p. 527). An additional reference for the term "dynamical" was to the abstract dynamics developed by Lagrange and elaborated by Hamilton.[25] Lagrangian methods had been used by both MacCullagh and Green[26] in their studies on the luminiferous aether. The approach consisted in postulating an appropriate potential-energy function for the aether, without specifying the details of its mechanical structure. Once that was done, a wave equation was derived that allowed the aether to be treated as a mechanical system without any specification of the machinery that gave rise to the characteristics exhibited by the potential-energy function. In contrast to the method of "On Physical Lines", where Maxwell attempted to describe "a particular kind of motion and a particular kind of strain so arranged as to account for the phenomena", in "A Dynamical Theory" he "avoid[ed] any hypothesis of [that] kind". He went on to point out that

in using such words as electric momentum and electric elasticity in reference to known phenomena of the induction of currents and the polarization of dielectrics, I wish to merely direct the mind of the reader to mechanical phenomena, which will assist him in understanding the electrical ones. All such phrases in the present paper are to be considered as illustrative and not as explanatory. (1965, vol. 1, p. 564)

It becomes clear, then, that the unifying power of the Lagrangian approach lay in the fact that it ignored the nature of the system and the details of its motion; one did not start with a set of variables that had immediate physical meanings. I shall say more about the technical details of that approach later, but for now it is useful to think about the unifying power as a function of the *generality* of the formulation. Because very little information is provided about the physical system, it becomes easier to bring together diverse phenomena under a common framework. Only their general features are accounted for, yielding a unification that, to some extent, is simply a formal analogy between two different kinds of phenomena. The Lagrangian emphasis on energetical properties of a system, rather than its internal structure, became especially important after establishment of the principle of conservation of energy, which also served as a unifying principle for different types of systems. In fact, as I claim later, the notion of "field energy" functioned in exactly that way for Maxwell. Energy conservation became the physical principle on which an otherwise abstract dynamics could rest.

As we saw earlier, Maxwell's mechanical model played a significant role in the formulation of the theory in "On Physical Lines", yet he remained agnostic about its value (at least with respect to the vortex hypothesis) as a true physical representation or theory about the structure of the medium and the mechanisms involved in displacement. His intention in "A Dynamical Theory" was to use some of the descriptive terminology from "On Physical Lines" simply as a way of providing some physical understanding of how one "might possibly" account for the phenomena, rather than as a true explanatory hypothesis. Maxwell did remark that only when speaking about the "energy" of the field did he wish to be understood literally. All energy was to be considered mechanical energy, whether it existed as motion, as elasticity or in any other form. Although the energy in electromagnetic phenomena was to be understood as mechanical energy, the crucial theoretical question concerned its location. According to action-at-a-distance theorists, the energy resides in electrical bodies, conducting circuits and magnets in the form of potential energy (the power to produce certain effects at a distance). In Maxwell's electromagnetic theory the energy resides in the electromagnetic field, that is, in the space surrounding the electrified and magnetic bodies, as well as in the bodies themselves. That energy has two different forms that, according to Maxwell, can be described without hypotheses as magnetic and electric polarization, or, with reference to what he terms a "very probable hypothesis" (1965, vol. 1, p. 564), as the motion and strain of one and the same medium.

The idea that the medium was *capable* of motion was supported by the phenomena of heat and light (1965, vol. 1, p. 525), especially the Faraday effect (magneto-optic rotation). In the latter case, Maxwell claims that

we have *warrantable grounds for inquiring* whether there may not be a motion of the aethereal medium going on wherever magnetic effects are observed, and we have some reason to suppose that this motion is one of rotation. (1965, vol. 1, p. 529; italics added)

He thought it possible that the motion of the medium (caused by electric currents and magnets) could be communicated from one part to another by forces arising from the connections of those parts: Because of the action of the forces there was a certain yielding that depended on the elasticity of the connections. Consequently, the energy of the medium was composed of the actual energy of the motions of the parts, and the potential energy was stored in the connections as elasticity (displacement). But aside from the very general claim about matter in motion, Maxwell offered no hypothesis about the nature of the medium, except that it must be subject to the general laws of dynamics. However, one ought to be able to work out all the consequences of its motion provided that one knew the form of the relation between the motions of the parts (Maxwell 1965, vol. 1, p. 532).

In the case of "On Physical Lines", the mechanical properties of the model provided the mechanisms for deducing electromagnetic effects, whereas in "A Dynamical Theory" the model played virtually no role; all consequences were derived from the Lagrangian formalism and experimental facts. In addition to that, the very *idea* of an electromagnetic medium was introduced in a rather different manner. On the basis of some work by Thomson (1854, p. 57) on the density of the luminiferous medium, Maxwell claimed that one could take the data from an independent domain (in that case, optics) and, given certain properties of an optical medium (especially elasticity and the velocity computation), draw conclusions about its identity with the electromagnetic medium. Although the *idea* of a medium seemed reasonable, and one could make certain assumptions about it, Maxwell did not consider that to be a basis for physical explanation.

On his view, an explanation was considered a true physical theory and therefore substantially different from an analogy or hypothesis. In "On Physical Lines" he postulated a mechanical electromagnetic medium and subsequently discovered that it could be identified with the optical aether, whereas in "A Dynamical Theory" he introduced the optical aether through a discussion of the phenomena of light and heat. That was followed by a discussion of the Faraday effect, which implied that magnetism was the result of the motion of an optical aether, and, finally, a discussion of the facts about electricity and magnetism that independently suggested the existence of an electromagnetic aether. Those three independent domains of evidence led to the general assumption that there was an aethereal medium pervading all space. But it was no more than an assumption; nothing could be concluded simply on the basis of evidence about effects that seemed to point to motion in the aether as their cause. According to Maxwell, legitimate inferences about the aether required more direct evidence for its existence than was provided by inferences from observable effects to unobservable causes.

In "On Physical Lines" Maxwell postulated the existence of an electromagnetic medium endowed with elasticity and then brought in the optical aether to lend credence to the idea:

The undulatory theory of light requires us to admit this kind of elasticity in the luminiferous medium in order to account for the transverse vibrations. We need not then be surprised if the magneto-electric medium possesses the same property. (1965, vol. 1, p. 489)

That physical analogy was not used in "A Dynamical Theory"; instead, Maxwell appealed to a kind of mathematical analogy based on a formal account of the motions of the parts of the medium provided by Lagrangian mechanics. That, together with what Maxwell termed "experimental facts of three kinds" (induction of currents, distribution of magnetic intensity according to variation of potential and induction of static electricity), provided the basis for the field equations (1965, vol. 1, p. 564). Those equations were then used to demonstrate the laws of mechanical force that governed electric currents, magnetic bodies and electrified bodies in the field. The important point is that the laws of mechanical action were *applicable* to the medium, but did not require the hypothesis of the medium for their introduction. [27]

Despite Maxwell's claim to provide deductions from experimental facts, his account still required the postulation of a displacement current, something that could neither be verified by nor "deduced" from experiment. But because Maxwell was not required to take account of the motions of the medium, all of the physical concepts in "A Dynamical Theory", except energy, were understood to be merely illustrative, rather than substantial. That enabled him to develop an account of the field that was consistent with mechanical principles but was at the same time free of the interpretative problems that plagued "On Physical Lines". Because the mechanical model had been abandoned, displacement was no longer seen as an elastic restoring force for the mechanical aether, and the negative value in $R = -4\pi E^2 h$ disappeared. In addition, the part of the theory that concerned the electrostatic equations was established without the use of any of the mechanical properties of the aether.[28]

Maxwell did retain a basic feature of his theory of charge as formulated in "On Physical Lines", that is, the identification of a charge with a discontinuity in displacement in a dielectric medium. The main difference between the accounts in "On Physical Lines" and "A Dynamical Theory" was that in the former case an equation connecting displacement with charge was not explicitly given, whereas in the latter that relation appeared as one of the basic equations.[29] Given Maxwell's claim about open currents, the notion of a displacement that gave rise to transient currents was essential; without it there would be no basis for the electromagnetic theory of light, because it represented the idea of electricity travelling in the field. So although displacement retained a prominent position in "A Dynamical Theory", its role was somewhat different from the role it played in "On Physical Lines". It was no longer associated with changes in positions of rolling particles; rather, Maxwell defined it simply as the motion of electricity, that is, in terms of a quantity of charge crossing a designated area (Bromberg 1968; Buchwald 1985).

The difficulty, of course, was that Maxwell's account of charge was itself problematic. Instead of interpreting it as a property accumulating in bodies, it was considered an epiphenomenon associated with the field – the result of a discontinuity in displacement (Maxwell 1873, vol.1, p. 3). But because there was no physical account of the field, and because the aether model had been abandoned, it was difficult to see just how charge could occur. The problem, of course, was traceable to the notion of displacement: If electricity was being displaced, how did this occur? Because of the lack of a mechanical foundation, the idea that there was a displacement of electricity in the field, though it was not associated with any mechanical source or body, became difficult to motivate theoretically.

Maxwell formulated 20 equations for the electromagnetic field containing 20 variable quantities. Again, in modern vector notation, using the same symbols as those in "On Physical Lines", with the exception of the addition of $\mathbf{p}'$ or (p', q', r') for total currents, we have the following:

(1) three equations for total currents

$$\mathbf{p}' = \mathbf{p} + \frac{d\mathbf{f}}{dt}$$

(2) three equations for magnetic force

$$\mu\boldsymbol{\alpha} = -\boldsymbol{\nabla} \times \mathbf{F}$$

(3) three equations for electric currents

$$-\boldsymbol{\nabla} \times \boldsymbol{\alpha} = 4\pi\mathbf{p}'$$

(4) three equations for electromotive force

$$\mathbf{P} = -\mu\left(\boldsymbol{\alpha} \times \frac{d\mathbf{x}}{dt}\right) - \frac{d\mathbf{F}}{dt} - \boldsymbol{\nabla}\psi$$

(5) three equations for electric elasticity

$$\mathbf{P} = k\mathbf{f}$$

(6) three equations for electric resistance

$$\mathbf{P} = -\rho\mathbf{p}$$

(7) one equation for free electricity

$$e = -\boldsymbol{\nabla} \cdot \mathbf{f}$$

(8) one equation for continuity

$$\frac{de}{dt} = -\nabla \cdot \mathbf{p}$$

(9) one equation for total field energy

$$\text{energy} = \frac{1}{8\pi} \iiint \left(\boldsymbol{\alpha} \cdot \mu\boldsymbol{\alpha} + \frac{1}{2}\mathbf{P} \cdot \mathbf{f} \right) dw^2$$

In modern form, these are written (where $\mathbf{I}$ is total current)

$$\mathbf{I} = \mathbf{J} + \frac{d\mathbf{D}}{dt}$$

$$\mathbf{B} = -\nabla \times \mathbf{A}$$

$$\nabla \times \mathbf{H} = 4\pi\mathbf{I}$$

$$E = -\left(\mathbf{B} \times \frac{d\mathbf{x}}{dt} \right) - \frac{\partial \mathbf{A}}{\partial t} - \nabla\psi$$

$$\epsilon_0 E = \mathbf{D}$$

$$E = -\rho\mathbf{J}$$

$$e = -\nabla \cdot \mathbf{D}$$

$$\frac{de}{dt} = -\nabla \cdot \mathbf{J}$$

The first three equations describe the relationships among electric displacement, true conduction and total current (which is a combination of the former two); the second three describe the relationship between lines of magnetic force and the induction coefficients of a circuit; the third three govern the relationship between the strength of a current and its magnetic effects; the fourth three give the value of the electromotive force in a body arising from the motion of the body, alteration of the field and variation in its electromagnetic potential; the fifth three describe the relationship between electric displacement and electromotive force that produces it; the sixth three represent the relationship between electric current and electromotive force; the seventh equation involves the relationship between the amount of free electricity at any point and the electric displacements in the neighbourhood; the eighth equation gives the relationship between the increase or decrease in free electricity and the electric currents in the surrounding area. Using those quantities, Maxwell then was able to formulate an equation that gave the intrinsic energy of the field as a function of its magnetic and electric polarizations at every point.

With these field equations in place, we now come to the significant feature of the theory: the derivation of the wave equation. Because the value for wave propagation emerged as a natural consequence of the model in "On Physical Lines", it is important to clarify its position within the more abstract dynamical framework of "A Dynamical Theory". Because the value of wave propagation for electromagnetic phenomena is equivalent to that for light, the wave equation functions as the manifestation of the unity of electromagnetism and optics. Hence, it is important to ensure that an abandonment of the aether model does not require any *ad hoc* steps in the new derivation; that is, we want to guarantee that the wave equation arises in a similarly natural way using Lagrangian dynamics. This issue is especially crucial for the question of unity, because the measure of a theory's unifying power should be linked to the number of free parameters required to encompass groups of diverse phenomena. That is, if one is required to add several free parameters in order to produce a unified theory, then one begins to question that the unity is simply the result of these additions, rather than a genuine theoretical or ontological unity.

In Part 6 of "A Dynamical Theory" Maxwell begins by asking if it is possible to explain the propagation of light using the field equations he has deduced from electromagnetic phenomena (1965, vol. 1, p. 577). Taking a plane wave propagating through a field with velocity V, he uses the equations of magnetic force to show that the direction of the magnetization must be in the plane of the wave. Because all electromagnetic functions will be functions of

$$w = lx + my + nz - Vt$$

where l, m and n are the direction cosines, we can multiply the equations of magnetic force,

$$\mu\alpha = \frac{dH}{dy} - \frac{dG}{dz}$$
$$\mu\beta = \frac{dF}{dz} - \frac{dH}{dx}$$
$$\mu\gamma = \frac{dG}{dx} - \frac{dF}{dy}$$

by l, m and n, which gives us

$$l\mu\alpha + m\mu\beta + n\mu\gamma = 0$$

Maxwell then combines the equations of magnetic force with those for electric currents to get

$$4\pi\mu p' = \frac{dJ}{dx} - \nabla^2 F$$
$$4\pi\mu q' = \frac{dJ}{dy} - \nabla^2 G$$
$$4\pi\mu r' = \frac{dJ}{dz} - \nabla^2 H$$

If the medium is a perfect dielectric, then there is no true conduction, and the currents are simply displacements caused by electromotive forces that are due to variations of either electromagnetic or electrostatic functions. By the equation for total currents, the displacement currents p', q' and r' become

$$p' = \frac{df}{dt}, \qquad q' = \frac{dg}{dt}, \qquad r' = \frac{dh}{dt}$$

The equations of electromotive force are

$$P = -\frac{dF}{dt} - \frac{d\Psi}{dx}$$
$$Q = -\frac{dG}{dt} - \frac{d\Psi}{dy}$$
$$R = -\frac{dH}{dt} - \frac{d\Psi}{dz}$$

The equations of elasticity

$$P = kf, \qquad Q = kg, \qquad R = kh$$

and electromotive force are then combined to yield

$$k\left(\frac{dJ}{dx} - \nabla^2 F\right) + 4\pi\mu\left(\frac{d^2 F}{dt^2} + \frac{d^2\Psi}{dx/dt}\right) = 0$$
$$k\left(\frac{dJ}{dy} - \nabla^2 G\right) + 4\pi\mu\left(\frac{d^2 G}{dt^2} + \frac{d^2\Psi}{dy/dt}\right) = 0$$
$$k\left(\frac{dJ}{dz} - \nabla^2 H\right) + 4\pi\mu\left(\frac{d^2 H}{dt^2} + \frac{d^2\Psi}{dz/dt}\right) = 0$$

If we differentiate the final equation with respect to y and the second with respect to z and subtract, then J and Ψ disappear, which, together with the equations for magnetic force, yields

$$k\nabla^2\mu\alpha = 4\pi\mu\frac{d^2}{dt^2}\mu\alpha$$
$$k\nabla^2\mu\beta = 4\pi\mu\frac{d^2}{dt^2}\mu\beta$$
$$k\nabla^2\mu\gamma = 4\pi\mu\frac{d^2}{dt^2}\mu\gamma$$

If we assume that α, β and γ are functions of $lx + my + nz - Vt = w$, the first equation becomes

$$k\mu \frac{d^2\alpha}{dw^2} = 4\pi\mu^2 V^2 \frac{d^2\alpha}{dw^2}$$

or

$$V = \pm\sqrt{k/4\pi\mu}$$

The other equations give the same value for V, so that the wave is propagated in either direction with a velocity V. The wave consists entirely of magnetic disturbances, and the disturbance at any point is transverse to the direction of propagation. Because k, the coefficient of electric elasticity, is related to v, the number of electrostatic units in one electromagnetic unit, by the equation $k = 4\pi v^2$, we have $V = v$. Hence these waves have all the properties of polarized light. So we see that here again displacement plays a central role in the derivation of the wave equation, and given its presence, the derivation is certainly no less straightforward than that provided in "On Physical Lines".

In summary, then, Maxwell's goal in "A Dynamical Theory" was to use the equations governing electromagnetic phenomena to present an account of the possible conditions that could exist in a mechanical medium responsible for the exchange and transmission of kinetic and potential energy.[30] The method allowed Maxwell to treat field variables as generalized mechanical variables interpreted within the context of the Lagrangian formalism. The implications for the physical interpretation of the theory will be further discussed later, but for now let me again stress that although the dynamical theory was applicable to mechanical phenomena, the equations were not connected to any specific mechanical model. Nor was the *motivation* for the theory based on mechanical hypotheses. Maxwell was anxious to give some physical interpretation to the formalism, but he was quick to point out that concepts like electric elasticity associated with the polarization of dielectrics (and hence displacement) should be understood as merely illustrative. The core of the theory was the Lagrangian formalism, a highly abstract and general method applicable to a wide range of different models of the data. Displacement retained its fundamental role as a necessary component for the field-theoretic view, but ultimately it was little more than a phenomenological equation.

3.3.2. *The Treatise on Electricity and Magnetism*

The methods used in "A Dynamical Theory" were extended and more fully developed in the *Treatise on Electricity and Magnetism* (hereafter, the *Treatise*), especially the role of energy conservation in providing a physical basis for the theory. Once Maxwell had formally expressed the fundamental laws of electromagnetism and the relations between the various quantities, he had to show how those laws could be deduced from the general dynamical laws applicable to any system of moving bodies. Although the connections between the motions of the medium and specific variables were eliminated from the equations (the variables being independent of any particular form of those connections), Maxwell did make use of mechanical *ideas* in the development of the theory. For example, in the chapter devoted to the Faraday effect, Maxwell once again appealed to the hypothesis of molecular vortices. But in applying his dynamical approach to magneto-optic rotation, Maxwell remained agnostic about the nature of those vortices and remarked that

the consideration of the action of magnetism on polarized light leads . . . to the conclusion that in a medium under the action of magnetic force *something* belonging to the same mathematical class as angular velocity, whose axis is in the direction of the magnetic force, forms part of the phenomena. (1873/1954, vol. 2, pp. 399–417, esp. 406, 408; emphasis added)

But there the vortex hypothesis was nothing more than a physical possibility. Although Maxwell claimed that he had "good evidence for the *opinion* that some phenomenon of rotation is going on in the magnetic field, and that this rotation is performed by a number of very small portions of matter" (1873/1954, vol. 2, p. 470; italics added), he stressed the importance of conclusions based on observed phenomena and claimed that results must be stated in language free from the assumptions of the undulatory theory:

Whatever light is, at each point in space there is something going on, whether displacement, or rotation, or something not yet imagined, but which is certainly of the nature of a vector or directed quantity, the direction of which is normal to the direction of the ray. This is completely proved by the phenomena of interference. (1873/1954, vol. 2, p. 460)

Toward the end of the chapter on the magnetic action of light (ch. 21) and the discussion of molecular vortices, Maxwell further emphasized that

the theory proposed in the preceding pages is evidently of a provisional kind, resting as it does on unproved hypotheses relating to the nature of molecular vortices and the mode in which they are affected by the displacement of the medium. We must therefore regard any coincidence with observed facts as of much less scientific value in the theory of magnetic rotation of the plane of polarization than in the electromagnetic theory of light, which, though it involves hypotheses about the electric properties of media, does not speculate as to the constitution of their molecules. (1873/1954, vol. 2, p. 468)

Because of the uncertainty about vortices, Maxwell thought it impossible to specify a formal law connecting the displacement of the medium with the variation of the vortices. Instead, he simply assumed that the variation caused by displacement was subject to the same conditions that Helmholtz had formulated for vortices in a perfect liquid.[31]

The method of dynamical explanation that Maxwell developed in the *Treatise* was one that had been used by Thomson and Tait in their *Treatise on Natural Philosophy* (1867). The method emphasized dynamical concepts, as opposed to the more abstract Lagrangian approach that focused strictly on mathematical formalism. To fully understand Maxwell's approach, we need to look briefly at his interpretation of the Thomson and Tait methods, as well as the connections among mechanical/dynamical concepts, energy physics and the Lagrangian programme.

Earlier I referred to the connection between the generality of Lagrangian dynamics and its unifying power. There we saw that the generality stemmed from ignoring the internal mechanical aspects of a system and focusing instead on its energy. The aim of Lagrange's *Mécanique Analytique* was to rid mechanics of Newtonian forces and the requirement that we must construct a separate acting force for each particle. Lagrange's equations of motion for a mechanical system were derived from the principle of virtual velocities and d'Alembert's principle. The method consisted of expressing the elementary dynamical relations in terms of the corresponding relations of pure algebraic quantities, which facilitated the deduction of the equations of motion.[32]

Maxwell's goal for the *Treatise* was to examine the consequences of the assumption that electric currents were simply moving systems whose motion was communicated to each of the parts by certain forces, the nature and laws of which "we do not even attempt to define, because we can eliminate [them] from the equations of motion by the method given by Lagrange for any connected system" (Maxwell 1954, sec. 552). Although Maxwell favoured that dynamical approach, his earlier papers indicate that he was especially critical of using purely abstract mathematics to describe physical phenomena, and the Lagrangian method was clearly of that sort, avoiding reference to concepts like momentum, velocity and even energy by replacing them with symbols in the generalized equations of motion. Lagrange's goal was to reduce all of physics to mathematics, and he was quite proud of the fact that one would not find a geometrical construction in all of *Mécanique Analytique*.[33] Maxwell, who was a strong proponent of a role for geometry and physical concepts in science, sought a middle ground in the dynamics of Thomson and Tait, a method that would enable him to "retranslate . . . the language of calculus into the language of dynamics so that our words may call up the mental image, not of some algebraic process, but of some property of moving bodies" (Maxwell 1873/1954, vol. 2, p. 200). That approach allowed one to "conceive of a moving system connected by means of an *imaginary mechanics used merely to assist the imagination in ascribing position, velocity and momentum to pure algebraic quantities*" (1873/1954, vol. 2, p. 200; italics added). Geometrical methods were especially appropriate, because dynamics was, after all, the science of motion.

The new physical dynamics of Thomson and Tait rested on the principle of conservation of energy (Thomson and Tait 1879–93, vol. 1, p. 265). In keeping with the goals of abstract dynamics, one could specify the energy of a system as determined by the configuration and motion of its parts, without any reference to the hidden mechanisms themselves. Thomson and Tait rejected the emphasis on d'Alembert's principle and the principle of virtual velocities and replaced them by reviving aspects of the Newtonian method, which emphasized motion as basic. They saw their programme as dynamical rather than mechanical, because it emphasized motion rather than machines,[34] with the generalized equations of motion (the Lagrangian and Hamiltonian versions) derivable from conservation laws. Conservation of energy was taken as basic, because d'Alembert's principle and the virtual-velocity principle, as well as conditions of equilibrium and motion for every case, could be derived from it.[35]

An important part of their dynamics was Thomson's minimum theorem, which was based on conservation laws and emphasized the role of impulsive forces. Because impulsive forces were thought to act in a vanishingly small period of time, they produced no change in the configuration or potential energy of the system and only a minimum change in the whole kinetic energy.[36] The result was that one could vary the system without taking into account its configuration or hidden mechanisms.

Because of the universal applicability of energy conservation, the concept of energy itself took on a substantial role in the new dynamics. As Tait remarked, "if we find anything else in the physical world whose quantity we cannot alter we are bound to admit it to have objective reality . . . however strongly our senses may predispose us against the concession" (Tait 1876, p. 347). Despite its "constant mutation", energy, unlike force, fulfilled that requirement.[37] That, of course, was in keeping with Maxwell's emphasis on a literal interpretation of field energy in "A Dynamical Theory", the only theoretical concept given a substantial interpretation.

In the *Treatise*, Maxwell followed the method of Thomson and Tait, reinterpreting Lagrange's symbols in terms of velocity, momentum, impulse and the energy principle. That allowed him to keep in mind "ideas that were appropriate for the science of dynamics while keeping out of view the mechanism by which the parts of the system are connected" (1873/1954, vol. 2, p. 198). The importance of both dynamical ideas and mathematical methods was evident throughout, with emphasis on the language of partial differential equations as the appropriate expression for the field and the action exerted between its contiguous parts. That integrated approach was also evident in Maxwell's preference for the Hamiltonian way of expressing kinetic energy, with emphasis on momentum and velocity, as opposed to the Lagrangian method, where momentum was expressed in terms of velocity. The physical emphasis on momentum in Hamilton's dynamics revealed the connection with Newtonian mechanics.[38]

We must remember, of course, that the Lagrangian formalism was understood quite differently by British field theorists than by modern physicists, and it is that difference that contributes in important ways to our overall understanding of the structure of nineteenth-century physical theory.[39] The basic idea inherent in the dynamical method practised by British physicists was that all processes could be fully described using continuous energy densities. By feeding different energy expressions through the dynamical equations, they got different boundary conditions and differential equations. In order to treat difficult or anomalous phenomena, they simply modified the energy expression to produce different consequences. In effect, all electromagnetic phenomena, including boundary conditions, could be obtained by applying Hamilton's principle to specified field energy densities. Because the world was thought to be a continuum, the variable in the equations represented properties of a continuum. Once the appropriate energy formula was specified for a particular problem, the field theorist was then committed to its consequences and boundary conditions. As Buchwald (1985) has pointed out, the concept of the "continuity of energy" (that it flowed continuously from one place to another) was one of the novel ideas required by the basic principles of Maxwell's theory: Although the values for properties of the medium (specific inductive capacity and magnetic permeability) changed in the presence of matter, the changes were continuous, with each material molecule effecting a continuous alteration in the properties of the medium. In contrast, modern theory employs energy densities only under limited conditions, and when these are unfulfilled one must resort to microstructures. Whereas modern physics appeals to the electron, British field theory simply formulated new energy expressions. Consequently, the basic equations were continually being changed, something that modern theory again gets around by introducing microstructures. Although the properties of the continuum may be dependent on microstructures, that assumption was not required by the dynamical method.

Despite the lack of commitment to specific mechanical models in that later work, we nonetheless see the importance of mechanical ideas and concepts in Maxwell's dynamical field theory, concepts that for him played the role of constructive images. In "A Dynamical Theory" (Maxwell 1965, vol. 1, pp. 533–4, 564) he speaks about determining the laws of mechanical forces acting on conductors, on magnetic poles and on electrified bodies. In the *Treatise*, Thomson's account of impulsive forces was used as a way of supplementing the abstract dynamics with mechanical concepts that would enable him to form a "mental image" (Maxwell 1873/1954, vol. 2, p. 200). But in keeping with his earlier work, Maxwell was explicit in his denial of any substantial interpretation of the internal forces used to explain the connections in the field. As a hypothesis it had no more certainty than action-at-a-distance accounts:

The observed action at a considerable distance is therefore explained by means of a great number of forces acting between bodies at very small distances, for which we are as little able to account as for the action at any distance however great.[40]

As he had in the earlier papers, Maxwell stressed the benefits not only of having a mental representation but also of having an alternative conception of the phenomena, for in "establishing the necessity of *assuming* these internal forces . . . we have advanced a step . . . which will not be lost, though we should fail in accounting for these internal forces, or in explaining the mechanism by which they can be maintained" (1873/1954, vol. 1, pp. 122–3, 127; italics added).

The emphasis on mental representation in Maxwell's early and later works was intimately connected with the importance he placed on geometric representation of the phenomena. It is the combination of these different approaches – abstract dynamics supplemented by dynamical/mechanical concepts and geometrical methods – that makes Maxwellian dynamics and methodology unique. Taken together, we have the benefits of a mathematically precise theory without the commitment to a specific mechanical account.

The use of abstract dynamics in the study of complex objects allows us to "fix our attention on those elements of it which we are able to observe and to cause to vary, and ignore those which we can neither observe not cause to vary" (Maxwell 1965, vol. 2, p. 776, n. 58). Maxwell uses the allegory of the belfry to illustrate this principle of inquiry:

In an ordinary belfry, each bell has one rope which comes down through a hole in the floor to the bell-ringer's room. But suppose that each rope, instead of acting on one bell contributes to the motion of many pieces of machinery, and that the motion of each piece is determined not by the motion of one rope alone, but by that of several and suppose, further, that all this machinery is silent and utterly unknown to the men at the ropes, who can only see as far as the holes in the floor above them.

Suppose all this, what is the scientific duty of the men below? They have full command of the ropes, but of nothing else. They can give each rope any position and any velocity, and they can estimate its momentum by stopping all the ropes at once, and feeling what sort of tug each rope gives. If they take the trouble to ascertain how much work they have to do in order to drag the ropes down to a given set of positions, and to express this in terms of these positions, they have found the potential energy of the system in terms of the known coordinates. If they then find the tug on any one rope arising from a velocity equal to unity communicated to itself or to any other rope, they can express the kinetic energy in terms of the coordinates and velocities.

These data are sufficient to determine the motion of every one of the ropes when it and all the others are acted on by any given forces. This is all that the men at the rope can ever know. If the machinery above has more degrees of freedom than there are ropes the coordinates which express these degrees of freedom must be ignored. There is no help for it. (Maxwell 1965, vol. 2, pp. 783–4)

Maxwell claimed that the attempt to imagine a working model of the mechanism connecting the different parts of the field ought to be seen as nothing more than a demonstration that it is *possible* to *imagine* a mechanism capable of producing a connection mechanically equivalent to the actual connection of the parts of the field, whatever it might be. The problem of determining the mechanism required to establish a given species of these connections was that there existed an infinite number of solutions or mechanical models that could account for the phenomena. A benefit of the Lagrangian method was that it enabled one to express all that could be deduced, within reasonable limits, from a given set of observational data. When applying the Lagrangian formalism to electromagnetism, the information derived from experiments on interacting electric currents was considered analogous to observations of the behaviour of ropes in the case of the belfry. The hidden mechanism was compared to motions that were assumed to exist in the medium surrounding the currents.

There are two relatively obvious suggestions as to why Maxwell adopted the method of dynamical theory. First of all, he had been unsuccessful in carrying out experiments intended to test the electromagnetic theory. As a result, his desire to move away from a method involving speculation about physical hypotheses in favour of a mathematical theory was further strengthened. Dynamical theory, as Maxwell understood it, involved no assumptions concerning the existence of specific bodies or their properties; his approach simply implied that all phenomena could be understood in terms of matter and motion. However, the reliance on a general dynamical approach was also bound up with Maxwell's philosophical views about the nature of scientific knowledge, physical theory and the role of models.

In his address to the mathematical and physical sections of the British Association in 1870, Maxwell was careful to distinguish between the structure of nature and the concepts we use to describe it:

> Molecules have laws of their own, some of which we select as most intelligible to us and most amenable to our calculation. We form a theory from these partial data and we ascribe any deviation of the actual phenomena from this theory to disturbing causes. At the same time we confess that what we call disturbing causes are simply those parts of the true circumstances which we do not know or have neglected, and we endeavour in future to take account of them. We thus acknowledge that the so-called disturbance is a mere figment of the mind, not a fact of nature, and that in natural action there is no disturbance. (Maxwell 1965, vol. 2, p. 228)

It was that distinction between nature and its representations that provided the basis for many of Maxwell's ideas about models and analogies. Because those modes of representation were thought to be heuristic devices or, at best, descriptions of what nature *might* be like, it was possible to propose several different and even contradictory models of the phenomena, without a firm commitment to any particular one. And indeed that approach fit nicely with the methodology suggested by the Lagrangian formalism. Just as there are many ways of expressing a proposition, a physical theory can be represented in a number of ways. It is also the case that different physical theories can be formulated in the same manner. For instance, insofar as the formal structure of the theory is concerned, analytical mechanics, electromagnetism and wave mechanics can all be deduced from a variational principle, the result being that each theory has a uniform Lagrangian appearance.[41] Recall that velocities, momenta and forces related to the coordinates in the equations of motion need not be interpreted literally in the fashion of their Newtonian counterparts. In the example of the belfry, the ropes correspond to the generalized coordinates that determine the configuration of the system, while the mechanism responsible for the bell-ringing (the connection between the movement of the ropes and the ringing) remains unknown. Similarly, we can think of the field as a connected mechanical system with currents, and integral currents, as well as generalized coordinates corresponding to the velocities and positions of the conductors. We can have a quantitative determination of the field without knowing the actual motion, location and nature of the system.

This degree of generality allows us to apply the Lagrangain formalism to a variety of phenomena, regardless of the specific nature of the physical system. This is particularly useful when the details of the system are unknown or when the system is simply assumed to lack a mechanism. Consider the following example. Suppose we want to find the path of a light ray between two given points in a medium of known refractive index. In the qualitative interpretation of this problem we employ an analogy that compares light rays and particles, that is, we describe light rays *as if* they were trajectories of particles or entities endowed with a mass. However, in the quantitative or formal treatment no mention of masses, motions or forces is made. Instead, a physically uninterpreted parameter t is introduced that allows us to specify optical length A as[42]

$$A = \int_{t_2}^{t_1} L \, dt$$

Nothing about the formalism or the use of dynamical analogies allows us to infer that light rays *are* particle trajectories, or that optics can be reduced to mechanics. The variables do not have an independent physical meaning, and because the Lagrangian formalism can be applied to such a variety of different systems, no conclusions about the nature of the system can be drawn. As was the case with Maxwell's theory, we cannot understand an analogy or formal identity as expressing an identity in kind or nature.

A rather different view is presented in some contemporary philosophical writing about the use of analogical argument as an appropriate method for establishing conclusions based on inductive evidence. For example, Salmon (1984) argues that if we can successfully extend an analogy into new domains, and provided the comparisons are relevant and close enough, then the analogical argument should give us good reason for accepting the conclusions we draw from it. He claims that the move from observables to unobservables can be sanctioned by analogical argument. The argument is based on the induction that if all effects $E_1, \ldots, E_n$ have been the results of causes $C_1, \ldots, C_n$, then we can infer that E_{n+1} is the result of C_{n+1}. Salmon cites some additional evidence for the legitimacy of analogical arguments, including the similarities between human research and animal research and the analogy between Newton's inverse-square law and Coulomb's theory of electrostatics.

This emphasis on analogy as a basis for evidential support is absent from Maxwell's work. Instead, the examples in "On Faraday's Lines" and "On Physical Lines" characterize an analogy between two different branches of physics by showing how the same mathematical formalism can be applied to each. Similarly, Thomson showed that there was an analogy between the theory of heat and electrostatics on the basis that both can be described by the same equation if one reads "temperature" for "potential" and "source of heat" for "positive electric charge". Consequently, the theory of heat was used as a model for the field theory of electrostatics. But because we don't know the mathematical structure of nature, and because nature can be described in a variety of mathematical ways, the most we can say is that there is a resemblance between the numerical consequences of our experimental outcomes and the numerical features of the model. In the case of atomic particles, we employ a model based on dynamics and electrostatics, because deductions from that type of model yield numbers comparable to those from experimental measurements. Salmon claims that a specific model may provide a better analogy than anything else with which we are acquainted and on that basis can lend plausibility to the conclusions of our inductive inferences. Although this appeals to our basic intuitions about induction, we need to be careful when linking analogy with inductive support. *Plausibility* (being such a loosely construed notion) is surely not enough when characterizing, as Salmon wants to, the evidential basis for theory acceptance.

According to Maxwell, the method for recognizing real analogies rests on experimental identification. If two apparently distinct properties of different systems are interchangeable in appropriately different physical contexts, they can be considered the same property. This is a species of what Mary Hesse has called substantial identification, which occurs when the same entity is found to be involved in apparently different systems. An example would be identification in "On Physical Lines" of the aetherial media of electromagnetism and light on the basis of the numerical value of the velocities of transmission of transverse electromagnetic waves and of light. However, as we saw earlier, Maxwell did not use that analogical argument as a way of underwriting his model of the aether. Experimental evidence of the kind he referred to had to be direct and independent, rather than the kind that produced an inferential basis for the medium. Hence, he was reluctant see the analogy as the foundation for a physical theory.

In contexts like the unification of the aethers, we use data from the domain of optics, a field supposedly independent from electromagnetism, as evidence for the reduction of the two media to one. A similar argument could be constructed for the displacement current. Although displacement played a pivotal role in facilitating the unification of electromagnetism and optics and, as such, could be seen as receiving confirmation from both domains, there was no theory-neutral or experimental evidence to suggest its existence. Hence its explanatory and unifying roles were not enough. The demand for independent evidence formed the basis for Maxwell's disregard for the method of the French molecularists, now commonly referred to as hypothetico-deductive:

In forming dynamical theories of the physical sciences it has been a too frequent practice to invent a particular dynamical hypothesis and then by means of the equations of motion to deduce certain results. The agreement of these results with real phenomena has been supposed to furnish a certain amount of evidence in favour of the hypothesis.

The true method of physical reasoning is to begin with the phenomena and to deduce the forces from them by a direct application of the equations of motion. The difficulty of doing so has hitherto been that we arrive, at least during the first stages of the investigation, at results which are so indefinite that we have no terms sufficiently general to express them without introducing some notion not strictly deducible from our premises.

It is therefore very desirable that men of science should invent some method of statement by which ideas, precise so far as they go, may be conveyed to the mind, and yet sufficiently general to avoid the introduction of unwarrantable details. (Maxwell 1965, vol. 2, p. 309)

The method of the molecularists violated the condition of independent evidence by assuming at the outset the existence of configurations of particles and specifying forces acting between them. In addition, because dynamical hypotheses admit of an infinite number of solutions, we require independent evidence as a way of determining the true solution. It was because that was unattainable that Maxwell employed the method of Lagrange:

When we have reason to believe that the phenomena which fall under our observation form but a very small part of what is really going on in the system, the question is not – what phenomena will result from the hypothesis that the system is of a certain specified kind? but – what is the most general specification of a material system consistent with the condition that the motions of those parts of the system which we can observe are what we find them to be? (Maxwell 1965, vol. 2, p. 781)

3.4. Realism and Dynamical Explanation

At the beginning of this chapter I claimed that my investigation would address two concerns: the methodological issue regarding the nature of theory unification and its relation to realism, and the historical question concerning Maxwell's theoretical commitments in the development of electromagnetism. Part of the answer to the methodological question is revealed in the historical discussion, and the remaining part requires that we draw some general philosophical conclusions from our historical case study. Let me address the historical question first, namely, what can we conclude with respect to the physical aspects of Maxwell's dynamical account of electromagnetism? Or, more specifically, what is the relationship between the mathematical structure of the electromagnetic theory and the physical concepts that Maxwell used to supplement the abstract dynamics of Lagrange?

Earlier we saw that Maxwell used the method of Thomson and Tait because he saw it as a way to employ dynamical methods while at the same time utilizing physical concepts. However, in order to answer the question, we must first understand the way in which Maxwell departed from the methods of Thomson and Tait, specifically the impact of his distinction between mathematical theory and physical theory. Although energy conservation occupied the primary place in Thomson and

Tait's *Treatise on Natural Philosophy* (1867), mechanical/dynamical concepts played a more substantial role for Tait and especially Thomson than for Maxwell. Thomson's ultimate goal was to provide an account of the hidden mechanisms that were ignored by abstract dynamics:[43]

The theory of energy cannot be completed until we are able to examine the physical influences which accompany loss of energy.... But it is only when the inscrutably minute motions among small parts, possibly the ultimate molecules of matter, which constitute light, heat, and magnetism; and the intermolecular forces of chemical affinity; are taken into account, along with the palpable motions and measurable forces of which we become cognizant by direct observation, that we can recognise the universally conservative character of all dynamic action. (Thomson and Tait 1879–93, vol. 1, pp. 194–5)

Maxwell, on the other hand, claimed that the true method of physical reasoning was to begin with the phenomena and to deduce forces from them by a direct application of the equations of motion (Maxwell 1965, vol. 2, p. 309):

It is important to the student to be able to trace the way in which the motion of each part is determined by that of the variables, but I think it is desirable that the final equations should be obtained independently of this process. That this can be done is evident from the fact that the symbols by which the dependence of the motion of the parts on that of the variables was expressed, are not found in the final equations.

Once a phenomenon had been completely described as a change in the configuration and motion of a material system, "the dynamical explanation of that phenomenon is said to be complete" (1965, vol. 2, p. 418). Maxwell goes on to say that we

cannot conceive any further explanation to be either necessary, desirable or possible, for as soon as we know what is meant by the words configuration, motion, mass, and force, we see that the ideas which they represent are so elementary that they cannot be explained by means of anything else. (1965, vol. 2, p. 418)

Although Maxwell followed Newton in his emphasis on deductions from the phenomena, notions like force were to be interpreted as mathematical entities distinct from and not the subject of ontological speculation. As Harman (1982a) points out, Newton had also declared that the meaning of the concept of force was defined by the mathematical formalism of the *Principia*. However, its intelligibility depended on a disjunction between the notion of force and inertia in Newton's theory of matter.[44] In the third rule, Newton claims that inertia is a universal quality known through experience to be essential. Because we perceive inertial qualities in macroscopic bodies, we can, using the "analogy of nature", ascribe it to all bodies, even those that are unobservable. Hence Newton's idea of force is inextricably tied to a substantial account of matter.

In contrast, Maxwell's views on the limitations of the analogical method and his remarks in the essay "Atom" indicate his rejection of that aspect of Newtonian methodology. In addition to the dynamical concepts like impulsive forces and momentum, matter had also been defined as a mathematical entity within Maxwell's dynamics. The notion of matter as the ultimate substratum had no place in the formalism of dynamics; instead, the dynamical interpretation of matter was grounded in the concept of mass:[45]

Whatever may be our opinion about the relation of mass, as defined in dynamics, to the matter which constitutes real bodies, the practical interest of the science arises from the fact that real bodies do behave in a manner strikingly analogous to that in which we have proved that the mass systems of abstract dynamics must behave. (Maxwell 1965, vol. 2, pp. 779–80).

For Maxwell, matter as a substantial entity was that "unknown substratum against which Berkeley directed his arguments" and was never directly perceived by the senses. It was, in a sense, a presupposition, the supposed cause of our perception; but as a thing in itself it was unknowable. Alternatively, as an abstract dynamical concept it enjoyed the same status as a straight line:

Why, then should we have a change of method when we pass on from kinematics to abstract dynamics? Why should we find it more difficult to endow moving figures with mass than to endow stationary figures with motion? The bodies we deal with in abstract dynamics are just as completely known to us as the figures in Euclid. They have no properties whatever except those which we explicitly assign to them. (Maxwell 1965, vol. 2, p. 799)

Maxwell remarks (1965, vol. 2, p. 781) that real bodies perhaps may not have substrata, but as long as their motions are related to one another according to the conditions laid down in dynamics, we call them dynamical or material systems.

This dynamical account of matter raises some important questions in connection with the energy of the field and the argument for the existence of an aetherial medium. Recall that in "A Dynamical Theory" Maxwell claimed that from the phenomena of heat and light we "had some reason to believe . . . that there is an aetherial medium filling space", a medium "capable of receiving and storing up two kinds of energy . . . actual . . . and potential" (Maxwell 1965, vol. 1, p. 528). In the same paper, Maxwell remarked that he wished to avoid any hypotheses of motion or strain in the medium, but "in speaking of the energy of the field I wish to be understood literally" (1965, vol. 1, p. 564). That energy could be described *without hypothesis* as magnetic and electric polarization, or, according to a very probable hypothesis, as the motion and strain of one and the same medium (1965, vol. 1, p. 564). Yet, among the key features of the *Treatise* are Maxwell's representations of the energy of the field and the stress in the medium. The kinetic or electromagnetic energy of the field was expressed in terms of the magnetic force and magnetic induction, or, alternatively, by the electrotonic state and electric current, both of which had the same observable consequences. Electrostatic or potential energy was expressed in terms of electric force and the displacement current.

Although the notion of stress in the medium was represented by electromagnetic quantities and was closely associated with the theory of polarization outlined in the *Treatise*, Maxwell was vague as to the nature of that stress or strain and the nature of polarization. The latter was defined in terms of a particulate theory, but no mechanical model was suggested.[46] As a result, the two questions that would naturally arise (What is the state of stress in the aether that will enable it to produce the observed electrostatic attractions and repulsions between charged bodies? What is the mechanical structure of the aether that will give rise to this state of stress?) were not sufficiently answered:

> It must be carefully borne in mind that we have made only one step in the theory of the action of the medium. We have supposed it to be in a state of stress, but have not in any way accounted for this stress, or explained how it is maintained.... I have not been able to make the next step, namely, to account by mechanical considerations for these stresses in the dielectric. I therefore leave the theory at this point. (Maxwell 1873, vol. 1, p. 132)

Indeed, in the *Treatise* (sec. 110–11, 645) Maxwell emphasized that while he postulated the existence of the stress, he was not making suggestions about its nature, but was merely showing that a *representation* of such a stress was possible.

Displacement, magnetic induction and electric and magnetic forces were all defined in the *Treatise* as vector quantities (Maxwell 1873, sec. 11, 12), together with the electrostatic state, which was termed the vector potential. All were fundamental quantities for expression of the energy of the field. Those vector quantities replaced lines of force and were conducive to Maxwell's preference for the geometrical approach. However, vectorial representation was more than just a mathematical method; it was a way of *thinking* that required one to form a mental image of the geometrical features represented by the symbols. Because it was a valuable method of representing directed quantities in space, Maxwell used those directed properties of vectors as a way of representing particulate polarization, where opposite parts of the particles differed in polarity.

Although Maxwell's emphasis on polarized particles was intended merely as an aid for understanding the theory, it raised several interpretive problems for others. The particulate representation of the medium made it difficult to distinguish between aether and matter. That, together with the fact that he had no account of the connection between matter and the aether, raised further difficulties for providing a theory of material dielectrics and the aether. Although all the electromagnetic energy or force was *in principle* located in the medium, Maxwell spoke of the aether as the limit of a dielectric of polarized particles, which further contributed to the confusion between the aether and material dielectrics. As a result, there was no independent variable representing the electromagnetic field when the theory was put in Lagrangian form. In addition, he considered certain quantities involved in the equations for material dielectrics to be fundamental quantities of electromagnetism, quantities that were identified as the vector potential of the electric current, electromagnetic momentum at a point and Faraday's electrotonic state (Doran 1974). Hence, Maxwell's theory of the electromagnetic field described the states of both the material and aetherial media. Although that confusion proved problematic for those trying to understand Maxwell's theory, it is possible to see why that may not have been the case for Maxwell himself. The reason has specifically to do with his dynamical account of matter and the role played by mechanical/dynamical concepts (including the vortex hypothesis) in the theory.

In the closing paragraph of the *Treatise* Maxwell remarked that energy could not be *conceived* independently of a medium or substance in which it existed. But he also claimed that his goal regarding the medium had been nothing more than the construction of a mental representation of its action (Maxwell 1873, vol. 2, p. 493). Although there seems to be some necessary connection between matter (the medium) and energy, commitment to the dynamical method and the importance of concepts and representations suggest that such a connection was required if a conception or "consistent representation" of the phenomena was to be possible. Yet, there was no substantial theoretical commitment to a medium, to molecular vortices or to the phenomena of rotation. The mathematical representation (or what Harman calls the functional representation) of matter was sufficient to account for its role in the theory. The same was true of the medium; because Maxwell did not intend to provide a substantial account of the medium, no problems arose for the theory.

In fact, there is substantial evidence to suggest that Maxwell's reluctance about the aether extended beyond the inability to provide a workable mechanical model. In his article "Ether" (Maxwell 1965, vol. 2, pp. 763–75) Maxwell discusses experiments he performed to detect the relative motion of the earth. However, faced with null results, he was reluctant to accept the Stokes drag hypothesis as a true explanation, claiming that "the whole question of the state of the luminiferous medium . . . is as yet very far from being settled by experiment" (1965, vol. 2, p. 770).[47] In addition to the experiments, Maxwell's insistence on the relativity of all physical phenomena suggests that perhaps the aether was not required as an absolute reference frame. In the *Treatise* (sec. 600, 601) he claims that for all phenomena relating to currents in closed circuits, the electromotive intensity is expressed by the same type of formula regardless of whether the motions of the conductors refer to fixed or moving axes. Similar ideas are presented in *Matter and Motion*, where he claims that acceleration, position, force and velocity are all relative terms and although "ordinary language about motion and rest does not exclude the notion of their being measured absolutely . . . this is because . . . we tacitly assume that the earth is at rest" (1952, p. 80):

There is nothing to distinguish one portion of time from another except the different events that occur in them, so there is nothing to distinguish one part of space from another except its relation to the place of material bodies. We cannot describe the time of an event except by reference to some other event, or the place of a body except by reference to some other body. (1952, p. 12)

What was important for Maxwell was knowledge of relations; hence the emphasis on analogy in theory construction.[48] The similarities expressed in Maxwell's analogies were between formal or mathematical relations, rather than between physical phenomena, and that approach was exemplified even in the *Treatise*. The fact that we can apply the Lagrangian approach to a variety of physical phenomena suggests that there is a formal similarity or analogy among them. So in using the dynamical method, Maxwell was again relying, in a sense, on mathematical analogies between electromagnetic and optical phenomena.[49] Matter is expressed dynamically in terms of mass, but the information we receive about it results solely from the mutual action between bodies. Although this mutual action (referred to as force or stress) is evidenced through the change of motion, it seems to require the medium to function as a vehicle through which this stress can act. However, the hypothesis of the medium simply provides a way of conceiving of the phenomena by allowing us to form a mental representation of how energy could be disseminated throughout space.

What does function in a substantial way is energy and the accompanying conservation law. Indeed, it provided for Maxwell, as well as for Thomson and Tait, the physical embodiment of the abstract dynamics of Lagrange. Energy was the capacity for performing work, and all energy was thought to be of the same kind (mechanical) regardless of its different forms. Because of its generality it could replace force as the concept that was common to observable motion, and it formed the only true *physical* basis for the unifying power of abstract dynamical theories. It is perhaps in *Matter and Motion* that Maxwell is most emphatic:

> The discussion of the various forms of energy – gravitational, electromagnetic, molecular, thermal, etc. – with the conditions of the transference of energy from one form to another, and the constant dissipation of the energy available for producing work, constitutes the whole of physical science, in so far as it has been developed in dynamical form under the various designations of Astronomy, Electricity, Magnetism, Optics, Theory of the Physical States of Bodies, Thermo-dynamics and Chemistry. (1952, p. 91)

Conservation of energy was the one generalized statement consistent with the facts in all physical sciences. In addition, it furnished the inquirer with a "principle on which he may hang every known law relating to physical actions, and by which he may be put in the way to discover the relations of such actions in new branches of science" (1952, p. 55).

In addition to its unifying power, conservation of energy exhibited the mark of what Maxwell earlier referred to as the true approach to physical theory, deduction from the phenomena:

> This doctrine, considered as a deduction from observation and experiment, can, of course, assert no more than that no instance of a non-conservative system has hitherto been discovered.

> As a scientific or science-producing doctrine, however, it is always acquiring additional credibility from the constantly increasing number of deductions which have been drawn from it, and which are found in all cases to be verified by experiment. (1952, p. 55)

What the history reveals, however, is the presence of a fundamental tension in trying to determine the basis for unification in Maxwell's work. Clearly the theory required displacement as a fundamental feature; it formed the basis for a field-theoretic approach, but lacked empirical support and did not emerge as a necessary consequence of the theoretical structure. Yet, as we saw earlier, it really did provide the fundamental parameter that functioned as the unifying mechanism for the theory. However, in terms of what could be given a realistic interpretation, the more legitimate basis for the unification of electromagnetism and optics was a general structural principle (energy conservation) together with the mathematical framework supplied by the Lagrangian formalism. A unified field theory of the kind Maxwell constructed would have been impossible without the Lagrangian framework, but what distinguishes the theory as truly unified is the relation between displacement and the velocity of light, that is, how displacement functions to yield a value for V.

3.4.1. Philosophical Conclusions

What methodological lessons can be learned from this historical analysis? At the very least it shows that theory unification is a rather complex process that integrates mathematical techniques and broad-ranging physical principles that govern material systems. In addition to the generality of the Lagrangian formalism, its deductive character displays the feature necessary for successful unification: the ability to derive equations of motion for a physical system with a minimum of information. But this mathematical framework provides little or no insight into specific physical details, leaving the problem of how to interpret the mathematical model and whether or not that physical interpretation should be construed realistically. Maxwell circumvented the issue by relying on a specific methodological assumption: Because one cannot be certain that hypotheses about the medium are true, models and concepts are to be understood as constructive images that either represent physical possibilities or are used for purely heuristic purposes. When there is independent experimental evidence available for these hypotheses or quantities, the debate about their realistic status can begin. My intention here is to highlight the ways in which unification relies on mathematical structures and how the generality and lack of detail in those structures enhance our ability to provide a unified account of diverse physical systems.

But my analysis would be incomplete without saying something about whether or not and how unification functions as a virtue in theory acceptance. What this historical picture shows is that the unifying process in fact failed to provide the basis for realism about theoretical structures like the aether model and displacement current. Although it is beyond the scope of this chapter, a more detailed history would reveal that unification also had no impact on the acceptance of the theory by Maxwell's contemporaries. However, the fact that displacement could not be theoretically justified in no way detracts from its status as the parameter that represents, in some sense, the unification. That is, as a theoretical concept we could not have a unified account of electromagnetism and optics were it not for the role of displacement in the formulation of field theory. This, of course, does not mean that this role justifies its acceptance as a real physical quantity or process. I am here simply commenting on the structural constraints required by the theory. Maxwell, rightly, saw the conservation of energy as the only legitimate physical basis of his theory. Undoubtedly it had a very important role to play, especially given its connection with the equations of motion generated by the Lagrangian formalism.

However, because displacement provides a necessary condition for formulating the field equations, it can enable us to distinguish between a truly unified theory that integrates or reduces various phenomena and one that simply incorporates more phenomena than its rivals.[50] In Maxwell's later versions of the theory, unity was achieved independently of specific theoretical hypotheses that functioned in a coherent explanatory framework. That is to say, the actual representation of the new theory cast in terms of the Lagrangian formalism lacked real explanatory power because of the absence of specific theoretical details. The field equations could explain both optical and electromagnetic processes as the results of waves travelling through space, but that very idea was given no theoretical foundation or understanding. Instead, the conceptualization that usually accompanies theoretical explanations was furnished by the mechanical images and concepts, which, according to Maxwell, lacked the status of explanatory hypotheses. In fact, many of Maxwell's critics, including Duhem and even Kelvin, remarked on the failure of the *Treatise* to provide a clear and consistent explanation of electromagnetic phenomena. Indeed, there is a sense in which Maxwell had shown only that a mechanical explanation of electricity and magnetism was possible.

The successful unification of electromagnetism and optics, despite its questionable explanatory merits, raises the possibility of distinguishing between unification and explanation. By focusing on the importance of mathematical structures and invariance principles like energy conservation, the difference between unified descriptive accounts of the phenomena and truly substantive explanatory theories becomes apparent. Although this distinction between descriptive and explanatory theories has been the focus of extended debate in the philosophy of science, the problem I am referring to here is significantly different. This case and the examples discussed in the following chapters show how mathematical structures are crucial to the unifying process. Because of that, it becomes possible, as in the case of Maxwell's theory, that the very mechanism that enables us to bring diverse phenomena together is one that need not provide a substantive account of the phenomena themselves. As some critics have remarked of Newton's unification of terrestrial and celestial mechanics, it permitted us to calculate everything but explained nothing. Not only is this distinction important for isolating the true nature of unification as something distinct from explanatory power, but also it enables us to see the exact ways in which phenomena are brought together in the unifying process. Of course, there is a sense in which Newton's theory explains the motions of a great many phenomena. The problem was that there was no explanation of the nature of the gravitational force that was cited as the cause, nor any explanation of how it acted except that it obeyed an inverse-square law. So, too, with Maxwell's theory. It provides very accurate descriptions of the behaviour of optical and electromagnetic processes using the field equations, yet there was no explanation of just what the field consisted in or how these waves could be propagated through space.

Although I have stressed the importance of a theoretical parameter in representing unification, one can still inquire about the *kind* of unity achieved by Maxwell. Earlier I claimed that the unity achieved in "On Physical Lines" reduced optical and electromagnetic processes to the mechanics of one aether. But what about the later abstract formulation of the theory? Did it consist in a reduction of two phenomena to one, or did it involve an integration of different phenomena under the same theory, that is, a synthetic unity in which two processes remained distinct but characterizable by the same theoretical framework? That is, did the theory show that electromagnetic and optical phenomena were in fact the same natural kind, hence achieving an ontological reduction? Or did the theory simply provide a framework for showing how those processes and phenomena were interrelated and connected? The answer is that it did both.

If we consider the structure of Maxwell's field equations, we cannot assume that the electromagnetic field reduces electricity and magnetism to one force. Instead, the electric and magnetic fields retain their independence, and the theory simply shows the interrelationship of the two – where a varying electric field exists, there is also a varying magnetic field induced at right angles, and vice versa. The two together form the electromagnetic field. In that sense the theory unites the two kinds of forces by integrating them in a systematic or synthetic way. Viewed this way, one calls into question the traditional relationship between unified theories and theoretical reduction. The former need not, even in the relatively straightforward case of Maxwell's theory, imply the latter. Yet there can be no doubt that the theory also reduced optical phenomena to their electromagnetic foundation. But without any substantial explanation of *how* that took place the reduction offered little in the way of true understanding.

Consideration of this historical case leads one to see that in at least some cases unity does not play a significant role in establishing theoretical realism or in providing a context for theory acceptance; some further condition is required. In the case of Maxwell's theory, its general acceptance followed the discovery of independent experimental evidence in the form of Hertz's electromagnetic waves. That occurred in 1888, some 15 years after publication of the *Treatise*. Up until that time, Maxwellian field theory had been popular only in Britain and had enjoyed no support on the Continent, where electromagnetism was dominated by action-at-a-distance accounts. Although Hertz's discoveries were interpreted as evidence for propagation of electromagnetic effects in time, as opposed to at a distance, several of Maxwell's physical ideas were either eliminated or reinterpreted by Hertz.[51] The field equations were recast in symmetric form, and unfortunately it was those phenomenological descriptions that Hertz took to be the basis of Maxwellian electrodynamics.

My claim here is not that we should be anti-realists about the electromagnetic field, nor that unification is not a desirable outcome. The moral is that theoretical unity is not always the sort of thing that should be taken to provide evidential support, even in such remarkable cases as Maxwell's electrodynamics. Nor am I suggesting that it would have been somehow irrational to have put any faith at all in the theory; rather, I am simply cautioning against seeing a successful phenomenological theory as evidence for an ontological interpretation of its theoretical parameters.

The example I want to consider next is the relationship between unity and mathematics in the electroweak theory. Although the cases bear important similarities, as we move into the domain of more abstract theory the division between the mathematical structure of the theory and the accompanying physical dynamics becomes increasingly blurred. As a result, the issue of unity in nature as the reflection of a unified theory becomes even more problematic.

4

Gauges, Symmetries and Forces: The Electroweak Unification

In Chapter 3 we saw how Maxwell's theory incorporates two different aspects of unity: the reductive and the synthetic. We also saw the different ontological and explanatory features associated with each. In this chapter we shall look more closely at a case of synthetic unity, one that integrates the electromagnetic and weak forces. The electroweak theory provides a unified structure within which electromagnetic and weak interactions can be integrated, and yet the particles that carry the forces remain distinct. Hence, at the level of particle ontology we do not have the kind of reduction that is present in Maxwell's unification of electromagnetism and optics, where the processes are identified with each other. Yet we do have a kind of theoretical unification in terms of the gauge theory SU(2) × U(1) that describes the mixing of the two fields. In this latter sense the history is remarkably like that of electrodynamics, insofar as the unification requires the presence of a mathematical structure and a particular theoretical parameter: specifically, gauge theory (symmetry) and the Weinberg angle, which represents the combination of the coupling constants for weak and electromagnetic processes.

What makes this case especially interesting is that if theoreticians had considered only the physical similarities between electromagnetic and weak interactions that might have made them candidates for unification, it is unlikely that any such unification attempt would ever have been undertaken. But the applicability of gauge theory and symmetry principles provided a powerful mathematical framework that created not only the context but also the mechanism for generating a synthetic rather than a reductive unity of these two processes. Two theories, each governed by independent symmetry groups, are combined under a larger symmetry structure, yielding a new hybrid theory that retains aspects of electrodynamics and the theory of weak interactions but introduces a new parameter necessary for presenting the separate theories in a unified form. Although this parameter (the Weinberg angle) represents the mixing of the weak and electromagnetic fields, its emergence was in some sense analogous to Maxwell's displacement current, in that it was a theoretical add-on; neither it nor its value was determined from within the larger theoretical structure.

This case is important philosophically because it shows how the structural constraints supplied by the mathematics can actually dictate the kinds of physical processes that the theory can accommodate. That is to say, the process of unifying these forces was driven by considerations grounded in the mathematics of gauge theory, rather than in the phenomenology of the physics. Hence, in this case, to an even greater extent than in the others, I consider unity to be a product of the mathematics, rather than a verification of a detailed causal hypothesis about relations between diverse phenomena or natural kinds.[1] The result is a unified theoretical framework that integrates forces that, at the level of phenomena, remain ontologically distinct.

In order to appreciate how electroweak unification was achieved, we must begin with the early work of Enrico Fermi on beta decay. Fermi's theory was constructed by analogy with quantum electrodynamics, so in that sense it represented the first attempt at bringing those two processes together. Tracing the history back to Fermi and the subsequent development of the V-A theory will also give us some indication of just how different in terms of phenomenology the weak and electromagnetic processes were, and how initially implausible it was that those two forces could be unified in a single theory. From there I shall discuss the evolution of the modern theory of weak and electromagnetic interactions, a process that spanned three decades. The development of the electroweak theory not only involved remarkable theoretical changes in the way physicists viewed the elementary forces of nature but also was associated with significant experimental discoveries, such as the discovery of weak neutral currents and the W and Z bosons. Although those experimental discoveries were significant for the theory, perhaps its most important feature was its renormalizability, something that proved to be the primary constraint on the theory's initial acceptance.

What is particularly interesting about the electroweak theory is the way unity and renormalizability are related within the theoretical structure. But, as with Maxwell's electrodynamics, at no point in the theory's development does unity emerge as an important feature in regard to its acceptance. Nevertheless, the theory does provide a unity of electromagnetism and the weak force; hence it is important to determine what that unity consists in, how it was achieved and what its implications are for an ontological unity in nature. Because the electroweak theory displays a synthetic unity, the comparison with electrodynamics becomes especially significant.

Because my emphasis is on theory unification, I shall focus mainly on the theoretical history; but a complete and historically accurate account of the developments would include a discussion of the ways in which theory and experiment merge to create what we take to be a well-confirmed theory. Others have focused on the experiments associated with the electroweak theory; Galison (1987) and Pickering (1984), in particular, have given rival accounts of the weak-neutral-current experiments and the role of theory in their interpretation. The discovery of the W and Z bosons is discussed by Watkins (1986).

4.1. Fermi and Beta Decay

Beta decay involves the emission of positrons and electrons from unstable nuclei, and it was one of the principal areas of research in the field of radioactivity in the

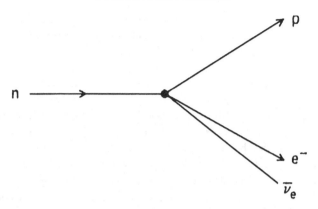

Figure 4.1. Weak beta decay of the neutron.

early twentieth century. No one seemed able to explain the fact that the energy spectrum for the electrons was continuous, rather than discrete as one might expect from a quantized system. It was Pauli who finally explained the phenomenon by suggesting that beta decay involved two particles, the electron and a neutral lepton called the neutrino.[2] The undetected massless neutrino was thought to carry away part of the energy emitted in the decay process, thereby giving the energy spectrum its continuous appearance. In 1934 Pauli's hypothesis was formalized by Enrico Fermi into a quantum field theory of weak interactions, the force responsible for beta decay. Fermi's account was modelled on electrodynamics; beta decay was considered analogous to the emission of a photon from an electromagnetic transition in an excited atom, and like photon emission, it took place at a single space-time point. The difference, however, was that the role of the emitted photon was played by an electron-antineutrino pair, rather than by the quantum of a new, weakly interacting field. So, in the beta decay of a free neutron, we have a proton, an electron and an antineutrino, four particles reacting at a single point: $n \rightarrow p + e^- + \bar{\nu}_e$ (Figure 4.1).

Although Fermi's theory of weak interactions was modelled on the vector currents of quantum electrodynamics, the analogy was far from perfect. The photon emitted in a radiative transition is the quantum of the electromagnetic force field; however, it was difficult to envision how the corresponding $e^- \bar{\nu}_e$ pair could be the weak-force quantum, especially because the effective mass of the pair varied from process to process. But because quantum electrodynamics (QED) had been reasonably successful, it seemed natural, at least as an initial step, to model other theories of interactions on QED. In that sense, Fermi was not explicitly trying to unify electrodynamics with the weak force; he was merely following basic principles of theory construction – using previously successful theories as models for new ones. Nevertheless, his work does represent the beginnings of the unifying process in much the same way that Faraday's analogies provided an important stepping-stone for Maxwell.

Fermi's account soon evolved into a more abstract theory based on the idea of interacting currents. Again, it was modelled on the vector currents of QED. Those new currents were generalizations of the charged currents found in the Dirac theory, $J_\gamma = e \bar{\psi} \gamma_\mu \psi$, where ψ is the four-component electron-spinor wave function, and γ_μ are the 4×4 Dirac matrices. In beta decay, the neutron and proton define a current J_μ that carries isospin and a baryon number, and the electron and its neutrino define another current J'_μ that has a well-defined lepton number. Together those currents replaced the fields originally used to represent particles. Weak interactions could then be described by coupling the currents together through a Hamiltonian, $H = (G/\sqrt{2}) J_\mu J'_\mu$, where G was the Fermi coupling constant, with a value of roughly 10^{-5} GeV^{-2}. The value for G was assumed to be a weak charge, analogous to the electric charge e; it governed the intrinsic strength of the interaction and hence the rate of beta decay. That process was known as a current-current interaction.

In 1957 it was found that parity (mirror symmetry) was violated in weak interactions. To accommodate that, it was suggested, on the basis of strong experimental evidence, that the interaction factors (contained in the weak currents J_μ and J'_μ) were a mixture of vector and axial-vector quantities.[3] Those factors had approximately equal strengths (exact for the lepton current), but opposite signs, hence the name V-A. To see how that solved the parity problem, consider that a vector quantity has well-defined properties under a Lorentz transformation. It will change its sign if rotated 180° and appear identical under a complete rotation of 360°. An axial vector transforms like a vector under rotations, except that it will transform with the opposite sign to a vector under an improper transformation like parity. Hence, if weak interactions consist of vector and axial-vector components, each one will look different under parity transformation.

Although the V-A theory was able to incorporate parity violation, there were other problems that needed to be corrected. Because it was a theory of point-like interactions, it gave unacceptable predictions for weak interactions at high energies. For instance, it predicted that cross sections for neutrino-electron scattering would rise linearly with the energy of the incoming neutrino. (A cross section is the basic measurement for the probability of particles interacting.)[4] But of course that prediction could not be true for arbitrarily high energies, because it would mean that neutrino collisions should be extremely common in bubble chambers and cosmic-ray photos, something that was not borne out empirically. In addition, the V-A theory, unlike QED, was non-renormalizable. In an effort to overcome those difficulties, it was suggested that the weak force be represented as mediated by particle exchange, rather than contact interaction between two currents. In order to retain the space-time structure of the V-A theory it was necessary that the exchanged particles that carried the weak force be spin-1 vector mesons, usually referred to as intermediate vector bosons (IVBs). The phenomenology of the V-A theory required that there be both positively and negatively charged particles; later those were identified with two charged states of the W boson.

Even though the W boson could be thought of as the weak analogue of the photon, it was there that the analogy with QED ended. There were significant differences between the photon and the W particle, differences that would seem to rule out any kind of unification between the weak force and electrodynamics. One dissimilarity was that the photon was electrically neutral, and weak processes like beta decay involved the exchange of electric charge between the currents J_μ and J'_μ. That required the W to exist in at least two charged states, $W^\pm$.[5] The second and most striking difference was that the W had to be massive (unlike the massless photon) in order to explain the extremely short range of the weak force.[6] The problem, however, was that the W-boson theory proved unsuccessful in overcoming the very problems it was designed to solve.[7]

Ironically, the difficulties stemmed from the fact that the bosons had spin 1 and non-zero mass, properties that were necessary to recover the phenomenology of the successful V-A theory. The spin of a particle is a concept used to characterize its properties under Lorentz transformations. A spin-1 particle transforms like a vector, with three components (two transverse, which are perpendicular to the direction of momentum, and one longitudinal, which is parallel) defining its orientation or polarization. For massless particles the longitudinal component has no physical significance; photons are always polarized in the plane transverse to their direction of motion. In the case of massive particles, the mass is described by a factor contained in the longitudinal component of the propagator.[8] It is this factor that creates an ever-increasing number ($\rightarrow \infty$) of contributions to the perturbation series, contributions that cannot be reabsorbed into a redefinition of masses and couplings. Consequently, the theory is not renormalizable and cannot provide realistic predictions. In addition, the presence of the mass factor in the W propagator causes the cross sections for some high-energy processes to rise with energy at a greater rate than is allowed by fundamental theorems. As a result, the high-energy behaviour and the renormalizability problem that plagued the V-A theory simply re-emerged in the boson case. Moreover, processes like π-meson decay, whose calculation was straightforward in the V-A theory, could not be easily explained in terms of W-boson exchange. So the attempt to construct a theory of weak interactions on the particle-exchange model of QED incorporated significant dissimilarities between the W and the photon. The very properties required by the W boson seemed to rule out any successful analogy between the photon exchange characteristic of electromagnetism and the W-boson exchange that supposedly was responsible for mediating weak interactions.

I mentioned earlier that Fermi's initial work on beta decay and weak interactions was not an explicit attempt to *unify* electromagnetism and the weak force; rather, it was an attempt to provide a theoretical framework for the process of beta decay that would be based on the emission of lightweight particles from the nucleus. At that time, relativistic theories were incapable of explaining how these particles could be bound in orbits of nuclear dimensions. Although Fermi agreed with Heisenberg's 1932 account, which stated that the nucleus consisted of only heavy particles (protons and neutrons), he nevertheless saw the electromagnetic model as the most plausible place to start. So although Fermi's work represented an important beginning, the first real attempt at unifying weak and electromagnetic processes was not undertaken until 1957, by Julian Schwinger.

In order to understand and appreciate the significance of Schwinger's approach, we first need to make a short detour into the domain of symmetry and its role in gauge theory. This is especially important, because one of the goals of this chapter is to show that it was the symmetry framework provided by gauge field theory that ultimately supplied the foundation for unification of the weak and electromagnetic interactions. That was achieved because of the power of the mathematics to generate a kind of particle dynamics that provided the form for various field interactions. As a result, the mathematical structure can, in some sense, be seen to dictate how the physics of the theory is fleshed out.

4.2. Symmetries, Objects and Laws

Many discussions of symmetries stress their global character, that is, the invariance (or covariance) of physical laws under a set of transformations described by a single parameter that applies regardless of location in space or time. When it comes to finding the appropriate symmetry for conservation of electric charge, we need to move beyond the geometric space-time symmetries like translation in time and rotation in space to a local symmetry where physical laws are invariant under a local transformation. This involves an infinite number of separate transformations that are different at every point in space and time.

In general, a theory that is globally invariant will not be invariant under locally varying transformations. But by introducing new force fields that transform in certain ways and interact with the original particles in the theory, a local invariance can be restored. In other words, these force fields are postulated simply to ensure that certain local invariances will hold. For example, the global form of charge conservation would necessitate instantaneous propagation of signals (i.e., charge cannot be created at one point and destroyed at another), an effect that would conflict with the special theory of relativity. Consequently, charge conservation must be thought of as locally invariant.[9] In order to achieve local gauge invariance, the field must be carried at the speed of light and interact with charges in a particular way. In the electromagnetic case this local symmetry is realizable only by introducing a new field called a gauge field. The new theory that incorporates the gauge field specifies interactions between the field and the conserved quantity of electric charge carried by the source current. In essence, the requirement of U(1), local gauge invariance, enables us to predict the existence of the photon and its properties. To put these ideas into proper context, let me describe some background to the notion of gauge invariance associated with the gauge field.

Gauge invariance had its beginnings in 1918 in the work of Hermann Weyl, who wanted to draw attention to the requirement that the laws of physics should remain the same if the scale of all length measurements were changed by a constant factor. Weyl wanted to extend that idea to a local invariance, where scale (or gauge) changes could vary at different points in space and time in a manner analogous to the curvilinear coordinate transformations of general relativity. Weyl thought he could show that from that gauge symmetry, one could, using Noether's theorem, derive the conservation law for electric charge. The attempt was unsuccessful, but the idea was resurrected in 1927 by Fritz London, who showed that the proper symmetry for electric charge was phase invariance rather than scale invariance, that the electromagnetic field in fact allowed for an arbitrary variation in the phase factor from point to point in space-time. Because Weyl was also involved in that reformulation, it continued to be labelled gauge invariance even though it has nothing to do with the notion of a gauge, which refers to a choice of length scale.

The original idea involved the introduction of a scaling factor $S(x)$. If we think of a vector at $x + dx$, the scale factor is given by $S(x + dx) = 1 + \partial_\mu S \, dx^\mu$.[10] The derivative $\partial_\mu S$ is the mathematical connection associated with the gauge change, which Weyl identified with the electromagnetic potential A_μ (because it transformed like a potential).[11] We can think of this connection as that which relates the scale or gauge of the vectors at different positions. The reinterpretation in light of quantum mechanics involved a change in phase of the wave function $\psi \rightarrow \psi^{-ie\lambda}$ (λ becomes a new local variable), with the gauge transformation for the potential A_μ becoming $A_\mu \rightarrow A_\mu - \partial_\mu \lambda$. If we write the non-relativistic Schrödinger equation

$$\left[\frac{1}{2} m (i\hbar\Delta - eA)^2 + e\Phi + V \right]\psi = i\hbar\frac{2\psi}{2t}$$

(where the canonical momentum operator $P_\mu - eA_\mu$ is replaced by $-i\hbar\Delta - eA$), then after the phase change an additional gradient term proportional to $e\Delta\lambda$ emerges, the result of the operator $i\hbar\Delta$ acting on the transformation wave function. This additional term spoils the local phase invariance, which can then be restored by introducing the new gauge field A_μ. The gauge transformation $A_\mu \rightarrow A_\mu - \partial_\mu \lambda$ cancels out the new term. This new gauge field is simply the vector potential defining the electromagnetic field, because the electromagnetic field is gauge-invariant in just this sense (Moriyasu 1983). A different choice of phase at each point can be accommodated by interpreting A_μ as the connection relating phases at different points. In other words, the choice of a phase function $\lambda(x)$ will not affect any observable quantity as long as the gauge transformation for A_μ has a form that allows the phase change and the change in potential to cancel each other. What this means is that we cannot distinguish between the effects of a local phase change and the effects of a new vector field.

The combination of the additional gradient term with the vector field A_μ prescribes the form of the interaction between matter and the field, because A_μ is what provides the connections between phase values at nearby points. The phase of a particle's wave function can be identified as a new physical degree of freedom that is dependent on space-time position. In fact, it is possible to show that from the conservation of electric charge one can, given Noether's theorem, choose a symmetry group, and the requirement that it be local forces us to introduce a gauge field, which turns out to be the electromagnetic field. The structure of this field, which is dictated by the requirement of local symmetry, in turn dictates, almost uniquely, the form of the interaction, that is, the precise form of the forces on the charged particle and the way in which the electric-charge current density serves as the source for the gauge field.[12]

The phase transformations of QED are one-parameter transformations and form what is called a one-dimensional Abelian group (meaning that any two transformations commute). This group is called the U(1) group of a U(1) gauge symmetry. Mathematical descriptions of all symmetry operations are dealt with in group theory, which uses algebra to describe operations like rotations.[13] These symmetry groups, however, are more than simply mathematizations of certain kinds of transformations, because in the non-Abelian case (non-commutative transformations) it is the mathematical structure of the symmetry group that determines the structure of the gauge field and the form of the interaction. In these more complicated situations there are several wave functions or fields transforming together, as in the case of SU(2) and SU(3) transformations, which involve unitary matrices acting on multiplets.[14] These symmetries are internal symmetries and typically are associated with families of identical particles. In each case the conserved quantities are simply the quantum numbers that label the members of the multiplets (such as isospin and colour), together with operators that induce transitions from one member of a multiplet to another. Hence the operators correspond, on the one hand, to the conserved dynamical variables (isospin, etc.) and, on the other hand, to the group of transformations of the symmetry group of the multiplets.[15]

The extension of gauge invariance beyond electromagnetism began with the work of Yang and Mills in the 1950s (Yang and Mills 1954). The connection between charge conservation and local gauge invariance, together with the fact that the latter could determine the structural features of electromagnetism, suggested that perhaps the argument could be generalized to other kinds of conservation laws. If so, gauge invariance would become a powerful tool for generating a theory of strong interactions. Yang and Mills began with the conserved quantity isospin (violated in electromagnetic and weak interactions), which allows the proton and neutron to be considered as two states of the same particle. Here, a local gauge invariance means that although we can, in one location, label the proton as the "up" state of isospin $\frac{1}{2}$, and the neutron as the "down" state, the up state need not be the same at another location. Because this assignment of isospin values is arbitrary at each point, some form of connection or rule is required to coordinate the various choices at different positions. In electromagnetism, phase values are connected by the potential A_μ; thus an analogous isospin potential field was introduced by Yang and Mills. But because the SU(2) symmetry group that governs isospin is also the group that governs rotations in a three-dimensional space, the potential field associated with this group is very different from the U(1) phase group of electromagnetism. In this SU(2) case the "phase" is replaced by a local variable that specifies the direction of the isospin.[16]

The interesting aspect of the Yang-Mills theory is the relation between the SU(2) group and the isospin connection. The components of isospin are transformed by the elements of the group, and consequently the connection (potential) is capable of performing the same isospin transformations as the transformations of the SU(2) group. In order to see how the potential generates a rotation in the isospin space, we first need to consider how the potential relates to a rotation.[17] A three-dimensional rotation of a wave function is $R(\theta)\psi = e^{i\theta L}\psi$, where θ is the angle of rotation, and L is the angular-momentum operator. It is important to note that the potential is not a rotation operator. In the electrodynamics case the amount of phase change has to be proportional to the potential to ensure that the Schrödinger equation remains invariant. In the Yang-Mills case the potential is considered to be a "generator" of a rotation and must be proportional to the angular-momentum operator to preserve gauge invariance.[18] Consequently, the potential must include three charge components corresponding to three angular-momentum operators, L_+, L_-, L_3. $A_\mu = \sum_i A_\mu^i(x)L_i$ is a linear combination of the angular-momentum operators, where the coefficients A_μ^i depend on space-time position. So the potential acts like a field in space-time, and an operator in the abstract isospin space. Each of the potential components behaves like an operator, as in the case of the raising operator L_+, which transforms a down state into an up state. This formal operation corresponds to the case of a neutron absorbing a unit of isospin from the gauge field and becoming a proton.[19]

We can see, then, how gauge invariance can dictate the form of the potential, because the gauge field must, unlike the electromagnetic potential, carry electric charge. In addition, the requirement of local invariance dictates that the mass of the potential field be zero. Unfortunately, in the Yang and Mills case the standard mass term required by the short range of the strong force destroys the gauge invariance of the Lagrangian. So although the Yang-Mills theory could not provide a complete account of strong interactions, it did furnish an important framework for dealing with the relations between conserved quantities, especially quantum numbers and symmetries. Yang and Mills showed that gauge invariance was more than just an "accidental" symmetry of electromagnetism or merely a property of phase changes. In light of their work, gauge symmetry became a powerful tool that could determine the form of the matter–field interaction and could define much of the dynamical content of a theory of strong interactions. Its ability to generate that kind of dynamics was a role it continued to play in constituting the core of the electroweak theory.

4.3. The First Steps to Unity: Symmetry and
Gauge Theory

As mentioned earlier, Schwinger's work represented one of the first attempts to unify electromagnetic and weak interactions. He wanted to find a description of the stock of elementary particles within the framework of the theory of quantized fields: The massive charged vector boson and the massless photon were to be gauge mesons. The two key components in his account were (1) that leptons carried a weak form of isospin analogous to the strongly interacting particles and (2) that the photon and the intermediate vector bosons (identified as the Z particle) were members of the same isospin multiplet. The interesting feature of Schwinger's paper is the central place of symmetry in constructing a field-theoretic model for elementary particles. On the basis of the spin values for Fermi-Dirac fields ($\frac{1}{2}$) and Bose-Einstein (B-E) fields (0), he began by describing the massive, strongly interacting particles. If spin values are limited in that way, then the origin of the diversity of particles must lie with the presence of internal degrees of freedom. (Although it is possible to have spin-1 B-E fields, they are assumed to refer to a different family of particles that includes the electromagnetic field.) Supposedly these degrees of freedom are dynamically exhibited by specific interactions, each with its own characteristic symmetry properties. So Schwinger's approach was to begin with some basic principles of symmetry and field theory and go on to develop a framework for fundamental interactions derived from that fixed structure. In strong interactions, the heavy particles were described using a four-dimensional internal symmetry space, and the weak-force leptons were represented by a three-dimensional rotation group from which one could then read off their characteristic features. And the symmetry properties of the lepton field could be generally stated at the same dynamical level as electromagnetic interactions.[20]

In the case of the bosons, symmetry arguments were once again fundamental. On the basis of the symmetry that existed between the isospin properties of heavy bosons and fermions, Schwinger wondered if there might also be a family of bosons that would be the realization of the three-dimensional rotation group. Because the electromagnetic field was assumed to be the third component of a three-dimensional isospin vector, it seemed likely that there were two other charged particles completing the triplet. Hence, Schwinger concluded that "we are led to the concept of a spin one family of bosons, comprising the massless, neutral photon and a pair of electrically charged (Z) particles that presumably carry mass in analogy with the leptons" (1957, p. 434). From the general suggestion that there existed a family of bosons that was the isotopic analogue of leptons, together with the identification of the photon as its neutral member, Schwinger developed the dynamics of a charged, unit-spin Z-particle field considered to be the "invisible instrument" of the whole class of weak interactions. There was, however, no experimental evidence regarding lepton polarization or the existence of the Z particle; moreover, the effects of various strong interactions obscured the predictions of the formalism. Nevertheless, Schwinger's "model" was important for several reasons. Not only did it provide the framework for future work, but also it showed how theory construction could proceed from a few general concepts to a dynamical theory of elementary particles. The fundamental and guiding assumption was the identification of the leptons and photon-Z-particle family as the physical realization of a three-dimensional internal symmetry space, in the way that the particles characterizing the strong interaction (the heavy fermions and bosons) were realizations of a four-dimensional symmetry space.

Schwinger's use of symmetry enabled him to construct powerful analogical arguments and to postulate the existence of a particle field that was the physical analogue of an imaginary mathematical symmetry space. However, a different kind of symmetry was required if electrodynamics and the weak interaction were to be unified. There were certain shared characteristics that gave rise to the idea that the two forces could be synthesized – parallels that emerged as a result of the supposition that weak interactions were mediated by bosons. The first was that both forces equally affected all forms of matter – leptons and hadrons. Both interactions individually possessed universal coupling strengths, and a single coupling constant sufficed to describe a wide class of phenomena. Finally, both forces were vectorial in character; that is, the interactions were generated by vectorial couplings of spin-1 fields. The principal difficulty was the difference between the masses of the charged bosons (large masses were required because of the short range of the weak force) and the massless photon. If there was only an isotopic triplet of leptons coupled to a triplet of vector bosons, the theory would have no internal symmetries; hence, a symmetry principle was required in order to relate the forms of weak and electromagnetic couplings.[21]

Because of the mass problem, it was thought that only partial symmetries – invariance of only part of the Lagrangian under a group of infinitesimal transformations – could relate the massive bosons to the massless photon. In order to achieve that it would be necessary to go beyond the idea of a triplet of bosons and introduce an additional neutral boson Z_s that would couple to its own neutral lepton current J_μ^s. That was proposed by Glashow in 1961. By properly choosing the mass terms to be inserted into the Lagrangian, Glashow was able to show that the singlet neutral boson and the neutral member of the triplet would mix in such a way as to produce a massive particle B (now identified as Z^0) and a massless particle that was identified with the photon. Those same masses guaranteed that the charged members of the triplet would become very massive. In addition, because leptons were characterized as representations of the SU(2) × U(1) symmetry group, the photon mediating electromagnetic interactions conserved parity, whereas the weak interactions mediated by the two charged and neutral vector mesons did not.

Glashow was led to the SU(2) × U(1) group by analogy with the approximate isospin-hypercharge group that characterized strong interactions. The idea of using gauge theory was becoming increasingly popular as a result of the work of Yang and Mills, but it seemed obvious that one needed a gauge group larger than

SU(2) to describe electroweak interactions. Moreover, because the universality (of coupling strengths) and the vector character of the forces were features of a gauge theory, those shared characteristics suggested that both the weak and electromagnetic forces might arise from a gauge principle. But perhaps the most striking feature of Glashow's model was that it predicted the existence of neutral currents, something for which there was no experimental evidence.

What is especially interesting about Glashow's work is that analogy with other successful ideas, rather than unity, seemed to be the driving force. In fact, Glashow himself claimed that the unity of the weak and electromagnetic forces as described in his 1961 paper was a nice "byproduct and NOT a direct stimulus".[22] What motivated him was the desire to find a theory that could explain the phenomena and be theoretically consistent. The conclusion of a paper surveying the use of gauge theories for weak, strong and electromagnetic interactions written with Murray Gell-Mann in 1961 is evidence that unification was not then "in the stars":

> In general, the weak and strong gauge symmetries will not be mutually compatible. There will also be conflicts with electromagnetic gauge symmetry. . . . We have not attempted to describe the three types of interaction together, but only to speculate about what the symmetries of each one may look like in an ideal limit where symmetry-breaking effects disappear.[23]

Undoubtedly he was influenced by Schwinger, under whose direction Glashow had completed his Ph.D. in 1957. But after the 1961 paper, and until 1970, he had more or less abandoned the idea of an electroweak gauge theory. Perhaps the main reason for that was the failure of renormalizability; there simply was no advantage in having a unified model of electroweak interactions if it was not renormalizable. Another consideration was the empirical success of the Gell-Mann and Ne'eman SU(3) scheme for strong interactions. That theory was originally presented as a gauge theory, making it difficult to see how one could describe both weak and strong interactions via gauge theories – there didn't seem to be enough room for commuting structures of weak and strong currents. But that was, in essence, a practical problem that perhaps could be overcome by finding a symmetry group large enough to accommodate both kinds of interactions. The renormalization problem was of a different order; regardless of how successful the theory might be, without renormalizability it could not be deemed acceptable.

A rather different account of electroweak interactions, without an additional neutral boson and neutral currents, had been published by Abdus Salam and J. C. Ward in 1958–9, just prior to Glashow's 1961 paper. They used the idea of a charge space and tried to show how gauge transformations in that space could generate weak and electromagnetic interactions. The three-dimensional gauge invariance of the charged space made it necessary to introduce into that space a triplet of fields consisting of two charged vector-boson fields, in addition to the electromagnetic field. And, supposedly, the form of the interaction of the charged fields would mediate the weak interactions. At that time, the gauge principle seemed to be the only guiding postulate for writing the fundamental interactions of fields. Whenever a symmetry property existed, the associated gauge transformation would lead, in a definite way, to the postulation of an interaction mediated by a number of particles. However, the difficulty with their theory was its inability to account, in a natural way, for parity violation in weak interactions. A few years later, in 1964, they introduced a model similar to Glashow's that incorporated neutral currents, something they considered to be the "minimum price one must pay to achieve the synthesis . . ." they were seeking (Salam and Ward 1964).

Regardless of the experimental status of the neutral currents, their introduction by no means solved the theoretical problems. Despite the similarities, there were profound differences between the weak and the electromagnetic interactions; the most troublesome was that their coupling strengths were significantly different. In order for the weak coupling constant to have a proper value, the boson mass needed to be extremely large, an assumption that could not be derived from within the theory itself. As a result, the boson masses were added to the theory by hand, making the models phenomenologically accurate but destroying the gauge invariance of the Lagrangian and ruling out the possibility of renormalization. Although gauge theory successfully generated an electroweak model, unlike the case with electromagnetism one could not reconcile the *physical* demands of the weak force for the existence of massive particles with the *structural* demands of gauge invariance. Both needed to be accommodated if there was to be a unified theory, yet they were mutually incompatible.

4.4. Unity through Symmetry-Breaking

Hopes of achieving a true synthesis of the weak and electromagnetic interactions were renewed a few years later, in 1967, with a paper by Steven Weinberg on leptons. Weinberg focused on leptons because they interact only with photons and the bosons that mediate the weak force. One of the most important assumptions of the gauge theory of weak interactions is that the lepton pairs can be considered weak isospin doublets like the neutron and proton. The core of the model is the description of the weak and electromagnetic interactions of leptons with the same type of coupling to the gauge field. Recall that one of the earlier problems was reconciling the different coupling strengths of the interactions. Part of the solution came, once again, from the structure of the SU(2) weak isospin symmetry group. In order to complete the group, a new gauge field was needed, and once that new field was in place, the weak and electromagnetic interactions could be unified under a larger gauge symmetry group given by the product of SU(2) × U(1). The problem, however, was that this new gauge field implied a new class of weak interactions, so a proper physical interpretation for the field was necessary. The solution offered by Weinberg was to assume the existence of a new weak interaction that involved a neutral current. Once the framework [the SU(2) × U(1) group structure] was in place for the mixing of the fields, the remaining problem was simply to generate the appropriate masses for the $W^{\pm}$ and the Z^0 particles/fields.

Weinberg's idea was that one could understand the mass problem and the coupling differences by supposing that the symmetries relating the two interactions were exact symmetries of the Lagrangian that were somehow broken by the vacuum. Those ideas originated in the early 1960s and were motivated by work done in solid-state physics on superconductivity. There it was common to use quasi-particles, which could not be directly mapped onto fundamental observable fields, to explain experimental results. Traditionally in field theory a direct correspondence was assumed between physical particles and terms in the Lagrangian. However, that idea was challenged on the grounds that when one takes interaction terms into account, the correspondence between fields and observable particles might be broken. In other words, one might not be able simply to read off from the Lagrangian an array of physical, observable particles – more might be involved. The characteristic feature of superconductivity was that an energy gap was produced between the ground state and the excited states of a superconductor. That fact had been confirmed experimentally, but the first successful field-theoretical account was given later in the now-famous BCS theory (Bardeen, Cooper, and Schrieffer 1957). That notion of an energy gap was introduced into the domain of particle physics by Nambu and Jona-Lasinio (1961), who suggested that the mechanism producing the gap was also responsible for the nucleon mass that arose largely because of the self-energy of a fermion field.[24]

The translation of those ideas to particle physics was based on an analogy between the physical vacuum of a quantum field theory and the ground state of an interacting many-body system – both of which are states of minimum energy. In quantum field theory the state of minimum energy need not be one in which all of the fields have zero average value. Because fluctuations are always possible, it may be the case that in the true ground state of the theory some field will have a non-zero equilibrium (expectation) value, producing an asymmetrical situation. One such example is the ferromagnet, where the basic interactions between the constituents of the system are rotationally symmetrical, even though its configuration is not. That is, the Hamiltonian describing the spin-spin interaction is rotationally invariant, but below the ferromagnetic transition temperature the ground state is not; instead, a particular direction in space is singled out as the preferred one, because of the alignment of the spins. This situation, where the ground-state configuration does not display the symmetry of the Hamiltonian, is referred to as a symmetry that is spontaneously broken. Technically, the symmetry is simply "hidden", because there are no non-symmetric terms added to the Hamiltonian. In a case in which, for example, an external-magnetic-field term is added, resulting in the selection of one specific direction, rotational invariance is explicitly broken. But in the case in which all directions are equally good, the symmetry still exists; it is merely hidden by the association of one of the possible directions with the ground state of the ferromagnet.

Although that was an attractive idea, it had one major drawback. Goldstone, working in high-energy physics, had shown that cases of spontaneous symmetry-breaking involved the production of a massless, spinless particle, now referred to as the Goldstone boson.[25] For each degree of freedom in which the symmetry is spontaneously broken, a massless scalar field always appears. Because it was thought that no such particles could exist, it was difficult to see how the idea of spontaneous symmetry-breaking could be applied to field theory. But once again a clue was forthcoming from the field of superconductivity. In that context, symmetry-breaking does not give rise to Goldstone excitations, a result that is directly traceable to the presence of the long-range Coulomb interaction between electrons (Anderson 1958). The first indication of a similar effect in relativistic theories was provided in 1963 by P. W. Anderson, who showed that the introduction of a long-range field like the electromagnetic field might serve to eliminate massless particles from the theory. Basically, what happens is that in cases of local symmetries that are spontaneously broken, the zero-mass Goldstone bosons can combine with the massless gauge bosons to form massive vector bosons, thereby removing the undesirable feature of massless particles. That result was fully developed for relativistic field theories by Peter Higgs and others, who showed that Goldstone bosons could be eliminated by a gauge transformation, so that they no longer appeared as physical particles.[26] Instead, they emerged as the longitudinal components of the vector bosons.

The Higgs model is based on the idea that even the vacuum state can fail to exhibit the full symmetry of the laws of physics. If the vacuum state is unique, meaning that there is only one state of lowest energy, then it must exhibit the full symmetry of the laws. However, the vacuum may be a degenerate (non-unique) state such that for each unsymmetrical vacuum state there are others of the same minimal energy that are related to the first by various symmetry transformations that preserve the invariance of physical laws. The phenomena observed within the framework of this unsymmetrical vacuum state will exhibit the broken symmetry even in the way that the physical laws appear to operate. The analogue of the vacuum state for the universe is the ground state of the ferromagnet. Although there is no evidence that the vacuum state for the electroweak theory is degenerate, it can be made so by introduction of the Higgs mechanism, which involves the artificial insertion of an additional field. This field has a definite but arbitrary orientation in the isospin vector space, as well as properties that would make the vacuum state degenerate. And not only is the symmetry-breaking Higgs mechanism able to rid the theory of the Goldstone boson, but in doing so it solves the mass problem for vector bosons while leaving the theory gauge-invariant.

Recall that in the discussion of the Yang-Mills gauge theory the gauge quanta were massless, with electric charges +1, −1 and 0. They could not be identified with the intermediate vector bosons responsible for the weak interactions, because those particles needed to be massive, and hence the masses had to be added to the theory by hand, resulting in a violation of gauge invariance. The Higgs mechanism solved the problem in the following way. Glashow's 1961 model was an SU(2) × U(1) theory (referring to the underlying symmetry group), which allowed for the possibility of adding scalar (spin-0) fields to the Lagrangian without spoiling the gauge invariance. The Higgs field (or its associated particle, the Higgs boson) was really a complex SU(2) doublet

$$\begin{pmatrix} \Phi^+ \\ \Phi^- \end{pmatrix}$$

consisting of four real fields that were required in order to transform the massless gauge fields into massive ones. So in addition to the four scalar quanta, we also have four vector gauge quanta, three from SU(2) and one from U(1). A massless gauge boson like the photon has two orthogonal spin components transverse to the direction of motion, whereas massive gauge bosons have three, including a longitudinal component in the direction of motion. Three of the four Higgs fields are absorbed by the $W^\pm$ and Z^0 and appear as the longitudinal polarization in each of the three quanta. Because longitudinal polarization in a vector field is equivalent to mass, we obtain three massive gauge quanta with charges positive, negative and zero. The fourth vector quantum remains massless and is identified as the photon. Of the four original scalar bosons, only the Higgs boson remains. The four gauge fields and the Higgs field undergo unique, non-linear interactions with each other, and because the Higgs field is not affected by the vector bosons, it should be observable as a particle in its own right.

The Higgs field breaks the symmetry of the vacuum by having a preferred direction in space (like the ferromagnet). When the W and Z particles acquire masses, the charges I_3 and Y (weak isospin and hypercharge) associated with the SU(2) × U(1) symmetry are not conserved in the weak interactions because the vacuum absorbs these quantum numbers.[27] The electric charge Q associated with the U(1) symmetry of electromagnetism remains conserved. The gauge theory associated with the electroweak unification predicts the existence of four gauge quanta, a neutral-photon-like object sometimes referred to as the X^0 or (B) associated with the U(1) symmetry, as well as a weak isospin triplet $W^\pm$ and W^0 associated with SU(2). As a result of the Higgs symmetry-breaking mechanisms, the X and the W^0 are mixed, so that the neutral particles one sees in nature are really two different linear combinations of these two. One of the neutral particles, the Z^0, has a mass, whereas the other, the photon, is massless. Because the masses of the $W^\pm$ and the Z^0 are governed by the structure of the Higgs field, they do not affect the basic gauge invariance of the theory. And the so-called weakness of the weak interaction, which is mediated by the $W^\pm$ and the Z^0, is understood as a consequence of the masses of these particles.

It was that idea of spontaneous symmetry-breaking and the Higgs mechanism that was used by Weinberg (1967) and independently by Salam (1968) to produce the first successful unification of electromagnetism and the weak force into a single gauge theory. Glashow and Schwinger had suggested in their earlier work that leptons should carry weak isospin like the other particles (baryons and mesons) involved in weak decays. And now that the problem of the gauge field masses seemed solved, it remained a matter of working out the appropriate mathematics required to formally express the weak interaction as a gauge theory. Here again it is important to note that Weinberg, like Glashow, did not set out to develop a unified theory of weak and electromagnetic interactions; instead, he simply wanted to understand weak interactions in a new way.[28] Unity was not the goal or even the motivating factor. The context is not unlike Maxwell's approach to electromagnetism. He was concerned to develop the work that Faraday had begun on a field-theoretic interpretation of electromagnetic phenomena. It isn't that he wasn't surprised or pleased when the theory yielded velocities that coincided with those of light waves, but that unity was not the basis for his development of the theory, nor was it seen as a significant reason for its acceptance by others. As we shall see, the same is true of the electroweak theory. Unity was simply not mentioned as a factor in the theory's acceptance, either by its founders or by their colleagues. The impetus to even carry out experimental tests for neutral currents and the massive bosons predicted by the theory was the result of the theory's renormalizability, rather than its unity. As mentioned earlier, unity and renormalizability are intimately connected in the electroweak case, but that is not because the former is a condition of the latter; Glashow's 1961 model was unified but not renormalizable.

But why was renormalization so important? Recall that one of the problems with the V-A account was that the mass factor for the W created an ever-increasing number ($\to \infty$) of contributions to the perturbation series, contributions that could not be reabsorbed into a redefinition of masses and couplings. It was that redefinition that constituted the basis of renormalizability. That same problem plagued the attempts of Glashow as well as Salam and Ward, because in all of those models the mass terms were added by hand, thereby ruling out renormalizability. In his 1967 paper, Weinberg suggested that his model might be renormalizable, but it wasn't until 1971 that Gerard 't Hooft developed a technique for the renormalization of spontaneously broken gauge theories. The issue of renormalizability was important not only for getting physically meaningful predictions but also because QED was renormalizable, and hence to produce a complete and coherent theory its unification with the weak force would have to exhibit that property as well.

But before moving on to the discussion of renormalization, let me briefly recap the situation with respect to the main topic: theory unification. Essentially, a unified theory of weak and electromagnetic interactions emerged not from the reduction of one force to another but through a combination or mixing of the gauge fields A_μ and W_μ^3. That mixing was possible through the identification of leptons with the SU(2) isospin symmetry group, which was then united with the U(1) group of electromagnetism to form the larger SU(2) × U(1). Employing gauge-theoretical constraints, one could then generate the dynamics of an electroweak model from the mathematical framework of gauge theory. However, nothing about the mixing of the two gauge fields follows from the mathematical structure of gauge theory alone. The actual unity was made possible via the classification of leptons with a particular symmetry group – the one governing weak isospin. Once that assumption was made, the mixing was then represented in the model by means of the Weinberg angle θ_w, which is a combination of the electromagnetic and weak coupling constants q and g into a single parameter.

Because of the neutral-current interactions, the old measure of electric charge given by Coulomb's law (which supposedly gives the total force between electrons) was no longer applicable. Owing to the contribution from the new weak interaction, the electromagnetic potential A_μ^{em} could not be just the gauge field A_μ, but had to be a linear combination of the U(1) gauge field and the W_μ^3 field of SU(2). Hence, the mixing was necessary if the electromagnetic potential was to have a physical interpretation in the new theory. One can see the sense in which the unification achieved by the Glashow-Weinberg-Salam (G-W-S) model emerges out of the formal constraints of gauge theory and the identification of particles with a particular symmetry group. The mixing of the gauge fields, which is really the essence of the unity, is represented by the Weinberg angle, a free parameter whose value is not determined by the structural constraints of gauge theory nor by the physical dynamics it generates. So although the two interactions are integrated under a framework that results from a combination of their independent symmetry groups, there is a genuine unity, not merely the conjunction of two theories. A reconceptualization of the electromagnetic potential and a new dynamics emerged from the mixing of the fields. As in Whewell's consilience of inductions, the phenomena are interpreted in a new way, an interpretation that results not from conjoining two theories but from a genuine synthesis – in this case, a synthesis that retains an element of independence for each domain but yields a broader theoretical framework within which their integration can be achieved.

From the discussion thus far we can begin to have some appreciation of the other similarities between the electroweak unity and Maxwell's electrodynamics. Although the latter clearly embodied a reductive unity, it was nevertheless made possible because of the power of the Lagrangian formalism, which provided a dynamics that could unite optical and electromagnetic processes under the same set of laws. Because no causal or mechanical details were needed, a description of the relevant processes via the field equations was sufficient. And, as with the electroweak theory, a fundamental parameter or theoretical quantity could be associated with the unification – a quantity that represented a new way of thinking about the previously separate phenomena. The idea that optical and electromagnetic processes were the results of waves propagating through the field required that electric current be capable of travelling across spatial boundaries. That was what the displacement current provided: the mechanism for a field-theoretic and hence unified account of those processes.

4.5. Renormalization: The Final Constraint

The notion of renormalization can be traced back to nineteenth-century hydrodynamics, where it was discovered that a large object moving slowly through a viscous fluid will behave, in some ways, as if it has an increased mass, because of the fluid particles that it drags along. As a result of interactions with a medium, the mass of such an object is renormalized away from the bare value that it has in isolation. The idea involves replacing many-body problems by simpler systems in which interactions are negligible. Hence, the complicated many-body effects are absorbed into redefinitions of masses and coupling constants.

In the 1930s, work on QED revealed several mathematical difficulties with the theory; calculations for many processes, when taken beyond first-order approximations, gave divergent results; that is, the theory gave infinite predictions for physically observable quantities. One problem, the ultraviolet divergences, was thought to result from the infinite self-energy of the electron. In addition, the possibility of electron-positron pair creation gave rise to infinite vacuum expectation values (vacuum fluctuations) and also implied an infinite self-energy for the photon. In 1947, experiments by Lamb and Retherford uncovered a shift in the 2s level of the hydrogen spectrum, a result that contradicted the well-established Dirac equation (Morrison 1986). That discrepancy was thought to result from the effects of the self-field of the electron. However, in order to show how those differences could be explained as radiative effects in QED, it was assumed that the masses and charges of the bare electrons and positrons that appeared in the formalism could not have their experimentally measured values. Because the electromagnetic field that accompanies the electron can never be "switched off", the inertia associated with the field contributes to the observed mass of the electron, and the bare mechanical mass is unobservable. Also, the electromagnetic field is always accompanied by a current of electrons and positrons that contribute to the measured values of charges as a result of their influence on the field. Consequently, the parameters of mass and charge need to be renormalized in order to express the theory in terms of observable quantities.

Although renormalization techniques had been successfully applied to QED, it was by no means obvious that they could also be applied to the electroweak theory. The criterion for renormalizability is gauge invariance, and, as we saw earlier, gauge invariance requires massless particles like the photon; any boson masses that were added by hand would destroy that invariance and render the theory unrenormalizable. However, if the masses could be introduced as a result of hidden or spontaneous symmetry-breaking, as they were in Weinberg's lepton model, the symmetry of the Lagrangian would be preserved, leaving open the possibility of renormalizability.

The basic idea in 't Hooft's proof was to show that the Feynman rules of the theory led to mathematical expressions for the W-boson propagators that avoided the problems associated with the use of massive spin-1 particles in perturbation theory. In perturbation theory, each contribution from a Feynman diagram should be a simple dimensionless number. But because the W-boson propagator has a dimensional mass factor, that spoils its contribution. In an effort to rectify the problem, each propagator (wavy line) must be multiplied by corresponding factors of momentum to give a dimensionless contribution. However, these momentum factors summed over all possible internal momenta will result in an infinite number of contributions to the perturbation series. What 't Hooft's proof showed was that for W-boson propagators with very large momentum, the mathematical expression for the propagator does not depend on mass, and as a result there is no need for corresponding momentum factors that were necessary to guarantee that the propagator was a dimensionless number. Because it is the extra momentum factors that lead to the infinite values (divergences) when summed over all possible internal momenta, their absence leads to finite results and renormalizability.

Without going through the complicated mathematics, one can give a brief sketch of the basic ideas. The Weinberg-Salam electroweak theory introduces a Z^0 gauge field whose coupling constant is related in a unique way to the couplings of the charged weak and electromagnetic interactions. In the case of neutrino and antineutrino scattering, where the outcome is a W^+ and W^-, there is an exchange of a Z^0. Consequently, this new term provides the requisite cancellation effect, so that the cross section no longer increases indefinitely with energy. In theories like G-W-S with spontaneous symmetry-breaking, the removal of the Goldstone boson is equivalent to choosing a particular gauge. (It is necessary to choose a gauge in order to write down the Feynman rules and do the appropriate calculations.) However, the mixture of the Goldstone boson and the W gauge particle that exists after the symmetry-breaking is equal to zero. 't Hooft retained that term, but left the choice of gauge open and added another term to the Lagrangian that would cancel the effects of the zero term. That, of course, was the term corresponding to the Z^0, whose gauge particle propagator was independent of the gauge boson mass. The factors in that propagator cancel the propagator representing a potentially real Goldstone boson.

The interesting issue here is that the renormalizability of the electroweak theory is intimately bound up with its unity, which, as we have seen, is the result of certain structural constraints that were imposed by gauge theory and symmetry. The most important of these is the fact that the electroweak theory is a non-Abelian gauge theory. To see how this all fits together, recall that the symmetry demands of the SU(2) group required that a third component of the gauge field corresponding to the isospin operator (τ_3) be present to complete the group. However, in order to have a unified gauge theory, that new component W_μ^3 had to be interpreted as a new gauge field responsible for a neutral-current interaction. Because W_μ^3 is not a purely weak field (it contains an electromagnetic component), the new "physical" weak field is associated with the Z^0, which can be defined from the neutral terms in the kinetic-energy Lagrangian:

$$Z_\mu^0 = \left(q\, A_\mu + g\, W_\mu^3\right) \big/ \sqrt{g^2 + q^2}$$

The coupling constant of the Z^0 gauge field is related in a distinctive way to the couplings of the charged weak and electromagnetic interactions. In the case of neutrino and antineutrino scattering,

$$\nu + \bar{\nu} \rightarrow W^+ + W^-$$

the exchange of a Z^0 provides a new contribution, a term that provides exactly the right cancellation effect necessary to prevent the cross section from increasing indefinitely with energy.

The crucial feature is the interaction between the $W^\pm$ and Z^0 fields. Figure 4.2 depicts the scattering of a neutrino and an antineutrino from Z^0 exchange, as described in the G-W-S model.

In any gauge theory, the conserved quantity serves as the source of the gauge field, in the same way that electric charge serves as the source of the electromagnetic field. In Abelian gauge theories like electromagnetism, the field itself is not charged, but in non-Abelian theories the field is charged with the same conserved

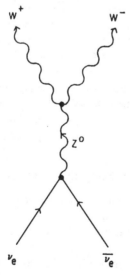

Figure 4.2. New contribution to the scattering of a neutrino and an antineutrino from Z^0 exchange.

quantity that generates the field. In other words, the non-Abelian gauge field generates itself, and as a result these fields interact with each other, unlike the situation in electrodynamics. The form of the interactions (also determined by the structural constraints of the symmetry group) leads to cancellations between different contributions to high-energy amplitudes. Thus, both the unity and renormalizability of the electroweak theory emerged as a result of the very powerful structural features imposed by the mathematical framework of the theory, rather than being derived from the phenomenology of weak and electromagnetic interactions. The Z^0 field that was necessary for the unification was also the feature that allowed the theory to be renormalizable. But that was not simply the result of using a gauge theory and symmetry; it was the non-Abelian structure of the gauge field that ultimately was responsible for this rather remarkable connection between unification and renormalization.

Although renormalizability was a crucial and final theoretical hurdle for the G-W-S electroweak model, it was by no means the last step in the journey to acceptance. In 1971 there was no experimental evidence that the new model was, to use Weinberg's phrase, "the one chosen by nature". There were several significant predictions and effects entailed by the model that needed to be confirmed, specifically the existence of the W and Z particles, as well as weak neutral currents. Recall that in the G-W-S model the weak force was mediated by massive vector bosons. The existence of the Z^0, an electrically neutral particle that functioned as the massive partner of the photon, was one of the crucial features of the model, but its very existence implied that there should also be neutral currents, a phenomenon assumed not to exist at that time. In recent years there have been some detailed studies analysing neutral-current experiments, most notably those by Galison, Pickering and Kim and associates.[29] They have different perspectives on the status of the results and the extent to which existing theory had an impact on the outcome of the experiments. Consequently, we shall not consider the experiments themselves, nor the differing accounts given by those authors. However, I would like to briefly draw attention to the importance of those experiments in legitimating the theory and the reasons why the neutral-current experiments were, in many ways, seen as a decisive test. Once again, as in the Maxwell case, unity was not seen as a persuasive argument for the theory's acceptance.

4.6. When Unity Isn't Enough

Recall that in the earlier discussion I mentioned that the electroweak interaction is also thought to be a current-current type of interaction that is mediated by an exchange current or particle, the W and Z together with the photon. There are two different kinds of weak interactions: charged-current interactions that are mediated by exchange of the W, and neutral-current interactions mediated by the exchanged Z (Figure 4.3). It was that latter type of interaction that formed a crucial part of the G-W-S model and added a new component to the theory of weak interactions.

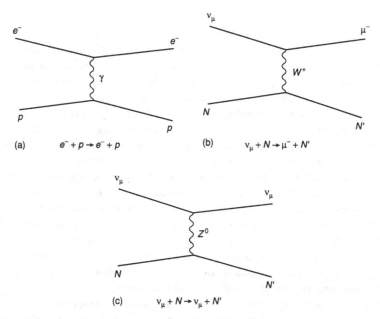

Figure 4.3. Various kinds of electroweak interactions: (a) electromagnetic interaction; (b) charged weak interaction; (c) weak neutral-current interaction.

Until the formulation of the G-W-S theory, all accounts held that weak interactions took place through charged-current processes, as described by the V-A theory. So in the case of a scattering event like the collision of a neutrino with an electron, the incoming neutrino turns into an outgoing electron, and an incoming electron turns into an outgoing neutrino. Because the neutrino is neutral and the electron carries a charge, the collision involves an exchange of charge. The problem, however, was that those kinds of neutrino-electron scattering experiments were very difficult to perform because neutrino-electron cross sections were extremely small. Hence, not only was there no empirical confirmation of neutral-current processes, but in 1970 the charged-current events had not been observed either. Both the neutrino and the electron, together with the muon, are species of leptons, with the latter two experiencing both weak and electromagnetic interactions, whereas the neutrino is involved only in weak interactions (if it is massless). Because experimental data on these leptons was unavailable, and because most of the information on weak interactions came from beta decay (which involved hadrons, i.e., neutrons and protons), that seemed to be the obvious place to look for evidence of neutral currents.

At the time that Weinberg and Salam produced their lepton model of electroweak interactions, the V-A theory provided a fairly good phenomenological account of various decay processes, without the added problem of a neutral boson associated with neutral currents. However, when dealing with, for example, K-meson (kaon) decay, in higher orders of perturbation theory neutral currents were predicted to arise as a result of the exchange of the W^+ and W^-, but were thought to be suppressed as a higher-order correction to the theory. And on the experimental side, kaon decay provided a context in which neutral currents could possibly manifest themselves in an unambiguous way. Kaons have a property known as strangeness, a quantum number analogous to electric charge or isotopic spin, that is thought to be conserved only in strong and electromagnetic interactions.[30] In weak interactions it may be violated, as in the case where a kaon decays into a pion plus a neutrino and an antineutrino. Because the pion and neutrinos are not strange particles, we have an instance where the strangeness is changed, and because in this case the kaon and the pion are both positively charged, the event also involves neutral currents. Experiments revealed, however, that the probability of a neutral-current weak interaction was less than a hundred-thousandth of the probability of a charged-current interaction. Hence, the neutral currents were either absent or extremely improbable in kaon decays. Similar results were found using just neutrino beams, without the overriding influence of electromagnetic interactions and without the complications of strangeness.

In short, all of the experimental results seemed to indicate that neutral currents simply did not exist, and although the V-A theory was able to suppress them, it could offer no reasonable explanation as to why that suppression might take place. So not only did the prospects for Weinberg's lepton model look dim, but also, as it stood, it was not a complete theory of weak interactions, because it covered only leptons. In order to be a truly unified theory, it needed to be extended to hadrons, but as we saw earlier, the experimental picture was no brighter on that front.

Interestingly enough, such an extension was already present in the SU(2) × U(1) model constructed by Glashow in 1960. At the time, he was motivated by the problem of beta decay, rather than interest in leptons; and he was convinced that because the charged current violated strangeness, the neutral current would need to do so as well. His model incorporated a neutral Z boson, but because it was thought that strangeness-changing neutral currents were either absent or strongly suppressed, he concluded that the Z^0 needed to be much heavier than the $W^{\pm}$. (Symmetry-breaking was unknown at the time.) Unfortunately, that resulted in all of the neutral currents being suppressed. Ultimately, in order to deal satisfactorily with the problem of strangeness-changing neutral currents, it was necessary to appeal to the existence of a fourth quark through a process that became known as the GIM mechanism.

Quarks are the fundamental constituents of hadrons, and in 1964 it was thought that there were only three quarks, with quantum numbers up, down and strange. Glashow's suggestion was that there existed a fourth "charmed" quark that provided a quantum mechanical cancellation effect that could be inserted into calculations involving neutral currents in such a way that it could suppress the currents. A significant part of the motivation for introducing the fourth quark was to strengthen the analogy between the weak lepton currents and weak hadron currents; because there were two weak doublets of leptons, it was thought that there should also be two weak doublets of quarks. The GIM mechanism (named after its inventors, Glashow, Iliopoulos and Maiani) cancelled the strangeness-changing neutral currents, in agreement with the data from kaon decay, but did not affect the strangeness-conserving neutral currents. Work on the GIM mechanism was done in 1970, three years after the electroweak models of Weinberg and Salam had been developed, but its importance for those models went unrecognized not only by Glashow and associates but also by Weinberg. However, once it became clear that the Weinberg-Salam account was renormalizable, the relevance of GIM as the final step in constructing a full theory of weak interactions was realized. Three years later, in 1973, neutral currents were discovered, providing what could be considered the most important piece of empirical evidence for the G-W-S electroweak theory.

The relationship between theory and experiment and the importance of unity in this context raise a number of interesting philosophical issues. The first has to do with the significance of renormalizability as the most important theoretical constraint on quantum field theories. We saw earlier that there was no interest in testing the prediction of the G-W-S model until it was proved to be renormalizable; hence, the fact that there existed a unified model for electroweak interactions did not have significant impact apart from its renormalizability. In fact, as many have noted in discussions of the development of the electroweak theory, Weinberg's seminal paper on leptons received only five citations in the literature between 1967 and 1971, compared with 226 citations in 1972–73. The fact that the G-W-S model successfully incorporated the properties of the weak force, as accounted for in the V-A theory, and unified them with the electromagnetic force, as described by QED, seemed to be of no significance in and of itself. It was 't Hooft's proof that marked the turning point; the *theoretical* superiority of the new electroweak model over the V-A theory became obvious, not for its unifying power but because it was, unlike the V-A account, renormalizable.

Ultimately, of course, it was the discovery of neutral currents that provided the crucial piece of missing evidence. But even here one can see the effects that renormalizability had on the relationship between theory and experiment. The electroweak model predicted the existence of neutral currents, and although all of the experimental evidence pointed to their non-existence, that need not necessarily have been seen as an obstacle. Weinberg's 1967–68 model predicted rates for the neutral-current processes that were significantly lower than the rates obtained from earlier studies, rates that were low enough to have escaped detection up until that time. As a result, it looked as though neutral currents would provide an important test of the new theory against the rival V-A account.

In the end, the experiments did successfully rule out the V-A theory, but it is important to recognize that it was not a matter of simply testing the predictions of one theory against those of another. Given the experimental climate at the time, there was no reason to look for neutral currents. And although the V-A theory offered no explanation for their suppression at higher-order corrections, the cross-section predictions of the electroweak model, which were in some sense consistent with the data, did not offer enough of an incentive for further searches. Those predictions became significant factors only after it was shown that the electroweak theory was renormalizable. In other words, if unification had been seen by the physics community as providing overwhelming evidence in support of the theory, then one would expect that there would have been sufficient interest among experimentalists to begin preliminary testing. In his account of neutral-current experiments, Peter Galison quotes Carlo Rubbia, one of the first people to be involved in the search for neutral currents: "Steven Weinberg was the one who, with rare insistence ... was chasing me and many other people to do the neutral current search" (Galison 1987, p. 213). But even after their discovery, Weinberg remained sceptical about the SU(2) × U(1) model, claiming that it was not sufficiently natural or realistic to win general acceptance and that parameters had to be carefully rigged to achieve even qualitative agreement with experiment (Weinberg 1974, pp. 258, 262). That pessimism, however, was directed at the specific model that had been constructed, not at the general idea that weak and electromagnetic interactions were associated with a spontaneously broken gauge symmetry. Nevertheless, his remarks are revealing about an attitude toward unity. The fact that that particular model had succeeded, to some extent, in producing a unification was not compelling in any sense of the word.

Much later, in 1983, there was another vital piece of experimental evidence: the discovery of the W and Z bosons. Although they were crucial elements in the theory, there was less urgency surrounding that search than the search for the neutral currents, simply because it was thought that the necessary technology required for detection of those particles was not available and was not expected to be available for some time. Because they were predicted to have very large masses, accelerators of very high energy were required. However, because of the work of Rubbia and a team of ingenious experimenters at the European Laboratory for Particle Physics (CERN), the discovery was made much sooner than anticipated, a discovery that verified not only the existence of the particles but also the very precise mass predictions given by the theory.

Throughout this discussion I have focused on the role played by formal, structural ideas about symmetry and gauge theory in developing a unified theory of weak and electromagnetic interactions. Now that that story has been told, I want to turn to a more specific analysis of the *kind* of unity that emerged. At the beginning of this chapter I claimed that the unity achieved in this context was structural rather than substantial and synthetic rather than reductive. In other words, although there is a unity at the level of theoretical structure, it is not an ontological or substantial unity that results from a reduction of the weak force and electromagnetic force to the same basic entity or natural kind. We have a synthesis or integration of these two forces produced by a combination of the symmetry groups governing each type of interaction. This is not to say that the electroweak theory is not a remarkable achievement, but only that in this case the mechanisms involved in producing unity say more about the kinds of mathematical models and structures at the theoreticians' disposal than about the ontological status of the phenomena themselves.

In the case of electrodynamics, the generality provided by the Lagrangian formalism allowed Maxwell to unify electromagnetism and optics without producing any specific details about how the electromagnetic waves were produced or how they were propagated through space. Yet the theory implied that they must be one and the same process. In Maxwell's early work, a mechanical model of the aether was introduced as a way of understanding how electromagnetic processes might take place. But he was always quick to distinguish between models that described possibilities and the more basic theoretical structure provided by the Lagrangian formalism. The SU(2) × U(1) gauge theory furnishes a similar kind of structure, by specifying the form of the interactions between the weak and electromagnetic forces. Although it describes the behaviour of the fields, no causal account is given as to why the fields must be unified. That fact emerges as a consequence of classifying leptons as weak isospin doublets. The Higgs mechanism is a *necessary* feature of this unity in that it allows for its possibility; nevertheless, the actual unification comes from constraints imposed by the isospin symmetry group and the non-Abelian structure of the fields. Although the unity is represented in the theory by the combination of the weak and electromagnetic coupling constants in the Weinberg angle, the idea of a unifying formal structure is crucial for understanding not only how the unity was produced but also how its nature should be understood.

What I want to show in the closing section is how the synthetic unity displayed by the electroweak theory is insufficient for drawing any conclusions about unity in nature. In fact, it recommends the opposite, by showing how it is possible to achieve a unity at the level of abstract theoretical/mathematical structure while allowing the forces themselves to remain distinct. At best it produces a kind of counterfactual unity. We know that the Higgs influence vanishes as the energy of the interacting particles increases above the rest-mass energy of the W and Z (100 billion electron volts); hence, these particles interacting at higher energies would obey a unified electroweak force in which the carrier bosons would have zero mass. But that is not now directly testable; rather, it is something we infer on the basis of the theory that describes lower-energy interactions.

4.7. Unified Theories and Disparate Things

Although the electromagnetic and weak forces remain essentially distinct, the unity that is achieved results from the rather unique way in which these forces

interact. As mentioned earlier, this interaction is not derivable from phenomenological aspects of the theory itself. By introducing the idea of broken symmetry, one
can accommodate the different *kinds* of particles associated with the electroweak
theory, particles whose properties are not only dissimilar but also incompatible
(e.g., massless and massive bosons). The incompatibilities were resolved by means
of the Higgs field. In addition, the U(1) and SU(2) symmetry groups governing
the electromagnetic and weak interactions were combined into a larger U(1) ×
SU(2) symmetry, as opposed to *reducing* the particles and forces to a common source
or form. In that sense the unity has been achieved not by simply conjoining the
two theories or symmetry groups but by introducing new components to the pre-
existing theories of weak and electromagnetic interactions. Two points are important here. The first is that one can more or less recover the entire structure of
QED from symmetry principles and the constraint of renormalizability. In the electroweak case, that kind of straightforward derivation is not possible. The Higgs
field has to be added in order for the theory to be phenomenologically accurate.
The second is a related point I mentioned earlier, one that deals with our interpretation of weak and electromagnetic processes. As with Whewell's consiliences,
there has been a reconceptualization of the phenomena – the physical dynamics –
that has resulted from the mechanisms involved in the unification. A different understanding of both weak interactions and the electromagnetic potential emerges.

This conceptual restructuring is a feature of all types of unification, and it can
occur for different reasons. In the case of reductive unity, the reconceptualization
results from seeing two seemingly distinct processes as one and the same kind.
In the case of synthetic unity, the phenomena are integrated under a single framework, but retain their distinct identities. As in the electroweak case, the framework
often introduces new elements into the synthesis. For example, the SU(2) × U(1)
gauge theory allows for the subsumption of the weak and electromagnetic forces
under a larger framework, but only by reconceptualizing what is involved in weak
interactions.

However, strictly with respect to the unifying mechanism itself, the core of the
electroweak theory is the representation of the interaction or mixing of the different
fields. Because the fields remain distinct, the theory retains two distinct coupling
constants: q, associated with the U(1) electromagnetic field, and g, for the SU(2)
gauge field. In order for the theory to make specific predictions for the masses of the
$W^{\pm}$ and Z^0 particles, one needs to know the value for the Higgs ground state $|\phi_0|$.
Unfortunately, this cannot be directly calculated, because the value of $|\phi_0|$ depends
explicitly on the parameters of the Higgs potential, and little is known about the
properties of the field. In order to rectify the problem, the coupling constants are
combined into a single parameter known as the Weinberg angle θ_w. The angle is
defined from the normalized forms of A^{em} and Z^0, which are respectively

$$A_{\mu}^{em} = \left(g A_{\mu} - q W_{\mu}^3 \right) / \sqrt{g^2 + q^2}$$

$$Z_{\mu}^0 = \left(q A_{\mu} - g W_{\mu}^3 \right) / \sqrt{g^2 + q^2}$$

The mixing of the A and W fields is interpreted as a rotation through θ (i.e., $\sin \theta_w = q/\sqrt{g^2 + q^2}$ $\cos \theta_w = q/\sqrt{g^2 + q^2}$). By relating the constant g to the Fermi coupling constant G, one obviates the need for the quantity $|\phi_0|$. The masses can now be defined in the following way:

$$M_W^2 = \frac{g^2}{2G} = \frac{e^2}{2G \sin^2 \theta_w} = \frac{(37.4 \text{ GeV})^2}{\sin^2 \theta_w}$$

$$M_Z^2 = \frac{M_W^2}{\cos^2 \theta_w}$$

In order to obtain a value for θ_w, one needs to know the relative signs and values of g and q; the problem, however, is that they are not directly measurable. Instead, one must measure the interaction rates for the $W^{\pm}$ and Z^0 exchange processes and then extract values for g, q and θ_w. What θ_w does is fix the ratio of U(1) and SU(2) couplings. In order for the theory to be unified, θ_w must be the same for all processes. But despite this rather restrictive condition, the theory itself doesn't provide a full account of how the fields are mixed by furnishing direct values for the Weinberg angle (i.e., the *degree* of mixing is not determined by the theory). More importantly, the theory doesn't determine in any strict sense that the fields must be mixed.

The latter remark was meant to indicate that there is nothing about gauge theory or symmetry per se that requires or specifies that a mixing of these gauge fields must occur. That condition ultimately depends on the assumption that leptons are in fact the weak isospin doublets that are governed by the SU(2) symmetry group. This assumption requires the introduction of a new neutral gauge field W^3 in order to complete the group generators (i.e., a field corresponding to the isospin operator τ_3). This is the field that combines with the neutral-photon-like object Z^0 necessary for the unity. We can see, then, that the use of symmetries to classify various kinds of particles and their interaction fields is more than simply a phenomenological classification constructed as a convenient way of representing groups of particles. Instead, it allows for a kind of particle dynamics to emerge; that is, the symmetry group provides the foundation for the locally gauge-invariant quantum field theory. Hence, given the assumption that leptons could be classified as weak isospin doublets, the formal restrictions of symmetry groups and gauge theory could be deployed in a way that would produce a formal model showing how these gauge fields could be unified. Interestingly, the particles that are the sources of the fields, and of course the forces themselves, remain distinct, yet the electroweak theory or model provides a unified *structure* that enables us to describe how these separate entities interact and combine.

In closing, let me review the important features of theory unification that emerge from the analysis of the electroweak theory. First, we saw that the unity produced in that context was a kind of non-reductive unity that resulted from a mixing and combining of various kinds of fields, rather than from reducing two processes to one. Second, although gauge theory facilitated the unification of the weak and electromagnetic interactions, a fully unified theory could not have emerged solely from the constraints of gauge theory and symmetry. Spontaneous symmetry-breaking via the Higgs mechanism was also required in order to preserve phenomenological aspects of the weak force. Finally, with respect to what is perhaps the decisive component of the unity, namely, the way in which the fields interact or combine, the ratio provided by the Weinberg angle cannot be derived from within the constraints imposed by the theory, and more importantly, the fact that the fields combine at all ultimately rests on the basic assumption that leptons are weak isospin doublets representable by the SU(2) symmetry group.

Setting aside the specifics of the electroweak theory for a moment, there is another important way in which gauge theory serves as a unifying tool; that is, it provides the form for all the force fields in nature. That is, the strong, weak, electromagnetic and gravitational fields are all thought to be gauge fields. In that sense it functions both in a global way to restrict the class of acceptable theories and in a local way to determine specific kinds of interactions, producing not only unified theories but a unified method as well.

Here again we can see that a number of important philosophical issues arise in the connection between explanation and unification. Although the electroweak theory is extremely successful in terms of its predictive power, its ability to explain crucial features of the theory itself is less obvious. The mixing of the gauge fields is not explained, either in the sense of being derived from within the theory or in the sense of showing why it must take place. Nor does the Weinberg angle and its integration of the two coupling constants *explain*, in any substantial way, the mixing of the fields. The Higgs field, on the other hand, explains how the particles can acquire a mass without destroying the gauge invariance of the Lagrangian. And in contrast to the case of electrodynamics, we do not have an understanding of electromagnetic processes in terms of weak interactions.

We saw earlier that unification was not a crucial factor in decisions to test the predictions of G-W-S. Yet one cannot simply conclude on the basis of empirical facts about unification and the way it functions in scientific contexts that it shouldn't be seen as an important criterion in determining a theory's future success, and therefore shouldn't be part of a larger group of reasons for acceptance. That is, the descriptive claim doesn't entail the normative one. But in order to see unification as having a confirmatory role in theory choice, we need a separate argument. One might want to claim that the unified theory is better simply because it treats more phenomena with fewer laws – it has a parsimonious quality that appeals to our desire for ease of calculation. But such reasons are functions not of the likelihood that the unified theory is true but of our desire for coherence and order – what van Fraassen (1980) has called pragmatic reasons for acceptance, not reasons for belief. It is quite a different matter to say, as Friedman and others do, that the unifying structure should be interpreted realistically simply in virtue of that function, or to claim that inference to the "unified explanation" is a legitimate methodological practice.

Given that there are good reasons for claiming that, at least in some theoretical contexts, unity and explanation are distinct, the argument that we can "infer the best explanation" fails on empirical grounds (by playing no role in decisions to accept or reject the theory) as well as philosophical grounds. Even if one accepts inference to the best explanation as a sound methodological principle, it is inapplicable in the cases I have discussed, for the simple reason that unity and explanatory power do not overlap in any significant way. Although I have not provided an account of the conditions required for a theoretical explanation, there are minimal expectations one can cite as basic features of an explanation. First, we should be able to appeal to existing theoretical structure as a way of answering at least *how* particular processes that fall within the domain of the theory take place – for example, how electromagnetic waves are propagated in space (something Maxwell's theory failed to do) and how the mixing of the gauge fields results from the constraints imposed by the theoretical structure. These are fundamental questions that deal with the explanatory power of the specific theories we have considered. One need not have a "theory" of explanation to recognize explanatory inadequacies at this level. This is not to deny that these theories *can* explain some crucial facts. But like displacement in Maxwell's theory, a good deal of the explanatory power of the electroweak theory comes from a mechanism (Higgs) for which there is no experimental evidence. My point here is simply to stress that at least in these two cases the core of the unification is not something that carries explanatory power as a fundamental feature. Although one might be tempted to cite the predictive success of unified theories as a good reason to prefer them in the future, this fails to constitute an epistemological defence of unification as a condition for realism.

In light of these conclusions, it appears as though the "unity argument" can be motivated only if one has an accompanying metaphysics that construes the world as a unified whole capable of being described in a systematic orderly way, that is, if one can use the success of natural science in constructing unified theories as evidence for an accompanying unity in nature. As we saw earlier, the synthetic unity provided by the electroweak theory provides no support for such a view, and the reductive unity present in the Maxwellian case does so only in a very limited way. The merits of unity as a metaphysical thesis applicable in scientific contexts have been called into question in the work of Cartwright (1983), Dupré (1993), Hacking (1996) and others. Those authors have attempted to undermine the idea of a unity of science, a unity of nature or any connection between the two. Science, they argue, fails to underscore any claims about unity. As I see it, however, the problem is that unity *in* science, as opposed to a unity *of* science, cannot be so easily dismissed. If the picture science presents is really one of fundamental disorder, as is sometimes suggested, how are we to understand the practice of theory unification that plays such a prominent role in the natural sciences? What I have tried to show in both this chapter and the one on Maxwell's electrodynamics is (1) how theory unification actually takes place, (2) the role that unification plays in the theoretical domain and (3) that one need not be committed to a metaphysics of unity in nature in order to uphold that unity as a feature of theory construction.

The tension present in the unity/disunity debate highlights the need for emphasizing the fundamental distinction already implicit in the practice of theorizing: a distinction between a metaphysical unity in nature and unified theories. The point illustrated by the discussion of the electroweak theory is that theory unification involves a specific process or methodology that need not commit one to a corresponding unity in nature. Consequently, we can make sense of the practice of unifying theories while embracing the persuasive arguments for disunity. This possibility exists in virtue of the role played by mathematical structures in bringing together several different phenomena under a common framework. Because the theoretical framework that facilitates unification functions at an abstract structural level, it may not directly support a metaphysical or reductivist picture of the unity present in natural phenomena. Hence, it is possible to have unity and disunity existing side by side not only within science but also within the more localized context of a single theory, where individual phenomena like fields and particles retain their independence within a unified theoretical structure.

The moral of the story is that there is no reason to deny or be sceptical about the scientific process of unifying theories. The arguments against unification are attempts to counter the ways in which philosophers use facts about unified theories in physics to motivate a metaphysical picture that extends to a unified natural order. This metaphysics is then used to argue for the epistemological priority of unified theories, without any investigation into how unified theories are constructed. As a response to this kind of metaphysical speculation, the disunity arguments are powerful and persuasive. However, once the structure and process of theory unification are fully articulated, one can see not only how the role of unity should be construed in particular cases but also that the link with metaphysics and epistemology is sometimes simply the result of an improper characterization of how theories become unified in the first place.

Appendix 4.1

Weinberg's Lepton Model

In order to get a sense of how the various ideas used by Weinberg fit together, let me briefly reconstruct the steps in the development of the model.[31] Weinberg focused on leptons because they interact only with photons and the intermediate bosons that mediate the weak force. The interaction of a lepton with a vector potential field is represented by the canonical momentum. The kinetic-energy term in the Lagrangian is $\bar{\psi}\gamma^{\mu}D_{\mu}\psi$, where ψ is a four-component Dirac spinor, and γ^{μ} are the 4×4 Dirac matrices. This kinetic-energy term can be further separated into a pure kinetic term and an interaction term:

$$\begin{aligned} \text{KE} &= \bar{\psi}\gamma^{\mu}\partial_{\mu}\psi - iq\,\bar{\psi}\gamma^{\mu}A_{\mu}\psi \\ &= \bar{\psi}\gamma^{\mu}\partial_{\mu}\psi - iq\,j^{\mu}A_{\mu} \end{aligned}$$

where the lepton current j^{μ} couples to the vector potential A_{μ}.

When A_μ is the electromagnetic potential, the corresponding current j^μ is given by $j_{em}^\mu = \bar\psi \gamma^\mu \psi$. For the weak-interaction potential W_μ, the charged weak current consists of the electron and its antineutrino. The current is

$$J_{wk}^\mu = \bar\nu \gamma^\mu (1 - \gamma_5) e$$

where the particle symbols $\bar\nu$ and e represent the appropriate Dirac spinors. The parity violation of weak interactions is built into the theory by the factor $(1 - \gamma_5)$, which projects out only the left-handed spin polarization state of the electron. The presence of the γ_5 matrix means that the weak current is not a pure vector like the electromagnetic current, but is a combination of vector and axial-vector contributions (similar to the V-A theory).

One of the most important assumptions in the gauge theory of weak interactions is that lepton pairs can be considered weak isospin doublets like the neutron and proton. Because the electromagnetic current is a vector, and the weak current has a V-A structure, the left and right polarization states are usually separated. The weak current is represented by $J_{wk}^\mu = \bar\psi \tau_\pm \gamma^\mu (1 - \gamma_5)\psi$, where ψ is the isospin doublet $\binom{\nu_e}{e}$. The neutrino is massless and always left-handed, so that

$$L = \begin{pmatrix} \nu_e \\ e \end{pmatrix}_L = \frac{1}{2}(1 - \gamma_5)\begin{pmatrix} \nu_e \\ e \end{pmatrix}$$

Hence the right-handed state contains only the electron

$$R = e_R = \frac{1}{2}(1 + \gamma_5)e$$

Hence the electron functions as a weak isospin singlet state, as well as part of an isospin doublet with the neutrino.

From here it is possible to describe the electromagnetic and weak interactions of leptons with the same type of coupling to the gauge field. Recall that one of the main difficulties with the early models was that there was no way of reconciling the different coupling strengths of the two kinds of interactions. In order to complete the SU(2) weak isospin group there needed to be a new component of the gauge field that corresponded to the isospin operator (τ_3). Because this field completed the set of generators for the SU(2) weak isospin group, both the weak and electromagnetic interactions could be unified under a larger gauge symmetry group that resulted from the product of the SU(2) and U(1). In other words, one could then construct a gauge theory for weak and electromagnetic interactions by adding the energy density terms for the A_μ and W_μ fields and the lepton masses. There remained, however, the question of providing a physical interpretation for this new gauge field: How should it be identified? One possibility was simply to identify it with the electromagnetic potential; however, that would not produce a unified theory, because the structure of the SU(2) group could not provide the correct coupling strengths for a conserved electric current and two charged W-boson fields. Moreover, the W_μ field would have to be massless, because the gauge symmetry could not be broken. Instead, the new gauge field W_μ^3 was assumed to have a neutral charge.

The assumption that this was simply a new physical field was by no means unproblematic, because its existence would entail a new set of weak interactions for both the neutrino and the electron. Because the W_μ^3 and the $W_\mu^\pm$ belong to the same set of generators for the SU(2) group, they would have the same coupling and would add an unwanted term to the

142   Unifying Scientific Theories

kinetic energy, which has the form $-ig\bar{L}\gamma^\mu W^3_\mu \tau_3 L$. In order to circumvent that problem, the Weinberg model predicts that this new interaction is a neutral-current interaction. Because the electron is in a weak isospin state, the new interaction will contribute a weak neutral force between electrons, in addition to the Coulomb force, and will also produce parity-violating effects.

The other important consideration resulting from this new weak field is that the definition of electric charge given by Coulomb's law is no longer applicable. Both the electric charge and the vector potential contain contributions from the weak interaction. Hence, one can no longer associate the electromagnetic potential A^{em}_μ with the gauge field A_μ of the U(1) group; instead, A^{em}_μ must be a linear combination of the U(1) gauge field and the W^3_μ field of the SU(2) group.

If we look at the purely neutral interaction terms in the kinetic-energy expression, we see that the one involving only the neutrino is

$$\bar{\nu}\gamma^\mu\Big[qA_\mu + g W^3_\mu\Big]\frac{1}{2}(1+\gamma_5)\nu$$

It is because the neutrino has no electromagnetic interaction that the potential A^{em}_μ can be defined as a linear combination of A_μ and W^3_μ, both of which are considered to be orthogonal unit vectors in the linear vector space specified by the group generators. The combination is then defined as $gA_\mu - q W^3_\mu$, which when normalized gives the proper electromagnetic potential $A^{em}_\mu = (gA_\mu - q W^3_\mu)/\sqrt{g^2+q^2}$. The field combination also defines a new weak field $Z^0_\mu = (qA_\mu + g W^3_\mu)/\sqrt{g^2+q^2}$, which is the true physical weak field, because W^3_μ contains an electromagnetic component. Charge is defined as the coupling of the electron to the A^{em}_μ field and can be calculated in terms of the coupling constants q and g and read off as the coefficient of A^{em}_μ. That is,

$$e = qg/\sqrt{g^2+q^2}$$

The coupling constant q associated with the U(1) gauge field is then defined as the weak hypercharge. A similar method applies in the case of the coupling of the Z^0_μ field to the neutrino and the electron.

The remaining and perhaps most important aspect of the theory is, of course, the symmetry-breaking. Because I outlined the basic workings of this earlier in the chapter, let me simply summarize the main points. The gauge field mass comes directly from the ground-state kinetic energy of the Higgs field. One can then calculate the kinetic energies of the $W^\pm$ and Z^0 using the broken form of the gauge potential. The Higgs field must carry weak isospin and hypercharge to interact with the $W^\pm$ and Z^0 fields, but its ground state must be electrically neutral in order to preserve the zero mass of the photon. The massless spin-1 fields (e.g., the electromagnetic field) have only two spin components, both of which must be transverse (i.e. perpendicular to the direction of momentum). In fact, Lorentz invariance and the requirements of special relativity show that a third longitudinal component has no physical significance for massless fields. However, massive vector fields have three polarization states, including a longitudinal component arising from the Higgs phase term. It is this component that endows the $W^\pm$ and Z^0 with their masses.

Appendix 4.2

Renormalization

One way of explaining the process of renormalization is through the use of Feynman diagrams. These diagrams are part of the propagator approach to QED, where the scattering of electrons and photons is described by a matrix (the scattering matrix) that is written as an infinite sum of terms corresponding to all the possible ways the particles can interact by the exchange of virtual electrons and photons. Each term can be represented by a space-time diagram known as a Feynman diagram. In each diagram the straight lines represent the space-time trajectories of non-interacting particles, and the wavy lines represent photons that transmit electromagnetic interaction. External lines at the bottom of each diagram represent incoming particles before interaction, and the lines at the top represent outgoing particles after interactions. The interactions themselves (between photons and electrons) occur at the vertices, where photon lines meet electron lines.

Each Feynman diagram represents the probability amplitude for the process depicted in the diagram. The amplitude can be calculated from the structure of the diagram according to specific rules that associate each diagram with a mathematical expression describing the wave function and propagator of the particles involved. The amplitude is a product of wave functions for external particles, a propagation function for each internal electron or photon line and a factor for each vertex that is proportional to the strength of the interaction, summed over all possible space-time locations of the vertices. The propagation function, corresponding to a line joining two vertices, is the amplitude for the probability that a particle starting at one vertex will arrive at another (Figure 4.4).

The set of all distinct Feynman diagrams with the same incoming and outgoing lines corresponds to the perturbation expansion of a matrix element of the scattering matrix in field theory. This correspondence can be used to formulate the rules for writing the amplitude associated with a particular diagram. To calculate the probability P of a physical event, one must first specify the initial $|i\rangle$ and final $|f\rangle$ states that are observed and then select all the Feynman diagrams that can connect the two. The wave functions of the quanta are multiplied together, giving the amplitude m for a number of subprocesses which, added together,

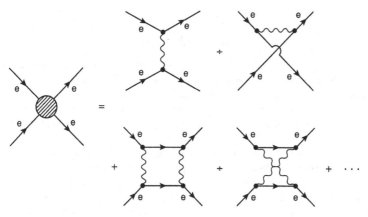

Figure 4.4. Feynman diagrams for electron-electron scattering.

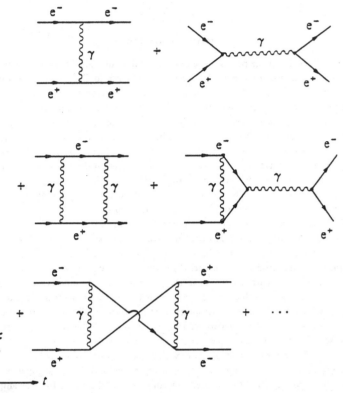

Figure 4.5. Perturbation series containing the various subprocesses possible in electron-positron scattering.

give the total amplitude M, which is then squared to give the probability of the event:

$$P = |\langle F | M | i \rangle|^2$$
$$M = m_1^{(1)} + m_2^{(1)}$$
$$m_1^{(2)} + m_2^{(2)} + m_3^{(2)} + \cdots$$

$m_i^{(1)}$ denotes "first-order" diagrams with two photon-electron vertices; $m_i^{(2)}$ are "second-order", with four photon-electron vertices, and so on. The diagrams in Figure 4.5 depict this process, where the first represents the exchange of a photon between an electron and a positron (amplitude $m_1^{(1)}$), the second ($m_2^{(1)}$) is the annihilation of the electron and positron into a photon and its subsequent reconversion and the third ($m_1^{(2)}$) is the exchange of two photons.

The perturbation expansion and Feynman diagrams are useful because only the first few in the infinite series must be considered. Because the strength of the electromagnetic force is relatively small, the lowest-order terms, or diagrams with the fewest vertices, give the main contribution to the matrix element.

Let us look at an example of how this works (Cooper and West 1988). Consider a typical scattering diagram (Figure 4.6). If we think of the quantum mechanical correction to

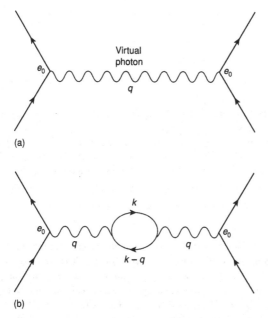

Figure 4.6. (a) Classical scattering of two particles of charge e_0; (b) quantum mechanical correction to (a) due to the creation of the virtual electron-positron pair.

the scattering of two particles with charge e_0, the exchanged photon can, in virtue of the uncertainty principle, create a virtual electron/positron pair represented by the loop in the Feynman diagram; k is the momentum carried around the loop by the two particles, and q is the momentum carried by the photon. There are many corrections that can modify the classic $1/q^2$ single-photon behaviour. If we take D_0 as a single multiplicative factor that includes all these corrections, then D_0/q^2 is the full photon propagator for all possible radiative corrections. D_0 is a dimensionless function that gives a measure of the polarization of the vacuum caused by the production of virtual particles.

The problem is that contributions from diagrams like Figure 4.6(b) are infinite, because there is no restriction on the magnitude of momentum k flowing in the loop. Hence, calculations lead to integrals that diverge logarithmically:

$$\int_0^\infty \frac{dk}{k^2 + aq^2}$$

There are different techniques for making the integrals finite, each of which involves a process called regularization, where a large mass parameter Λ is introduced.[32] The standard (Pauli-Villars) technique involves a factor $\Lambda^2/(k^2 + \Lambda^2)$ introduced into the integrand with the understanding that Λ is to be taken to infinity at the end of the calculation. Hence, the foregoing integral is replaced by

$$\lim_{\Lambda^2 \to \infty} \int_0^\infty \frac{dk^2 \Lambda^2}{(k^2 + aq^2)(k^2 + \Lambda^2)} = \ln \frac{\Lambda^2}{aq^2}$$

whose divergence can be expressed in terms of the infinite mass parameter Λ. That generates the following series:

$$D_0(q, e_0) \approx 1 + a_1 e_0^2 \left(\ln \frac{\Lambda^2}{q^2} + \cdots \right) + e_0^4 \left[a_2 \left(\ln \frac{\Lambda^2}{q^2} \right)^2 + b_2 \ln \frac{\Lambda^2}{q^2} + \cdots \right] + \cdots$$

Consequently, the structures of the infinite divergences in the theory are parameterized in terms of Λ, which serves as a finite cutoff in the integrals over internal virtual momenta. All of the divergences are absorbed by an infinite rescaling of the fields and coupling constants, rather than imposing an arbitrary cutoff. This allows for a finite propagator D (which does not depend on Λ) to be derived from D_0 by rescaling, as long as one rescales the charge at the same time. The rescalings take the form

$$D = Z_D D_0 \quad \text{and} \quad e = Z_e e^0$$

The important point is that the scaling factors are independent of the physical momenta (q), but depend on Λ in such a way that when the cutoff is removed, D and e remain finite. So when $\Lambda \to \infty$, Z_D and Z_e develop infinities of their own that compensate for the infinities of D_0 and e_0. These original bare parameters calculated from the Lagrangian have no physical meaning; only the renormalized parameters D and e do.[33]

Because the Z's are scale factors, they are dimensionless. However, they do depend on Λ, and because a dimensionless function cannot depend on a single mass parameter, a finite mass scale μ must be introduced such that $Z = Z(\Lambda^2/\mu^2, e_0)$. What this, in effect, means is that a mass scale not manifest in the Lagrangian is inserted as a consequence of renormalization. This provides the possibility of generating mass even though no mass parameter appears in the Lagrangian. The finite charge e that is the result of an infinite rescaling of the bare charge e_0 is given by

$$e = \lim_{\Lambda \to \infty} Z_e \left(\frac{\mu^2}{\Lambda^2}, e_0 \right) e_0$$

and depends implicitly on the renormalization scale parameter μ, with the infinities absorbed into the definition of the bare quantities. Once the divergences have been absorbed by the bare parameters, we are left with the physical renormalized parameters. Similar considerations hold for the dimensionless D's, where $D_0 = D_0(q^2/\Lambda^2, e_0)$ and $D = D(q^2/\mu^2, e)$. D, the finite renormalized propagator, is related to its bare divergent counterpart by an infinite rescaling:

$$D \left(\frac{q^2}{\mu^2}, e \right) = \lim_{\Lambda \to \infty} Z_D \left(\frac{\mu^2}{\Lambda^2}, e_0 \right) D_0 \left(\frac{q^2}{\Lambda^2}, e_0 \right)$$

But because there is only a finite number of such physical parameters, only a finite number of redefinitions is allowed. Any quantum field theory could be renormalizable if an infinite number of redefinitions could be used to render all orders finite.

5

◁══════════════════════════════════════▷

Special Relativity and the Unity
of Physics

Although Maxwell's unification of electromagnetism and optics was, in most respects, an unqualified success, electricity and magnetism remained distinct forces within the framework of the theory. Despite the fact that they were integrated in a way similar to the synthesis of the electromagnetic and weak forces, it was not until Einstein's formulation of the special theory of relativity (STR) in 1905 that the two forces could be said to be truly unified. Indeed, it was that lack of theoretical unification that was responsible for the famous "asymmetries which do not appear to be inherent in the phenomena" (Einstein 1952a, p. 37), one of the key features that prompted Einstein's thoughts on relativity.

But the unification of two types of phenomena, electric and magnetic, in the STR was not, in and of itself, where the real power of the theory lay. In other words, the unity of electricity and magnetism was, in that case, indicative of something deeper and more pervasive – specifically, a unification of two domains of physics: mechanics and electrodynamics. In that sense, the STR, in particular the relativity principle, can be seen as both unifying empirical phenomena and unifying other theories; it functions as a kind of unifying meta-principle that extends to the whole of physics. As Einstein noted in his autobiographical remarks, "the universal principle of the special theory of relativity is contained in the postulate" that the laws of physics are invariant under Lorentz transformations from any inertial system to any other (1949, p. 57). Special relativity was what Einstein called a "theory of principle", one that furnished constraints to which other theories had to adhere. The unification of electrodynamics and mechanics was a realization of the requirement that the laws of physics must assume the same form in all inertial frames. Hence the STR's status as a meta-theory: It specified what form the laws of physics should have in order to investigate the nature of matter.

Because of the different ways in which the STR functions as a unifying theory, one of my goals is to specify its structure and function in a way that will highlight both the physical nature of the theory and the role that mathematics played in achieving unification. The latter, of course, involves not only the significance of the Lorentz transformations in the development of relativity but also the unification of space and time produced by Minkowski's mathematization of the event structure of the theory. The Lorentz transformations are especially important because they take on a different role and meaning in the STR than that initially attributed to them by Lorentz, who considered them as merely an "aid to calculation". Consequently, their unifying power is different as well.

In an analysis of the "physical" basis for the theory, two issues emerge as pivotal with respect to unification. The first concerns the relationship of the two so-called postulates to each other, that is, the two axioms that form the foundation of the theory. What connection, if any, is there between the claims (1) that the laws of physics have the same form in all inertial frames of reference and (2) that light is always propagated in empty space with a definite velocity c that is independent of the state of motion of the emitting body? Some commentators, Zahar (1989) and possibly Miller (1981), consider the two postulates/axioms as heterogeneous or independent, a claim that, if true, would violate Einstein's methodological dictum that physical theories ought to be unified homogeneous wholes whose primitive notions are closely interconnected. Einstein himself claimed that the axioms were "only apparently irreconcilable". In what follows I want to show how an analysis of the ways in which unity emerges from the foundations of the STR indeed bears out Einstein's view. The two postulates are in fact quite closely intertwined and together form a unified theoretical core.

It is at this point in the story that things become more complicated. The second issue concerning the physical interpretation of the theory relates to the claim sometimes made that the two postulates composing the STR do not have physical content in and of themselves; that is, they do not make claims about the physical nature of the world. Instead, they express constraints or conditions on making certain kinds of measurements (position, time and velocity) in the effort to gain knowledge about the world (e.g., Pais 1982). Einstein himself refers to the two postulates, taken together, as constituting merely a heuristic principle that, considered by itself, contains only statements about rigid bodies, clocks and light signals (1989, p. 236). The theory supplies additional information only in the sense that it requires connections between laws that would otherwise appear independent. In other words, if the postulates of the STR are interpreted solely as providing a basis for measurement procedures, it becomes questionable whether the theory itself represents a unity in nature or whether it simply provides a framework in which unity can be realized. That is, perhaps relativity, as a theory of principle, functions in a transcendental way to structure our view of nature so that it becomes possible to rid physics of the incompatibilities that exist at the level of phenomena. If so, then the STR would become simply a condition for the possibility of unity at the empirical level. What it would yield would be an abstract structural unity governed by symmetry conditions – the requirement that the laws of physics (mechanics and electrodynamics) be Lorentz-covariant. Obviously the second point relates to the first. If the two postulates are in fact interconnected, it would seem that they must yield more than just measurement procedures, especially because the constancy of the speed of light indeed makes an empirical claim distinct from those concerning clocks and light signals.

My task, then, is to uncover the different ways in which the STR functions as a unifying theory and identify the source of that unity (empirical or formal/ structural) with respect to the theory's foundations, both mathematical and physical. Because unity can have both formal and material dimensions, the relationships between unified theories and a unified physical world are, as we have seen, not always straightforward. The intricate structure of relativity exemplifies these rather complex relationships.

My discussion begins with the requisite background for understanding the significance of the STR: Lorentz's electromagnetic theory. It is only from this perspective that one can appreciate the novelty of Einstein's approach to the problems of physics in the late nineteenth and early twentieth centuries. I shall then discuss the genesis of relativity itself, the relationship between the physics and mathematics and how the theory not only enables the unification of electricity and magnetism, mechanics and electrodynamics, but also how it connects two seemingly independent laws (the conservation of energy and the conservation of mass) in the relation $E = mc^2$.

Finally, we shall look briefly at the development of relativity in the hands of Minkowski (1909). Instead of the requisite four numbers to specify an event in space and time (three spatial dimensions and one temporal one), Minkowski produced a mathematical unification of those dimensions into a four-dimensional space-time. This space-time structure can be seen as a framework within which the electric and magnetic fields become unified. In fact, Minkowski's formalism showed that the mathematical form of a law itself could guarantee its invariance under Lorentz transformations. The unification provided by this formalism raises additional questions about the relationship between physical phenomena and the mathematical frameworks within which they are embedded. Although my focus is not specifically whether or not the unifying power of Minkowski space-time has implications for realism about space-time structure, it is nevertheless important to determine the extent to which the space-time of the STR reflects or even extends the unity presented in Einstein's original formulation of the theory.

5.1. Electrodynamics circa 1892–1904: Lorentz and the Origins of Relativity

Lorentz's interest in the electrodynamics of moving bodies began with an article published in 1887 that dealt with the influence of the earth's motion on optical phenomena. There he turned his attention to A. A. Michelson's first interferometer experiment performed in 1881, the results of which Michelson took to be confirmation of Stokes's hypothesis that the aether was completely dragged along by the earth. Lorentz, a proponent of Fresnel's theory of partial drag, argued that Michelson had miscalculated the fringe shift, and consequently the experiment had not refuted the Fresnel hypothesis. However, by 1887 a significantly more precise version of the experiment had been performed by Michelson and E. W. Morley. No fringe shift was detected, a result that raised difficulties for the predictions of Fresnel drag. What was needed was a way of modifying the aether structure that would reconcile it with the experimental result. Lorentz, however, failed to mention the Michelson-Morley experiment in his paper (Lorentz 1892a), referring only to their 1886 repetition of Fizeau's earlier (1851) measurement of the velocity of light in moving water, a result that confirmed Fresnel's prediction for the drag coefficient in a quantitatively accurate way.

The goal of Lorentz's paper (1892a) was a theoretical deduction of Fresnel's coefficient of entrainment, or the convection coefficient. In order to derive that result, he assumed that the sources of electromagnetic disturbances were microscopic charged particles ("ions") moving in an all-pervasive aether at absolute rest. (By contrast, Fresnel's aether permitted relative motions of its parts.) The state of the aether at every point was described by Maxwell's equations:

$$\nabla \times \mathbf{E} = -\frac{1}{c}\frac{\partial \mathbf{B}}{\partial t} \tag{5.1}$$

$$\nabla \times \mathbf{B} = \frac{1}{c}\frac{\partial \mathbf{E}}{\partial t} + \frac{4\pi}{c}\rho\omega \tag{5.2}$$

$$\nabla \cdot \mathbf{E} = 4\pi\rho \tag{5.3}$$

$$\nabla \cdot \mathbf{B} = 0 \tag{5.4}$$

where $\mathbf{E}$ and $\mathbf{B}$ are the electric and magnetic fields, ρ is the particle's charge density, ω is the particle's velocity relative to the aether and $(4\pi/c)\rho\omega$ is the convection current. Conservation of charge was guaranteed on the ground that the particles were rigid.

Lorentz then used these field equations to treat the electromagnetic properties of matter, as well as the dispersion and propagation of light in dielectrics at rest and in motion. His explanation involved the assumption that charged particles oscillated in response to incident light. He was then able to obtain the optical and electromagnetic properties of matter by averaging over the electromagnetic fields that arose from the ions. In the absence of charge, where $\rho = 0$, the field equations took the form

$$\nabla \times \mathbf{E} = \frac{1}{c}\frac{\partial \mathbf{B}}{\partial t} \tag{5.5}$$

$$\nabla \times \mathbf{B} = \frac{1}{c}\frac{\partial \mathbf{E}}{\partial t} \tag{5.6}$$

$$\nabla \cdot \mathbf{E} = 0 \tag{5.7}$$

$$\nabla \cdot \mathbf{B} = 0 \tag{5.8}$$

These yield a wave equation

$$\left(\nabla^2 - \frac{1}{c^2}\frac{\partial^2}{\partial t^2}\right)\mathbf{E} = 0 \tag{5.9}$$

describing an electric or magnetic disturbance that is independent of the source, because neither its velocity nor acceleration appears in the equation itself. What this entails is that relative to the aether the velocity of light is constant and independent of the motion of the source.

To treat the optics of moving bodies, he used transformation equations that were equivalent to Galilean transformations, where the system S_r designates an inertial reference system.

$$x_r = x - vt \tag{5.10}$$
$$y_r = y \tag{5.11}$$
$$z_r = z \tag{5.12}$$
$$t_r = t \tag{5.13}$$

The convective derivative for linear translation was

$$(\partial/\partial t)_{S_r} = (\partial/\partial t)_S + \mathbf{v} \cdot \nabla_r \tag{5.14}$$

In the untransformed system S at rest, the wave equation for radiation, with its sources (derived from the Maxwellian field equations), described an electromagnetic wave travelling through the aether with a velocity c that was independent of the motion of the source. However, when transformed to an inertial system, the resulting equation did not properly describe wave motion. Consequently, Lorentz proposed an additional transformation of the inertial coordinates into a new system S':

$$x' = \gamma x_r \tag{5.15}$$
$$y' = y_r \tag{5.16}$$
$$z' = z_r \tag{5.17}$$
$$t' = t - (v/c^2)\gamma^2 x_r \tag{5.18}$$

where $\gamma = 1/\sqrt{1 - v^2/c^2}$. These transformation equations yielded a proper wave equation, where an electromagnetic disturbance was propagated with velocity $c\sqrt{1 - v^2/c^2}$, but it came at a cost: It relied on the motion of the source, a clear violation of an aether-based theory.[1] Without offering an explanation of that difficulty, Lorentz simply concluded that his calculations were accurate only to first order, v/c. He did claim that because the approximation gave rise to further calculations it led to a *théorème générale*. To first order in v/c, the equations for the electromagnetic field quantities of molecules constituting matter had the same form in S as in a system connected with S_r by the equations

$$x' = x_r \qquad (5.19)$$
$$y' = y_r \qquad (5.20)$$
$$z' = z_r \qquad (5.21)$$
$$t' = t - (v/c^2)x_r \qquad (5.22)$$

One thing that seems clear, however, is that this additional transformation had only mathematical significance for Lorentz: He introduced both t' and x' as new independent variables. He successfully derived the Fresnel drag coefficient

$$\kappa = 1 - \frac{1}{N^2} \qquad (5.23)$$

(where N is the absolute refraction index of the medium at rest) from the wave equation in the inertial system, attributing the cause to the interaction between light and the oscillating particles that made up the moving medium.

Although Lorentz failed to mention the Michelson-Morley experiment in his paper (1892a), a letter to Rayleigh on August 18, 1892, indicates that he was clearly bothered by the result:[2]

Fresnel's hypothesis taken conjointly with his coefficient $1 - 1/N^2$, would serve admirably to account for all the observed phenomena were it not for the interferential experiment of Mr. Michelson, which has, as you know, been repeated after I published my remarks on its original form, and which seems decidedly to contradict Fresnel's views. I am totally at a loss to clear away this contradiction, and yet I believe if we were to abandon Fresnel's theory, we should have no adequate theory at all, the conditions which Mr. Stokes has imposed on the movement of the aether being irreconcilable to each other.

Can there be some point in the theory of Mr. Michelson's experiment which has as yet been overlooked?[3]

As the quotation suggests, the problem was made more difficult by the fact that the absence of a second-order effect of the motion of the earth through the aether seemed to point to an aether being totally dragged by the earth. This contradicted Lorentz's (1892a) results regarding the drag coefficient that supported a Fresnel-type aether. Not only was there no obvious way to account for the null result, but also it was clearly at odds with his previous work on the electrodynamics of moving bodies.

With hindsight we can see that the problem stemmed from a fundamental incompatibility between his aether-based ontology and his use of the transformation equations. An aether-based physics could not provide the right sort of wave equation, but a further transformation to an inertial system contradicted the fact that the velocity of c had to be independent of its source. However, at that time a denial of the aether was not a viable option, and even had it been, it was not one that

Lorentz would have been prepared to accept![4] His account of ions was, in his view, a way of reconciling the problem of matter–aether interaction in Maxwell's work. By introducing an ontology of particles and fields, Lorentz saw himself as well on the way to explaining mechanics in terms of electrodynamics.

Lorentz spelled out what he saw as a plausible solution to the Michelson-Morley result in a second paper published in 1892, entitled "The Relative Motion of the Earth and the Aether" (1892b). Using the Newtonian velocity-addition rule, he calculated that if v was the earth's velocity relative to the aether, then to second order in v/c the difference in times for the light to traverse the interferometer's arms (both parallel and normal to its motion) should be exactly the measure of the expected fringe shift that was not detected.

$$\delta t = t_p - t_n = lv^2/c^3 \tag{5.24}$$

Lorentz's proposal was as follows: Consider the line joining two points of a solid body; the one that is at first parallel to the direction of the earth's motion does not keep the same length when turned through 90 degrees. If the length in the rotated position is l, then one can retain the Fresnel drag hypothesis if the length in the parallel position is

$$l' = l(1 - v^2/2c^2) \tag{5.25}$$

This shrinking effect in the direction of motion became known as the Lorentz-FitzGerald contraction hypothesis.[5]

In order to explain the contraction effect, he assumed that the molecular forces that determined the shape of a body were affected by motion through the aether. That motion, in turn, influenced the shape of a moving body by a factor of second order in v/c. Because the nature of molecular forces was unknown, the hypothesis could not be directly tested. Instead, Lorentz argued by analogy, claiming that the influences of the motion of ponderable matter on electric and magnetic forces could be extended to molecular forces. He then went on to claim that they transformed in the same way as electromagnetic force, but only if the dimensions of a body in the direction of motion in the inertial system were shortened by the relevant factor.[6] In other words, the molecular-forces hypothesis and the contraction effect mutually supported each other, in that the required numerical value for contraction could be derived from the transformation of molecular forces. Appropriately, Lorentz was somewhat sceptical about the result, but his dissatisfaction seemed to rest more with the theoretical *explanation* of the contraction effect than with the supposed *existence* of the effect itself. He claimed that the application to molecular forces of what was to hold for electric forces was too venturesome to attach much importance to his result; moreover, other deformations could answer the problem equally well (1892b).

Here again it becomes obvious how the Newtonian picture of an aether at absolute rest created the kinds of difficulties that relativity theory would eventually solve. Recall that in his electromagnetic theory Lorentz needed to introduce an additional transformation S' on the inertial coordinates in order to get the proper form for a wave equation, a strategy that violated the foundations of an aether-based theory of light. Then, in calculating the expected fringe shift for Fresnel drag, Lorentz used Newton's velocity-addition rule. However, in extending the argument to the Michelson-Morley experiment, a more complicated story was required. In order to justify the contraction hypothesis, Lorentz claimed that when the molecular forces were transformed from the inertial system S_r to the new S' system, a correction factor corresponding exactly to the contraction length was necessary if the transformation was to be successful. We can see, then, how the argument had come full circle: The desire to retain the aether in the face of a null result by Michelson-Morley required the move to a reference frame to which no physical meaning could be attached, a frame that required that the velocity c be dependent on the motion of the source – a clear violation of Maxwellian electrodynamics and an aether-based theory of light.

The contraction hypothesis was and still is the focus of criticism both by Lorentz's contemporaries and by present-day historians and philosophers of science.[7] The complaint is primarily that it was hopelessly *ad hoc*, simply cooked up for the narrow purpose it was to serve – as a response to the Michelson-Morley experiment. However, if we consider the entire argument, it becomes evident that the contraction hypothesis was in fact well integrated into Lorentz's electromagnetic theory. As we saw earlier, the argument for the molecular-forces hypothesis that formed the theoretical basis for the contraction effect was connected with the transformation scheme for electromagnetic phenomena. What made the connection plausible was the fact that the numerical value for the contraction effect could be calculated as a direct consequence of the transformation of molecular forces.

Moreover, it is reasonable to suppose that that kind of argument was not developed in an *ad hoc* fashion to account for the null result of Michelson-Morley, especially given the way in which Lorentz referred in his second paper (1892b) to the origins of the contraction hypothesis presented earlier (1892a). He claimed that from the equations he developed in the first paper it was possible to deduce the forces that the particles in a system B moving through the aether in the direction of the x axis with a velocity p would exert on one another. To do that, one must first suppose that there is a system A of material points carrying electrical charges, and it is at rest with respect to the aether. One must then introduce a third system that is also at rest but differs from A with respect to the location of the points. This new system C can be obtained from A by an extension that results from multiplying all the dimensions in the direction of the x axis by a factor of $1 + p^2/2V^2$, while dimensions perpendicular to it remain unchanged. The connection between the forces in B and C can then be thought of in the following way: The x components in C are equal to those in B, whereas the components at right angles to the x axis are larger than those in B by a factor of $1 + p^2/2V^2$, where $V = c = $ speed of light, and p is the speed of the moving frame.

Despite Lorentz's initial concern, as discussed earlier, that the electromagnetic argument could be carried over and applied to molecular forces, by 1895 he seems to have been more assured that both the contraction hypothesis and the transformation of molecular forces were indeed legitimate options for explaining the Michelson-Morley experiment.[8] Because one could definitely assume that electric and magnetic forces were transmitted through the aether, it seemed reasonable to assume that molecular forces were as well. And if so, then the translation would most likely affect the action between two molecules or atoms in matter in a way that would resemble the attraction or repulsion between charged particles. So, because molecular forces were transmitted via the aether and hence shared the same substratum, they should behave and transform like electromagnetic forces. Rather than having the character of an *ad hoc* hypothesis, that work was considered by Lorentz as providing the foundation for a systematic theory of electrodynamics. That was especially evident given the way he went on to develop the theory in *Versuch einer Theorie der elektrischen und optischen Erscheinungen in bewegten Körpern*, published in 1895.

An important aspect of that 1895 discussion arises from an issue that figures significantly in the 1892 papers. Recall that when Lorentz first introduced the transformation equations for inertial coordinates to the new system S', he claimed that they were merely aids to calculation. The reason for that was clear. Their use entailed that the velocity of light was dependent on the motion of the source, a conclusion that clearly contradicted the aether theory to which most, including Lorentz, subscribed. Yet from that purely mathematical foundation he was eventually able to articulate and motivate a dynamical explanation of the Michelson-Morley experiment, one that embodied an aether at rest but also relied on the transformation equations to systematically yield the correct value for the contraction effect. And, as we saw earlier, those two theoretical moves were clearly at odds. That internal conflict, however, revealed the significance of the relationship between the mathematical and physical structures of Lorentz's electromagnetic theory. Not only did it expose the root of the dynamical problems that would eventually lead Einstein to reformulate an entirely new kinematics in the STR, but also it showed the power of mathematics (in this case, the transformation equations) in suggesting the kind of physical/dynamical theory that sometimes proves difficult to motivate initially on purely physical grounds.

As we saw in Chapter 3, Maxwell's Lagrangian formulation of electrodynamics failed to provide a theoretical explanation of *how* electromagnetic waves were propagated or how matter interacted with the aether, but it did provide a physical interpretation of electromagnetic phenomena as field processes. In that sense Maxwell's theory resulted in a conceptual shift regarding electromagnetic phenomena by isolating the field as primary. Charge was no longer associated with material sources, but became a kind of epiphenomenon of the field. But just as

Maxwell could not reconcile his field theory with a mechanical account of matter and the aether, Lorentz faced problems harmonizing his transformation equations with his theory of the aether and its constituent particles. In 1892 Lorentz had simply ignored the theoretical problem. However, even the 1895 attempt to put his electrodynamics on a firmer foundation via the theory of corresponding states failed to provide the kind of physical interpretation that could justifiably raise the status of the transformation equations beyond that of a mathematical tool.

5.1.1. 1895: Versuch *and the Theory of Corresponding States*

The version of the electromagnetic theory developed in 1895 was especially significant for a number of reasons. Not only was it the last of Lorentz's work that Einstein claims to have read before the development of the STR, but also it stood as the most systematic exposition thus far of both the transformation equations and the physical basis for the contraction hypothesis. Yet the theory was especially problematic in a number of respects. First, it is tempting to claim that the theorem of corresponding states formulated in 1895 can be construed as a physical interpretation of the transformation equations, but that interpretation seems not quite right for a couple of reasons.[9] First, the dynamics used to explain the transformation equations provides no physical interpretation for the variable t_L used to represent local time. Because this is an essential component in the transformation of physical systems, we are left with no account of the temporal features of empirical phenomena in the newly transformed system.[10] Second, because Lorentz's account of the electron was essentially a theory of rigid particles, it could not consistently accommodate the physical effect of contraction. Another, perhaps more serious, difficulty with Lorentz's theory was the *disunity* it introduced into the Maxwellian framework for electrodynamics; instead of presenting a systematic way of transforming problems for bodies in motion to problems for bodies at rest, two different transformation equations were used for different phenomena (optical and electrodynamical).[11] That resulted in a more complex theoretical structure and highlighted the fundamental problem in bringing together electrodynamics and mechanics. To see how those problems provide a context from which to evaluate the unity provided by the STR, let us take a closer look at the theoretical picture that emerges from Lorentz's *Versuch*.

One of the key features of the *Versuch* is the procedure for transforming electrodynamical problems into electrostatic ones. That had the advantage of allowing one to treat problems related to moving bodies as though they belonged to a reference system having the properties of being at rest in the aether. In order to appreciate the full implications of that procedure and why it proved unsuccessful, we must keep in mind the fact that Lorentz was operating within an aether-based ontology. Maxwell and his successors, especially FitzGerald and Hertz, recognized the difficulties in using mechanics to explain electromagnetic phenomena. Lorentz tried instead to explain mechanics using electrodynamics, but in the end fared no better. Despite their presence in his earlier work, the first systematic explanation of the transformation equations was presented in the *Versuch*. There Lorentz used the notion of local time, introduced in 1892, to obtain the theorem of corresponding states. That allowed him to conclude that the earth's motion through the aether would have no first-order effects on experiments using terrestrial light sources. Hence he was able to explain from within his theoretical framework the absence of any evidence for a stationary aether. The absence of first-order effects was derived from his electrodynamics (he could derive the partial-drag hypothesis), and he could then invoke the contraction hypothesis to account for second-order effects, as he had done previously. Those pieces of Lorentz's electrodynamics fit together in the account presented in the *Versuch*. A brief exposition of his approach will not only reveal its deficiencies but also allow us to see both how and why it was left to Einstein to succeed in completing and extending the unification that Maxwell had begun 50 years earlier.

Lorentz began with an inertial system in which the sources of the electromagnetic field (ions) were at rest, with their properties constant. In such cases the convective derivative of the ion's electromagnetic quantities would vanish. Using Galilean transformations, Lorentz transformed the wave equations for an aether-fixed system to those for an inertial system. All quantities were expressed in the inertial system, except for the electric and magnetic fields, which were treated from the aether-based system. Lorentz then introduced a further coordinate transformation from the inertial system S_r to a new system S'' obtained from S_r by increasing all dimensions along the x_r axis (including the dimensions of the ions) by a ratio of $\sqrt{1 - v^2/c^2}$ to c. Those transformation equations were similar to the ones used in 1892 in the transition from S_r to S':

$$x_r = x''\sqrt{1 - v^2/c^2} \tag{5.26}$$

$$y_r = y'' \tag{5.27}$$

$$z_r = z'' \tag{5.28}$$

$$t_r = t'' \tag{5.29}$$

However, that new reference system S'' could only be interpreted as a purely mathematical move resulting from extending the dimensions along the x axes that were common to S'' and S_r. No physical meaning could be associated with the extension, because Lorentz's ions were thought to be rigid bodies. By using those transformation equations, Lorentz was able to reduce the wave equation for electromagnetic processes to Poisson's equation, thereby enabling him to treat electrodynamical problems electrostatically.

Lorentz then attempted to devise a method for solving problems in the optics of moving bodies. He transformed a wave equation into one that could describe a wave travelling with velocity c by introducing a new independent variable that he referred to as the local time coordinate:

$$t_L = t - \frac{v}{c^2} \cdot \mathbf{r}_r \tag{5.30}$$

That was to be distinguished from the universal time t, which was relative to the aether, and from the time as defined relative to the inertial reference system $S_r = (x_r, y_r, z_r, t_r = t)$, either of which Lorentz was prepared to acknowledge as "true" time.[12] In the new wave equation, c was independent of the source's motion provided that electromagnetic quantities were functions of a set of modified Galilean transformations of the sort introduced for the inertial system in 1892:

$$x_r = x - vt \tag{5.31}$$
$$y_r = y \tag{5.32}$$
$$z_r = z \tag{5.33}$$
$$t_L = t - \frac{v}{c^2}x_r \tag{5.34}$$

However, it was only in a charge-free aether that the field equations were covariant in the transformation between the rest and inertial systems. Lorentz also introduced two new quantities into the field equations, $\mathbf{E}$ and $\mathbf{B}$, which he simply referred to as vectors:

$$\mathbf{E}_r = \mathbf{E} + \frac{v}{c} \times \mathbf{B} \tag{5.35}$$

$$\mathbf{B}_r = \mathbf{B} - \frac{v}{c} \times \mathbf{E} \tag{5.36}$$

and so

$$\nabla_r \times \mathbf{E}_r = -\frac{1}{c}\frac{\partial \mathbf{B}_r}{\partial t_L} \tag{5.37}$$

$$\nabla_r \times \mathbf{B}_r = \frac{1}{c}\frac{\partial \mathbf{E}_r}{\partial t_L} \tag{5.38}$$

$$\nabla_r \cdot \mathbf{E}_r = 0 \tag{5.39}$$

$$\nabla_r \cdot \mathbf{B}_r = 0 \tag{5.40}$$

Using the new local time coordinate, Lorentz was able to show that the laws of optics were the same to order v/c in the inertial system as in the rest system. Because the transformation equations for electrodynamics and for optics differed in form, they failed to provide a unified way of dealing with Maxwellian-type problems.

The limited form of covariance achieved by the equations was explained using Lorentz's "theorem of corresponding states", which said that for the state of a system characterized by $\mathbf{E}$ and $\mathbf{B}$ as functions of (x, y, z, t) in an aether-fixed frame, there was a corresponding state in an inertial system that was characterized by (x_r, y_r, z_r, t_L). The result was that one could treat electromagnetic radiation in the inertial system as if it were at rest relative to the source; that is, phenomena in a moving system were analysed as though they were in a reference frame that was at absolute rest. It is tempting to conclude that the theorem of corresponding states provides a general physical interpretation of the Lorentz transformations. In other words, instead of seeing it as simply part of a calculational scheme, we consider the spatial transformation as associated with a shrinking of the moving system by a factor $[1 - (v^2/c^2)]^{-1/2}$ relative to the moving system. In the optics case we are dealing with covariance in a charge-free aether, so we can ignore the difficulties associated with expansion of supposedly rigid particles. Instead, we can think in terms of the molecular-forces hypothesis. As Lorentz himself claimed, if molecular forces were transmitted through the aether in the same way as electromagnetic forces, their translation most likely would affect the action between two molecules or in a manner resembling the attraction or repulsion between charged particles. Because the form and dimensions would ultimately be conditioned by the intensity of molecular actions, a change in dimension would be inevitable. However, in dealing with optical problems, a transformation of the wave equation into a form that would make c independent of the source would require the local time coordinate, but no physical interpretation had been given. To that extent it becomes difficult to see the theorem of corresponding states as the physical basis for the transformation equations when the fundamental quantity t_L responsible for their success is left uninterpreted.

When dealing with electrodynamical problems, it was possible to systematically ignore the difficulties surrounding the interpretation of the local time coordinate. One could simply consider a system of particles at rest in the inertial system, that is, a system of particles all moving with the same velocity through the aether. One could then characterize the field as depending only on x, y and z or x_r, y_r and z_r. Corresponding to the moving system S_r was a system S'', at rest in the aether, that was obtained by expanding S'' by the factor $\gamma = 1\sqrt{1 - v^2/c^2}$ along the x axis. Having computed the forces acting at corresponding points in the two systems, we would find that each component of a force in S'' would be proportional to the corresponding component of the same force in S_r. But there we would have to assume that the aether was not charge-free; hence we would need an account of how the ions that composed the aether expanded and contracted. But that would seem impossible if ions were rigid bodies.

The difficulty in trying to establish the theorem of corresponding states as a general interpretation for the transformation equations required that one deal with both the problem of the local time coordinate and the lack of a deformable electron. Each of those could be accommodated, but within different contexts. One set of transformation equations was used for optical problems, and another for electrodynamical ones. In the former case the theory of corresponding states enabled Lorentz to explain all optical phenomena to an accuracy of first order in v/c. It is perhaps no accident that Lorentz introduced the theorem of corresponding states and the local time coordinate in the section of the *Versuch* that dealt with optical problems. There he could ignore the difficulties raised by rigid ions, making the theorem seem physically more plausible, and leaving only the problem of the local time coordinate to be solved.[13]

The fundamental difficulty raised by Lorentz's electrodynamics is that it proposes two rather different physical and mathematical theories to deal with problems that Maxwell's theory had shown to be one and the same. What Lorentz demonstrated was that it was possible to have a limited version of covariance for the field equations by reducing problems of moving bodies to problems of bodies at rest, but because that reduction relied on two different sets of transformation equations, it was significantly less powerful as a result. Lorentz's attempt to derive covariance from his transformation equations and to reduce mechanics to electrodynamics was plagued by the classical assumptions that formed the foundation of his theory. Just as the success of Maxwellian electrodynamics had depended to a great extent on its departure from classical accounts of the relationships between electromagnetic field processes and material objects, the failure of Lorentz's programme can be traced to its retention of classical ideas (in particular, the aether) that could not be accommodated within a truly covariant account of electrodynamics.

In 1904 Lorentz introduced a deformable electron and a slightly different variation of the transformation equations in an effort to provide a more comprehensive theory of electrodynamics. That was motivated in part by his own theoretical concerns, by his desire to answer criticisms levelled by Poincaré concerning the accumulation of hypotheses and by his need to account for further negative experimental results on second-order aether-drift detection. However, just prior to that, in 1899, he published a paper attempting to generalize the results of the *Versuch*. There he gave a simpler explanation of his theorem of corresponding states, but also introduced transformation equations that he claimed would hold for second-order quantities of v/c:

If, in a body or system of bodies, without a translation, a system of vibrations be given, in which the displacements of the ions and the components of F' and H' are certain functions of the coordinates and time, then, if a translation be given to the system, there can exist vibrations, in which the displacements and the components of F' and H' are the same functions of the coordinates and the *local* time. This is the theorem ... by which most of the phenomena, belonging to the theory of aberration may be explained.[14]

F' and H' refer to the electric and magnetic forces, respectively. The new transformation equations involved an indeterminate coefficient of the scale factor that differed from unity by a quantity of the order v^2/c^2.

$$x = (\epsilon/k)x'' \tag{5.41}$$

$$y = \epsilon y'' \tag{5.42}$$

$$z = \epsilon z'' \qquad (5.43)$$

$$t' = k\epsilon t'' \qquad (5.44)$$

where t'' is modified local time.[15] Lorentz claimed that those were precisely the transformations required in his explanation of the Michelson-Morley experiment. In that context the value of the scale factor ϵ could be left indeterminate, but for the real transformation produced by translatory motion the factor should have a definite value. Unfortunately, he saw no means for determining it.

It was in that 1899 paper that Lorentz introduced ideas about the effects of vibration and translatory motion on the electron's (ion's) mass. Some such account was needed if his second-order transformation equations for electric and molecular forces were to be successful. In other words, in the transformation from the rest system to the moving system, a given electron must have different masses for vibrations perpendicular to and parallel to the velocity of translation.

By 1903, Trouton and Nobel had performed experiments on the motion of the aether on a charged condenser that, according to Lorentz's electron theory, should have yielded positive results. Similarly, in 1902 an experiment by Rayleigh on moving liquids, which was repeated in 1904 by Brace, failed to yield the double refraction of the order of v^2/c^2 as predicted by the Lorentz contraction hypothesis. The null results from those experiments suggested a fundamental problem with the Lorentz electron. The 1904 Lorentz paper provided the fundamentals for a contractible electron theory, a generalized version of his theory of corresponding states that supposedly could account for the null experiments, and a slightly revised version of the transformation equations was introduced in 1899 (with the factor ϵ set at unity). I shall not discuss the particulars of those new transformation equations, partly because even with them Lorentz failed to achieve full covariance, but also because they are in some sense incidental to the remainder of the story. Einstein claimed not to have read any of Lorentz's work after 1895; hence nothing in the presentation of Lorentz's 1904 equations was significant to Einstein's reformulation of them in his 1905 relativity paper. The 1904 paper is of interest, however, for its patterns for solving problems in electrodynamics, patterns that reveal the fundamental inability of the classic approach to attain the required solutions.

Lorentz's starting point was electrodynamics. On the basis of dynamical considerations he formulated a set of transformation equations that, in the face of anomalous experimental results, led him to introduce new dynamical hypotheses that governed the behaviour of the electron. The strategy, however, gives us the wrong kinematics for understanding the electrodynamics of moving bodies. In the pre-1904 formulation, the transformation equations were non-reciprocal for two observers at rest in frames moving relative to each other; the contraction effect would occur in whatever frame was moving with the greater absolute velocity. Similarly for the case of local time. Time would slow down in the frame of reference with the faster absolute velocity through the aether. That non-reciprocity also affected the transformation equations for intermolecular forces, charge density and so forth. Although he did introduce a limited form of reciprocity in 1904, ponderomotive force remained non-reciprocal. The difficulty resulted from the fact that Lorentz retained the aether as a privileged frame of reference, a frame in which systems at rest were dilated as compared with those in motion. Ultimately, problems with his electron theory can also be traced to the transformation equations. He determined that the longitudinal and transverse masses of the electron would have values that would depend on velocity, the consequence being that the mass of the electron would be completely electromagnetic in origin. However, at no point in that calculation was the energy of the electron taken into account. As Abraham (1905) pointed out, the mass calculated from the electron's momentum differed from the mass calculated from its energy; electromagnetic forces alone were insufficient to account for the entire energy of Lorentz's electron. In that sense, the Lorentz transformations fail to produce unity with respect to integration of the electric and magnetic fields. Not only do the equations suffer from formal difficulties (non-reciprocity), but no systematic dynamics can emerge because of interpretive problems with respect to the coordinate transformations.

It was Poincaré (1906) who pointed out that Lorentz covariance could not be achieved in a theory in which particle masses came from electromagnetic self-fields. He also addressed other fundamental problems with Lorentz's theory, specifically the lack of covariance for Maxwell's equations and the instability of the deformable electron due to the Coulomb repulsion of its constituents (a problem related to the difficulties surrounding mass).[16] In fact, as Zahar (1989) points out, most of the content of relativity theory can be found in Poincaré's work. Curiously, Einstein made little mention of Poincaré; by 1905 he had read only the main philosophical work *Science and Hypothesis*. Instead, he chose to focus on Lorentz's contributions as the precursors to the STR. Because my goal in this chapter is to follow, as closely as possible, Einstein's own path to relativity and unity, I shall also move directly from Lorentz to Einstein circa 1905.

5.2. Einstein circa 1905: Lorentz Transformed

Einstein began the famous relativity paper with an observation about electromagnetic induction. He claimed that Maxwell's electrodynamics, as it was then understood, when applied to moving bodies would lead to "asymmetries that do not appear to be inherent in the phenomena". For example, if one took the reciprocal actions of a magnet and a conductor, the observable phenomenon depended only on the relative motion of the conductor and the magnet; yet the theoretical interpretation involved a sharp distinction between the two cases in which either the magnet or the conductor was in motion. If the magnet was in motion, and the conductor at rest, an electric field with a definite energy was produced, which in turn produced a current in certain areas of the conductor. However, if the magnet was stationary and the conductor was in motion, no electric field was produced in the vicinity of the magnet. Instead, one found an electromotive force to which there existed no corresponding energy in the conductor, yet it produced electric currents of the same direction and intensity as those arising from the electric forces in the case of the moving magnet.

What this means, essentially, is that from the point of view of the phenomena, there should be no difference whether the conductor or the magnet is in motion; that is, according to the Maxwell *equations*, the phenomena behave as though only relative motion is significant. Nonetheless, their theoretical *interpretation* involved a different account of the induction effect for each case. Why should that have been so? Einstein went on to claim that an example of that sort suggested that no properties corresponding to the idea of absolute rest were to be found in mechanics or electrodynamics. Instead, it appeared to confirm what Lorentz had already shown, namely, that to first order, for small quantities, the laws of electrodynamics and optics would be valid for all frames of reference in which the laws of mechanics also held. But the argument there was not immediately obvious. That is, it was not at all clear how electromagnetic induction was connected with the notion of absolute rest. In order to clarify the point, let us retrace the steps that would have led Einstein to that conclusion.

Electromagnetic induction is a phenomenon that exhibits both electrodynamical and mechanical properties. It was generally accepted at the time that the laws of mechanics were Galilean-invariant. That is, they obeyed the classical relativity principle, which states that mechanical laws should be the same in any two coordinate systems. Maxwell's equations, however, were not Galilean-invariant and displayed only a limited form of Lorentz covariance. If one attempts to impose Galilean invariance on all physical laws, then difficulties immediately arise regarding the aether. Recall that such was the problem Lorentz faced. In his attempts to achieve covariance, Lorentz had been forced to use an uninterpreted local time coordinate that allowed him to derive the proper form for an equation for a wave travelling with velocity c. However, unless some physical meaning could be given to that notion of time, it seemed unlikely that full covariance could be achieved within his theory of electrodynamics.

The fundamental difficulty can be traced to a kind of duality between electric and magnetic phenomena that pervaded Lorentz's theory. Not only was there an ontology of particles and fields that attempted to answer questions about the relationship between matter and the field (i.e., the material basis for electromagnetic phenomena), but also the electromagnetic field itself consisted of two distinct entities: an electric field and a magnetic field, related mathematically by an angle of intersection. To that extent the Lorentz theory certainly provided no more unity than Maxwell's equations alone; in fact, it introduced an element of disunity into the theoretical framework via the use of different forms for the transformation equations. Indeed, one of the points I want to stress is the way in which that disunity ultimately gave rise to the asymmetries problem. If one assumes that the electric and magnetic fields are separate entities, then the differences described by the induction effect are real. However, if one assumes full covariance, then the separation of the two fields simply becomes a frame-dependent phenomenon. Judged from the *perspective* of the magnet, there were no electric fields, but one did exist when considered from the point of view of the conduction circuit. It was that assumption of full covariance that constituted the principle of relativity, the first postulate of Einstein's new theory stating that the laws of physics take the same form in all inertial frames.

We can see, then, how the relativity principle seemed to point toward a solution to the asymmetries problem. But how, exactly, did it relate to the issue of absolute rest? At the beginning of the relativity paper, Einstein cited, in addition to the induction effect, the "unsuccessful attempts" to discover any motion of the earth relative to the light medium as further evidence for an absence of anything corresponding to absolute rest, and consequently as evidence for the covariance of physical laws. Some of those experiments had been discussed by Lorentz in the *Versuch*, where he had also shown that to order v/c electromagnetic induction was unaffected by the aether's motion. As mentioned earlier, although Lorentz could successfully account for the absence of first- and second-order effects, the theory nevertheless contained a fundamental flaw. The dualism present in Lorentz's theory had its mathematical origins in the local time coordinate required by the transformation equations. Lorentz had no means for interpreting time transformed in that way, because its use was at odds with the existence of an absolute reference frame in which the speed of light was constant.[17] As we shall see later, Einstein himself suggested that there was a seeming incompatibility between the principle of relativity and the claim that light was always propagated in empty space with a definite velocity c that was independent of the motion of the emitting source. In other words, on the old theory the constancy of the speed of light seemed to require an absolute-rest frame of the sort ruled out by the relativity principle. Maxwell's equations were not Galilean-invariant, and if one insisted on retaining the Galilean relativity principle, then the speed of light should vary when measured by two sets of observers in relative motion.

Ironically, once it was possible to abolish the aether as an absolute-rest frame, the issue of interpreting local time emerged as the key for solving electrodynamical problems. For Lorentz, there had been only two "real" times, as it were, defined relative to the aether frame and the moving frame. If the absolute-rest frame no longer figured in the theory, then the notion of time became a phenomenon defined purely relative to an (any) inertial frame. Because there was no evidence for the existence of an aether at absolute rest, one could simply dispense with it. But how would that affect the assumption regarding the constancy of c? Until then, the constancy of c had been thought to presuppose an aether rest frame. However, if one simply raised that assumption to the status of an axiom or postulate, as Einstein did, then using the principle of relativity one would be able to formulate a set of transformation equations that would provide the kind of full covariance that had eluded Lorentz. Here we can begin to see the connections between the two postulates of the STR. It was precisely Einstein's relativity postulate that made the velocity of c frame-independent. If Maxwell's equations were to be covariant, as specified by relativity, then it had to be the case that c was constant. Because there was no longer any need for an "aether mechanics" in the STR, the dualism inherent in Lorentz's theory between electrodynamics and mechanics was finally overcome.

The new electrodynamics of moving bodies began, then, with a reformulated kinematics in which time occupied the pivotal role. Lorentz had tried, in some sense, to reduce mechanics to electrodynamics. By taking as his starting point dynamical problems and assumptions, he attempted to develop a kinematics that would allow him to transform problems concerning bodies in motion to problems of bodies at rest. The recognition that electrodynamical problems suggested the non-existence of an absolute-rest frame allowed Einstein to approach the problem kinematically. Instead of deriving the transformation equations for specific problems as Lorentz had done, Einstein's strategy was to ask how one should transform physical systems so that the asymmetries and disunity in electrodynamics could be made to disappear. Because the asymmetries problem was a problem in the electrodynamics of moving bodies, and because any theory of electrodynamics would first have to consider the kinematics of rigid bodies, the relativity paper began by developing the appropriate kinematics.

Before going on, let me make a brief editorial comment. As anyone familiar with relativity theory knows, there is no shortage of books, both complex and simple, on the topic. Many of them are concerned with giving an explication of the theory by clarifying the basic concepts through the use of examples and providing a derivation of the Lorentz transformations in one form or another. In addition, most of them discuss various consequences of relativity, such as the twin paradox, the Doppler effect and so on. Because my project is to clarify the kind of unity produced by special relativity and how it emerged, this discussion will be limited to only the relevant issues. No mention will be made of philosophical problems in the foundations of relativity, problems such as the conventionality of simultaneity and the twin paradox. The exposition of the theory will closely follow the text of Einstein's 1905 paper in an attempt to reconstruct the remarkable interconnectedness that the argument displays.

5.2.1. Synchronizing Clocks: The Here and Now

If we want to understand the motion of a material point, then we must give the values of its coordinates as functions of time; but in order to do that, we must provide some physical interpretation of what we mean by "time". Because all judgements in which time plays a role are judgements of simultaneous events, Einstein began by providing a definition of simultaneity. From there he went on to discuss the relativity of lengths and times, followed by presentation of the transformation equations and their physical interpretation. Once the kinematical theory required by the two principles was in place, he showed how it could be applied to the Maxwell equations and in particular to the forces in a moving magnetic field.

Earlier I suggested that the conjunction of the relativity principle with the constancy of the speed of light led to a new conception of time, which in turn provided the foundation for the transformation equations. The apparent incompatibility of the two postulates, as initially mentioned by Einstein, is evident from the difficulties surrounding the notion of simultaneity. For example, consider a light source S moving relative to an observer O with velocity v; there is a second observer O' at rest with respect to S. Both observers must see as wave fronts spheres whose centres are at rest relative to O and O'; that is, they see different spheres. The difficulty resolves itself if one assumes that the space points reached by the light simultaneously for O are not reached simultaneously for O', which is, in essence, the relativity of simultaneity. Einstein's definition of simultaneity centres on the synchronization of two clocks at different places. The definition is given in terms of the propagation of light rays: If a light ray is emitted from point P at time t, is reflected at Q at time t' and returns to P at t'', then the clock at Q is considered synchronized with the one at P if $t' = (t + t'')/2$. In other words, one can define a common time for P and Q by the equation $t' - t = t'' - t'$, which is tantamount to claiming that the velocity of light travelling backward from Q to P is equal to the velocity travelling from P to Q. Light is used as a method for regulating the clocks because the two postulates of the STR enable one to make definite claims about the propagation of light signals. This point is significant for the measurement issue I raised at the beginning of this chapter and the question whether or not it is possible to attribute physical content to the two postulates of the STR independent of the constraints they place on certain kinds of measurements. Although claims about rods and clocks are undoubtedly bound up with the two postulates, it is possible to isolate these measurement constraints as providing a physical basis for the appropriate kind of kinematics. That is, the relativity principle clearly has certain physical *implications* for electrodynamics and optics; specifying the covariance of its laws not only involves an assertion about their form but also implies, in some sense, that the velocity of light can no longer be frame-dependent. In other words, once the two postulates are in place, we can then make claims about the synchronization of clocks.

By defining what is meant by synchronous clocks, Einstein contends that he has defined not just simultaneity but also the notion of time itself. "The time of an event is that which is given simultaneously with the event by a stationary clock located at the place of the event, this clock being synchronous ... with a specified stationary clock" (1952a, p. 40). Here again we can see how the unity in Einstein's account begins to emerge. Lorentz's theory embodied two separate notions of time, one physical, the other mathematical, whereas in the STR there are as many times as there are frames of reference. The local time, which has only mathematical significance in Lorentz's transformation equations, becomes "true" or physical time for Einstein – it is simply what is registered on a clock at the spatial position where the event takes place.

Einstein claims that the notion of a "common time" that embodies the value for the one-way velocity of light $\epsilon = \frac{1}{2}$ is established by definition, but on closer inspection we can see that it is intimately connected with the two postulates of the STR. It functions, in a sense, as an instantiation of both the homogeneity and isotropy of space for the inertial reference systems in which light travels. The homogeneity of space rules out any particular point as preferential to any other, and isotropy guarantees that all directions are equivalent. These latter claims lie at the core of the transformation equations, especially the homogeneity of space, which guarantees that there is no preferred reference frame. Indeed, the linearity of the transformation equations follows from the assumption of homogeneity, which in essence means that linearity is a consequence of the relativity principle. Thus far we have defined time and simultaneity only for the stationary system, so before looking at Einstein's formulation of the transformation equations, let me first mention the implications of the two postulates for inertial systems, the relativity of lengths and times.

Because the notion of a time interval had already been specified, Einstein was able to define the velocity of light as

$$\text{velocity} = \text{light path/time interval}$$

What needs to be shown is how the synchronization of clocks by observers in two different inertial systems can be monitored by both observers. Like the notions of simultaneity and synchronization presented for the rest frame, the new definitions of length and time are expressed operationally for observers in different reference systems. For instance, suppose there are two inertial systems k and K that are at rest, with their spatial axes aligned. In each frame there is an observer and a stationary rod r of length l, as measured by another rod r' that is also stationary. Now assume that the system k is accelerated with velocity v relative to the system K. The length of the moving rod can be determined using two operations:

1. The observer moves together with the two rods and measures the length by superimposing one on the other, as though all three were at rest.
2. Using stationary clocks set up in the stationary system and synchronized in the appropriate way, the observer at rest in K determines at what points in the stationary system the two ends of r are located at a definite time. The distance between these two points measured by r' is called the length of the rod.

Einstein claims that in accordance with the relativity principle the length l determined in operation 1 (the length of the rod in the moving system) is equal to the length of the stationary rod. However, the observer at rest in K who measures the length of the moving rod in the rest system finds a different value r_{ab}. But now suppose that in k a clock is placed at each end, A and B, of the rod and that these clocks are synchronized with the clocks in the stationary system, so that their readings correspond at any instant in time. We say, then, that the clocks in k are synchronous with those in the stationary system k. We also assume that there are two observers, O and O', in k with each of the clocks. They synchronize the clocks according to the specified procedure, and the observer O in K observes events with clocks synchronized with A and B in k. A light ray is then sent from A at time t, is reflected at B at t' and reaches A again at t''. Given postulate two of the STR, the constancy of the velocity of light, the relation between these events is

$$c(t' - t) = r_{ab} + v(t' - t) \quad \text{and} \quad c(t'' - t') = r_{ab} - v(t'' - t') \qquad (5.45)$$

from which it follows that

$$t' - t = r_{ab}/c - v \quad \text{and} \quad t'' - t' = r_{ab}/c + v \qquad (5.46)$$

What this means is that from the point of view of the observer O in K, the distance travelled between A and B is r_{ab}. And if the time at which the light reaches B according to the clock in K is t', then if the clocks on the moving rod remain in synchrony with those in K, they will not be synchronous for observers in k, the moving system, for whom the length of the rod is l.

From that Einstein concluded that "we cannot attach any absolute signification to the concept of simultaneity. . . . two events which, viewed from a system of co-ordinates, are simultaneous, can no longer be looked upon as simultaneous events when envisaged from a system which is in motion relative to that system" (1952a, p. 42). Many notable issues arise from this analysis of length and time. One that is certainly pivotal is the difference between the geometrical notion of length and the kinematical notion. The received view of kinematics in 1905 assumed that the lengths determined by the two operations defined earlier would be precisely equal, that a moving rigid body identified at time t could in geometrical respects be represented by the same body at rest in a definite position. What Einstein's argument shows is that given the procedure specified for synchronizing clocks at different locations, the relativity of length and the relativity of simultaneity follow as immediate consequences. The observer in k measures the geometrical shape of the length of the rod in the moving system, and the observer in K measures the kinematical shape of the moving rod in the stationary system. This latter value, Einstein claims, is determined on the basis of the two postulates or principles of the STR. In other words, the definition of synchronicity is in some sense determined by the two principles.

The other significant matter to arise out of this analysis is the operational definition of an "event" that figures importantly throughout the relativity paper. An "event" is an occurrence at a specific spatial position that is measured relative to an inertial frame by a rigid measuring rod and whose time is registered by a clock located at the same point. In fact, all our judgements of time are judgements of simultaneous events, such as the arrival of a train and the reading of a clock or watch. But when we evaluate the temporal connection of a series of events occurring at different places that are remote from the watch, we need a more elaborate account of simultaneity involving the synchronization of clocks using light rays. Hence, the notion of simultaneity is inextricably linked to the causal order of events. Similarly, the definitions of time and length are operational in that they involve synchronized clocks and rigid bodies acting as systems of coordinates. These operational definitions extend the content of the two principles by specifying a relationship with physical objects, that is, by making explicit the implications of the principles in the domain of kinematics. As Einstein himself notes, the theory of electrodynamics he develops in the 1905 paper (1952a) is founded on the kinematics that emerges from the two principles. Later I shall have more to say about the relationships between these operational definitions and the postulates, but first let us continue with the remainder of the kinematical argument. Once it has been presented in full, the interpretive issues should become clearer.

Taken together, the two principles and the definitions that follow form the core of the STR. But before those foundations can be applied to electrodynamical problems, such as specifying the electromotive force in a moving magnetic field, we need some way of determining the relationship between what is observed in k and what in K, that is, defining exactly how l and r_{ab} are related. Because time and length are determined relative to an observer in an inertial frame (i.e., there is nothing corresponding to the notion of true time or length), one requires transformation equations that can relate spatial and temporal coordinates in k and K. Recall that Lorentz formulated the transformation equations in response to a particular problem, whereas in Einstein's paper we see how they have a kind of axiomatic status. And in the same fashion as the definitions, they can be seen to follow from the two postulates, forming a remarkably coherent and unified theoretical structure.

5.2.2. Deriving the Transformation Equations: The Interconnection of Space and Time

The derivation of the transformation equations begins with the assumption of two inertial reference frames k and K that result from three rigid lines perpendicular to one another and emanating from a common point. Each system has a rigid measuring rod and several clocks that are all similar to one another. Reference frame k is moving at a constant velocity. The spatial coordinates for k are ξ, η and ζ and are measured by the rod moving with the system. Those in K space are measured using the stationary measuring rod, and the coordinates x, y and z are obtained. In order to fully characterize an event, the temporal coordinate needs to be specified using a synchronized clock at the requisite spatial position. So time t of the stationary system is determined by means of light signals for all points where there are clocks. Similarly, time τ in the moving system is determined by clocks at rest relative to that system. We can now guess as to how the argument will unfold.

The definition of clock synchronization in an inertial system yields the notion of simultaneity:

$$t' = \tfrac{1}{2}(t + t'') \qquad (5.47)$$

If we assume that $x' = x - vt$, then it follows that a point at rest in k must have the coordinates x', y and z independent of time. A consequence of the two postulates or principles of the theory is the homogeneity of space and time, which implies that the transformation equations must be linear; in other words, x' must be expressible in the form $ax + by + cz + dt$, where the coefficients a, b, c and d depend only on the relative velocity V of the moving system. If the equations were not linear, the transformation would depend on the choice of the origin, which would be unacceptable given the constraints of homogeneity.[18] At that point in Einstein's derivation these spatial coordinates transform according to Galilean relativity. To relate the two systems k and K, one must first show (in equations) that τ is simply an accumulation of the data from clocks at rest in k. In keeping with his usual methodology, Einstein developed a further thought experiment based on that operational definition of τ.

Assume that a light ray is emitted from A (the origin of system k) at time τ_0 along the x axis to x', and then at τ_1 it is reflected back to the origin, arriving at τ_2. Given the definition for clock synchronization in k, we get

$$\tau_1 = \frac{1}{2}(\tau_0 + \tau) \qquad (5.48)$$

In order to arrive at the proper set of transformation equations, Einstein needed to determine the way in which τ_0, τ_1 and τ_2 were interrelated with the coordinates x', y, z and t in K. If the light ray leaves A at time τ_0 as registered by the clock at A, and at time t when A is the position of an observer on the x axis of K, then the time of this event E_1 is specified as $\tau_0 = \tau_0(0, 0, 0, t)$. Because the times τ are functions of x', y, z and t, the value 0 means that $x' = y = z = 0$. The light ray then arrives at B at time τ_2; this is signified at E_2. If the distance between A and B is x', then, assuming the constancy of c, we can calculate when the light ray arrived at the spatial position of E_2 in k as witnessed by an observer in K. That is, a distance x was covered in the time $t_1 = x/c$. Because $x = x' + vt_1$, we get $t_1 = x'/c - v$. Hence the time between the light ray leaving A and arriving at B for the observer in K is $t + t_1 = t + x'/c - v$. This gives us a value for τ_1 that is

$$\tau_1 = \tau_1[x', 0, 0, t + x'/(c - v)] \qquad (5.49)$$

From the point of view of K, the time taken for the ray to travel from B back to A is $t_2 = x'/c + v$, making the round-trip time $t + t_1 + t_2 = t + x'/(c - v) + x'/(c + v)$, and making the value for τ_2

$$\tau_2 = \tau_2[0, 0, 0, t + x'/(c - v) + x'/(c + v)] \qquad (5.50)$$

The definition of clock synchronization in k with which we began, equation (5.48), relates the values for τ_0, τ_1 and τ_2; now, together with the values for t, t_1 and t_2 as measured in K, we have a new definition of synchronization for k, which is

$$\tau\left(x', 0, 0, t + \frac{x'}{c - v}\right)$$

$$= \frac{1}{2}\left[\tau(0, 0, 0, t) + \tau\left(0, 0, 0, t + \frac{x'}{c - v} + \frac{x''}{c + v}\right)\right] \qquad (5.51)$$

Assuming x' to be infinitely small, Einstein expanded (5.51) to get

$$\frac{1}{2}\left(\frac{1}{c - v} + \frac{1}{c + v}\right)\frac{\partial\tau}{\partial t} = \frac{\partial\tau}{\partial x'} + \frac{1}{c - v}\frac{\partial\tau}{\partial t} \qquad (5.52)$$

or

$$\frac{\partial\tau}{\partial x'} + \frac{v}{c^2 - v^2}\frac{\partial\tau}{\partial t} = 0$$

The solution to equation (5.52) gives the functional dependence of τ on x' and t. An analogous argument can be applied to the axes y and z, keeping in mind that from the perspective of the stationary system light is always propagated along those axes with velocity $\sqrt{c^2 - v^2}$.[19] That will yield the following result:

$$\frac{\partial\tau}{\partial y} = 0, \qquad \frac{\partial\tau}{\partial z} = 0 \qquad (5.53)$$

where τ is independent of y and z. Because τ is linear, this gives us

$$\tau = a\left(t - \frac{v}{c^2 - v^2}x'\right) \qquad (5.54)$$

where a is a function of ϕv.

Einstein then went on to determine ξ, η and ζ given the constraint that light travelling in a moving system has velocity c. So for a light ray emitted at $\tau = 0$ in the direction of increasing ξ, we get

$$\xi = c\tau \quad \text{or} \quad \xi = ac\left(t - \frac{v}{c^2 - v^2}x'\right) \qquad (5.55)$$

where t is substituted for τ. Recall that $x' = x - vt$. Relative to the stationary system K, the light ray moving relative to the origin of emission in k travels with velocity $c - v$ and travels a distance $x = ct$, which yields

$$x'/c - v = t \qquad (5.56)$$

Substituting this value of t for ξ gives us

$$\xi = a\frac{c^2}{c^2 - v^2}x' \qquad (5.57)$$

Hence, we see that the Galilean coordinate x' and the relativistic coordinate ξ are related by a scale factor. If we then eliminate x', we have

$$\xi = \frac{a}{1 - v^2/c^2}(x - vt) \qquad (5.58)$$

Einstein then gave analogous results for η and ζ. If

$$\eta = c\tau = ac\left(t - \frac{v}{c^2 - v^2}x'\right) \qquad (5.59)$$

then when

$$t = \frac{y}{\sqrt{c^2 - v^2}}, \qquad x' = 0 \qquad (5.60)$$

we get

$$\eta = a\frac{c}{\sqrt{c^2 - v^2}}y \quad \text{and} \quad \zeta = a\frac{c}{\sqrt{c^2 - v^2}}z \qquad (5.61)$$

Einstein then replaced the Galilean coordinate x' with its value $(x - vt)$, and a with $\phi\beta$, and obtained the following relations:

$$\tau = \phi(v)\beta(t - vx/c^2) \qquad (5.62)$$
$$\xi = \phi(v)\beta(x - vt) \qquad (5.63)$$
$$\eta = \phi(v)y \qquad (5.64)$$
$$\zeta = \phi(v)z \qquad (5.65)$$

where

$$\beta = \frac{1}{1 - \sqrt{v^2/c^2}} \qquad (5.66)$$

and ϕ is an unknown function of v.[20]

The insertion of β is of crucial importance here. Not only does t need to be transformed in light of the two postulates, but the x parameter requires a similar transformation. Although initially it may not be obvious that the spatial coordinate needs to be transformed in this way, a little reflection on what has been established thus far will reveal why. Before one can convey an understanding of the motion of a material point as a function of time, we need the notion of a rigid body. This body must be identifiable in terms of a system of coordinates; its spatial position must be defined by specific standards of measurement, including the methods of Euclidean geometry. Motion is then described using the values of the coordinates as functions of time. Because spatial position and time are interconnected in this way, the relativity of one will necessitate the relativity of the other. Moreover, we saw that Einstein began his discussion in the 1905 paper (1952a) by citing the asymmetries problem and the unsuccessful attempts to detect an aether drag – phenomena that suggested that neither electrodynamics nor mechanics embodied any properties that corresponded to absolute rest. Hence, there was no privileged frame that identified the spatial coordinate in a non-relativistic way. Here we see the beginnings of what, in Minkowski's work, will be the unification of space and time in one framework, a structure that will *formally* represent the unity of electric and magnetic fields achieved by the STR. Prior to that point, however, more work needed to be done. Recall that ϕ was still an unknown function of v, and there was no proof that the constancy of c was in fact consistent with the principle of relativity (i.e., the transformation equations).

Again, recall that Einstein initially referred to the two principles as only apparently irreconcilable. The conflict stemmed from the fact that the principle of relativity asserted the physical equivalence of all inertial frames, and the Maxwell-Lorentz version of electrodynamics seemed to imply the existence of a privileged frame where c was constant. Also, if we adopt the constancy of c as a postulate in the way that Einstein did, then we seem to have a related problem: If the principle of relativity is true, then why is the velocity of light constant in all inertial frames? The fact that that conflicted with the Newtonian law for addition of velocities meant that drastic revisions were required in the kinematical foundations of electrodynamics and the Galilean transformations. Earlier I discussed some reasons why the two principles appear interconnected. Einstein, however, provided a simple proof. It began with the assumption that when the origins of k and K coincide (when $t = \tau = 0$), a spherical light wave is emitted and propagated with velocity c in K. If (x, y, z, t) is a point reached by the wave, then that point satisfies the equation

$$x^2 + y^2 + z^2 = c^2 t^2 \qquad (5.67)$$

Using the transformation equations, we can replace the K coordinates with those for k, where

$$\xi^2 + \eta^2 + \zeta^2 = c^2 \tau^2 \qquad (5.68)$$

As it happens, this wave is also spherical, and it travels with velocity c, a consequence that shows the compatibility of the two principles by showing that the transformation equations satisfy them.

It now remains to determine the value for ϕ that is also consistent with the relativity principle. To do this, a third system of coordinates K' is introduced that, relative to k, is in translatory motion parallel to the x axis such that its origin moves along that axis with velocity $-v$. Or, more simply, K' moves relative to k with velocity $-v$. When $t = 0$ and all three origins coincide ($t = x = y = z = 0$), then t' in K' also equals 0. The coordinates x', y' and z' of K' can be related to K by two applications of the transformation equations. We first transform from K' to k to get

$$t' = \phi(-v)\beta(-v)(\tau + v\xi/c^2) \qquad (5.69)$$
$$x' = \phi(-v)\beta(-v)(\xi + vt) \qquad (5.70)$$
$$y' = \phi(-v)\eta \qquad (5.71)$$
$$z' = \phi(-v)\zeta \qquad (5.72)$$

then from k to K, which yields

$$\tau + (v\xi/c^2) = \phi(v)\beta(t) \qquad (5.73)$$
$$\xi + v\tau = \phi(v)\beta(x) \qquad (5.74)$$
$$\eta = \phi(v)y \qquad (5.75)$$
$$\zeta = \phi(v)z \qquad (5.76)$$

Finally, with the appropriate substitutions, we have

$$t' = \phi(v)\phi(-v)t \qquad (5.77)$$
$$x' = \phi(v)\phi(-v)x \qquad (5.78)$$
$$y' = \phi(v)\phi(-v)y \qquad (5.79)$$
$$z' = \phi(v)\phi(-v)z \qquad (5.80)$$

Because the transformations between x', y' and z' and x, y and z do not depend on time, K and K' are at rest relative to each other, which makes them identical with each other. Hence $\phi(v)\phi(-v) = 1$.

Given this identity transformation between K and K', what is the relation between $\phi(v)$ and $\phi(-v)$? Einstein asks us to imagine a rod of length l measured by observers at rest in k that lies along the y axis and moves perpendicularly with velocity v relative to K. Its coordinates in K are $y_1 = l/\phi(v)$ and $y_2 = 0$, giving us a length $l/\phi(v)$. Hence, the length as measured from K depends only on the relative velocity, and not on direction. Because the length of the moving rod measured in the stationary system does not change with a change in the direction of motion, we can interchange v and $-v$, allowing us to conclude that $\phi(v) = \phi(-v)$ and that $\phi(v) = 1$. This reciprocity is also a necessary consequence of the isotropy of space. Given the value 1 for $\phi(v)$, the transformation equations now become

$$\tau = \beta(t - vx/c^2) \tag{5.81}$$

$$\xi = \beta(x - vt) \tag{5.82}$$

$$\eta = y \tag{5.83}$$

$$\zeta = z \tag{5.84}$$

where

$$\beta = \frac{1}{\sqrt{1 - v^2/c^2}}$$

One of the things that clearly differentiates these transformation equations from those given by Lorentz is not simply the reciprocity condition but also the fact that they embody a "physical meaning". That is, not only does t refer to a real time coordinate, but the implications of these equations for the kinematical foundations of physical theory involve an entirely new way of conceptualizing experience, one that has its origins in the data received from rigid rods and clocks and how they transform in the shift from a moving system to a stationary system.[21] This physical meaning involves the relativity of length and time mentioned in conjunction with the derivation of the equations. One of the consequences Einstein demonstrates from the notion of length contraction is that a rigid sphere at rest relative to k will appear as an ellipsoid of revolution for an observer in the stationary system K. This results because the sphere's dimension along the common direction of the x axes of k and K appears contracted in the ratio $1 : \sqrt{1 - v^2/c^2}$. The greater the value of v, the greater the contraction; hence length is a relative quantity. The difference has to do with what I referred to earlier as the geometrical shape (the sphere) versus the kinematical shape (the ellipsoid), two concepts defined by Einstein in 1907 (Einstein 1989). Here again we see the advance over Lorentz's version, in which the contraction effect occurred, but the inertial observer was simply unable to detect any change in shape.

An analogous result was proved for the relativity of time. We can compare the slowing down of a moving clock by comparing it with two others. If K's x axis contains several synchronized clocks and at k's origin there is one clock, then if the origin of k moves a distance $x = vt$ relative to K's origin, the clock in k will register a time $\tau = t\sqrt{1 - v^2/c^2}$ as observed from K. τ is the elapsed time of k's clock, and t is the time on a clock located at $x = vt$ on K's x axis. The result $\tau = t - (1 - \sqrt{1 - v^2/c^2})t$ implies that the time interval in k is slower than the one in K by the value $1 - \sqrt{1 - v^2/c^2}$.

Einstein concluded the kinematical portion of the paper by deriving an equation for the addition of velocities where c is a constant. That was necessary because as a postulate it conflicted with the Newtonian velocity-addition law. He showed that the velocity of light could not be altered by composition with a velocity less than c, giving us a value for V:

$$V = (c + w)/1 + (w/c) = c \qquad (5.85)$$

However, before going on to apply the new kinematics to electrodynamical problems, Einstein had an interesting observation. In cases where v and w have the same direction, it is possible to transform these coordinates using one transformation, rather than two. A consequence of the initial derivation was that the relativistic transformations each have both an inverse and an identity; hence if V is replaced with $(v + w)/1 + (vw/c^2)$, the result is that the parallel transformations form a group. That would have significant implications for the mathematization of the STR by Minkowski, but what is perhaps more important at this point is what it reveals about the structure and content of Einstein's formulation.[22]

5.2.3. Synthetic Unity: From Form to Content

We know that the problem of asymmetries in the phenomena requires the relativity principle for its solution, that is, the claim that the laws of electrodynamics and optics will be valid for all frames of reference in which the laws of mechanics hold good. From that postulate, together with the constancy of the velocity of light for all inertial frames, it follows that space and time must be homogeneous. So two principles that seem to be independent, especially with respect to their content, furnish what can be seen as a structural constraint on the nature of space and time. That homogeneity, in turn, implies that the transformation equations must be linear, a precondition for the group-theoretic requirements that emerge at the end of the kinematic analysis. From the simple observation of an asymmetry in the phenomena, Einstein was able to show that the solution in fact points to a mathematical symmetry in the equations. The desire to eliminate the disunity in electricity and magnetism produced both a formal unity and a material unity that applied not only in those specific fields but to physics in general.

The application of the new kinematics to electrodynamics included (1) a demonstration of covariance for the Maxwell-Hertz equations, (2) the relativistic expression for the aberration from the zenith, (3) the transformation law for light frequencies and (4) the dynamics for a slowly accelerated electron.[23] Without going through the details for each of those cases, suffice it to say that the broad range of applications shows the power of relativity in accounting for electrodynamic and optical phenomena in a truly unified way. Before going on to discuss one of the most important consequences of the STR, the equivalence of mass and energy, let us briefly review the differences between Lorentz's account of the transformation equations and Einstein's. This kind of comparison will enable us to appreciate the ways in which the new kinematics produced a non-reductive yet purely simple theory that displays both a mathematical unity and unity of a more qualitative kind in the description of nature itself. The ultimate simplicity of the theory reveals the interconnectedness of physical phenomena achieved via a synthesis of physics itself with electrodynamics as a primary participant.

The most significant difference, of course, was that for Einstein, all coordinates in the transformation equations had physical meaning. In the Lorentz theory the transformations implied the contraction hypothesis and the slowing down of time, modifications that arose as a result of the undetectable but real velocity through the aether. Because that velocity would be undetectable for an observer in a moving frame, Einstein eliminated it as a theoretical parameter. Instead, the modifications arose solely as a result of the relative motion of one frame with respect to another. But Einstein's transformation equations were not simply mathematical-coordinate transformations that related the data gathered on rods and clocks of observers in two different inertial frames. Because Lorentz had a preferred frame – the system at rest in the aether – only the x, y, z and t coordinates in that frame could be interpreted physically. For Einstein, coordinates describing all inertial frames, both at rest and in motion, were equally real, enabling him to move from k to K by simply changing v (the velocity of one frame relative to the other) to $-v$ and substituting the appropriate Greek and Roman letters. Because K was fixed in the aether, that kind of transformation was meaningless for Lorentz, for whom v referred only to velocity through the aether. Hence, the limited reciprocity of the Lorentz equations was merely a mathematical property, with no real physical implications.

Similarly for the case of time. We saw earlier that the local time t, uninterpreted in the Lorentz transformations, became the real time in Einstein's formulation. Because for Einstein there was no privileged observer situated in an aether rest frame, there was no significance attached to the notion of true time in that frame, contrary to the view of Lorentz. Rather, the local times in all inertial frames were true times, and consequently the descriptions of all phenomena could be treated in an equivalent way. What that meant was that the transformation equations were universally applicable to all phenomena governed by the laws of physics. Moreover, the effects produced by relative motion were general and were not dependent on specific attributes of particular bodies. For example, the explanation of contraction in Lorentz's theory involved assumptions about the electronic and atomistic constitution of matter together with elastic forces. In Einstein's view, none of that needed to be taken into account; contraction was the result of modifications of our space and time measurements due to relative motion and was indifferent to the composition of the body itself. Hence, one did not require knowledge of the microscopic nature of the body in order to explain or predict relativistic changes.

We can see, then, how the general character of Einstein's transformation equations reveals both the unity and simplicity of the theory. Lorentz's account was not symmetric between two sets of coordinates; in fact, it did not really relate two inertial frames at all, but rather expressed a mathematical relation between a physical frame (the aether) and a non-physical set of coordinates. In that sense it was not really a coordinate transformation, because it could not be used to move from one physical frame to another. The equations were arrived at by taking account of particular features of physical phenomena. Lorentz assumed that the failure to detect the earth's motion through the aether was something that needed to be deduced from electrodynamic equations. For Lorentz, the transformation equations were a way of explaining anomalous behaviour within the existing framework; hence he derived them by beginning with particular problems in the field. That was why, in the end, they lacked the unifying power possessed by Einstein's formulation – equations that were formulated on the basis of two fundamental principles that applied to the whole of physics. Physics moved from simple assumptions that were themselves interconnected to a system of equations for connecting physical systems to each other. That, in turn, enabled us to see how the electric and magnetic fields were unified.

The theory can be divided into four levels: the two postulates; the kinematical definitions deduced from the two postulates; the transformation equations; the applications of those results to problems in the electrodynamics and optics of moving bodies (i.e., the derivation of equations of motion for a charged particle) or to any physical theory. The two principles are interrelated in that the speed of light must be constant if the relativity principle is to be applicable to Maxwell's electrodynamics. Or, put slightly differently, if Maxwell's equations are to be covariant, then for all inertial frames, the speed of light must be a constant. The simplest assumption is to take such a constant to be also an invariant (i.e., not changing from one frame to the next). But the true unity produced by the postulates comes in the form of their consequences, the transformation equations. Einstein himself pointed out that the postulates did not constitute a "closed system" in which individual laws would be implicitly contained or from which they could be deduced, but only a "heuristic principle, which considered by itself alone contains only assertions about rigid bodies, clocks and light signals".[24] In that sense the two postulates taken together as the foundation of the theory are akin to a universal formal principle that imposes constraints on all physical theories. But, as we saw earlier, those postulates and hence the theory are not without content, a content that is more than what is simply derived from the transformation equations as applied to specific phenomena. From the operational definitions we learn something about the physical behaviour of measuring rods and clocks, and consequently about the nature of time and motion. Because the coordinates x, y, z and t are the results of measurements obtained by means of these clocks and rods, they not only furnish methods of measurement applicable to all physical systems but also provide information about the physical world.

In the Introduction, I mentioned some possible options for thinking about the unity exemplified by the postulates of the STR. One option was that they, together with the Lorentz transformations, provided a theoretical unity, where that is understood in terms of the theory's ability to connect phenomena like the electric and magnetic fields, mass and energy, and to show how the postulates form a coherent system. Another possibility was that the theory produced a kind of unified method for doing physics; it described how to define time, simultaneity, motion and so forth. Despite the interconnection of the postulates, the important question is whether or not the formal unity provided by the transformation equations in bringing together the electric and magnetic fields implies a unity in nature. Or is this situation akin to the electroweak unification discussed in Chapter 4? There we saw a unity produced by the theory's equations, accompanied by a physical independence with respect to the particles carrying the two forces. In order to fully answer this question, we need to look at the mathematical unity of space and time produced in the work of Minkowski. Ultimately I want to claim that despite Minkowski's formalism, it is a mistake to say that space and time are, in essence, the same. Similarly with the electric and magnetic fields. Although the theory combines them into a single entity, they are capable of being isolated in a frame-dependent way. Nevertheless, the STR does satisfy both of the previously mentioned options for thinking about unity. The postulates, definitions and transformation equations together embody a unity of theoretical content, method and form, producing a synthetic unity with far-reaching implications, specifically for theoretical concepts and laws that otherwise appear independent. Perhaps the best example is the equivalence of mass and energy. So before going on to discuss the Minkowski formalism, let us look briefly at the relationship between these two concepts to see if this unification qualifies as an ontological reduction.

5.2.4. The Most Famous Equation in the World

A few months after publishing the relativity paper, Einstein wrote to Conrad Habitch:

One more consequence of the electrodynamical paper has also occurred to me. The principle of relativity, together with Maxwell's equations, requires that mass be a direct measure of the energy contained in a body; light transfers mass. A noticeable decrease of mass should occur in the case of radium. The argument is amusing and attractive; but I can't tell whether the Lord isn't laughing about it and playing a trick on me.[25]

That relationship was not new. Before 1905, the possibility that inertial mass might be associated with electromagnetic energy in special cases had been commonly discussed. Indeed, it had been suggested that all mechanical concepts could be derived from electromagnetism; in particular, there had been attempts to derive the inertial mass of the electron from the energy associated with the electromagnetic field. At around the same time it was demonstrated that a container (a black body) filled with radiation manifested an apparent inertial mass that was proportional to the energy of the enclosed radiation. But again, what Einstein did was to show that a particular relationship known to be valid in specific cases was in fact a universal principle.

The argument in Einstein's 1905 paper on the mass–energy relation began by considering the radiation from a body as seen in two different frames:[26] In the rest frame, equal amounts of energy (Einstein actually referred to light waves) are radiated in opposite directions (total energy $= L$). If the initial energy in the rest frame is E_0, and if we assume a conservation principle, then the final energy E_1 with respect to the frame will differ from E_0 by the amount L, such that

$$E_0 = E_1 + L \qquad (5.86)$$

If we now consider similar emissions from a second inertial frame in which a body is moving at velocity v and, given conservation, H_0 and H_1 are the initial and final energies with respect to the second frame, they should differ by the amount $L/\sqrt{1 - v^2/c^2}$, so that

$$H_0 = H_1 + L/\sqrt{1 - v^2/c^2} \qquad (5.87)$$

This follows from the transformation law for the energy of a light wave from one inertial frame to another. If we then subtract (5.86) from (5.87), we get

$$(H_0 - E_0) - (H_1 - E_1) = L[1/\sqrt{1 - v^2/c^2} - 1] \qquad (5.88)$$

Einstein then claims that H and E are energy values for the same body referred to two coordinate systems in motion relative to each other, in one of which the body is at rest. He goes on to say that it is "therefore clear" that the difference $H - E$ can differ from the kinetic energy K of the body with respect to the other system only by an additive constant C. C's value will depend on the choice of the arbitrary additive constants in the energies H and E. As a result, we can put

$$H_0 - E_0 = K_0 + C \quad \text{and} \quad H_1 - E_1 = K_1 + C \qquad (5.89)$$

because C does not change during emission of light. By substitution, we can now get

$$K_0 - K_1 = L[1/\sqrt{1 - v^2/c^2} - 1] \qquad (5.90)$$

where K_0 and K_1 are the initial and final kinetic energies of the body with respect to the inertial frame in which it has velocity v.[27] What this means is that the kinetic energy of the body decreases as a consequence of the emission of light, and we ignore any assumptions about the nature of the body, the type of radiation and the interactions between those two factors and any measuring instruments. The difference depends solely on the velocity, in the same way that the kinetic energy of the electron does, something Einstein showed in the relativity paper.

If we neglect quantities of the fourth order and higher, we can then put

$$K_0 - K_1 = \frac{1}{2}(L/c^2)v^2 \qquad (5.91)$$

From this it follows that if a body emits energy L in the form of radiation, its mass will diminish by L/c^2. If we use the Newtonian limit and define the inertial mass of a body in translational motion as

$$m = \lim_{v \to 0} \frac{K}{v^2/2} \qquad (5.92)$$

then it follows from equation (5.91) that

$$\lim_{v \to 0} \frac{K_0 - K_1}{v^2/2} = \frac{L}{c^2} \qquad (5.93)$$

Given the definition of mass, (5.93) can be written as

$$m_0 - m_1 = L/c^2 \qquad (5.94)$$

where m_0 is the mass of the body before radiation, m_1 is its mass after radiation and L is the total amount of energy radiated. Hence, we can conclude that a body at rest emitting radiation must lose inertial mass in amounts equal to the energy lost divided by c^2 – the conclusion being that the mass of a body is a measure of its energy content.

Again, the beauty of this mass–energy relation is that it applies to any kind of energy, in the same way the relativity principle applies to any physical system. It is also important to note that the assumption of small velocities in no way limits the generality of the conclusion, nor is it suggestive of a merely "approximate" result. The quantities in equation (5.93) are measured in the body's rest frame; hence the relation between them does not depend on the velocity of the frame in which the kinetic energy is expressed. Does this mean that mass and energy are the same thing? Does their equivalence mean that they are somehow reducible to each other? This is a rather complex question that technically will take us beyond the boundaries of relativity; but because it is crucial to an overall assessment of the unifying power of the STR, we need to know what the implications of $E = mc^2$ actually are. The formula tells us that the mass of a body is equal to the total energy content divided by the square of the speed of light. But the question whether or not mass is always equivalent to energy depends in part on what we mean by "mass". Einstein's concern in the original paper was with the rest energy of the system and cases in which the rest energy could be simply associated with the rest mass. Let us take particle collisions as an example: In a case in which one or more particles is destroyed, others will take their places. If the total rest mass of the particles decreases, then the lost mass corresponds exactly to the overall increase in kinetic energy. A similar relation holds for relativistic mass. Consider the example of a photon whose rest mass is zero: The particle still experiences a finite gravitational pull of E/c^2 because of its relativistic mass. Similarly, in the case of a nucleus containing particles bound together by internal forces, the total energy contains the relativistic energy of the particles (rest energy and kinetic energy) and the potential energy associated with the various force fields. The important point is that the observed mass of the nucleus is related to the total energy by $E = mc^2$, and if the nucleus is at rest, then m represents its observed rest mass. If it loses energy by emitting radiation, then its mass will be reduced by the corresponding amount.

But in order for this relation to imply a complete identity, it seems that a total conversion of mass into energy should be possible. However, such a conversion is not possible, because of the conservation of baryon number, which says that the total number of these particles cannot change; they can be converted into other baryons, but cannot disappear entirely. Baryons include, among other things, protons and neutrons, which make up most of the mass of ordinary matter, and therefore most of the mass of ordinary matter is not available for conversion into energy. The energy that results from fission and fusion and from cases of ordinary combustion comes from small differences in the binding energies of different nuclei, not from the destruction of particles.[28] So we are left in the rather odd position of asserting an equivalence between mass and energy, but not an identity that would allow one to be fully transformed into the other. To that extent, the unity resulting from this relationship is perhaps best expressed using Einstein's own words:

The most important result of a general character to which the special theory has led is concerned with the conception of mass. Before the advent of relativity, physics recognized two conservation laws of fundamental importance, namely, the law of conservation of energy and the law of the conservation of mass; these two fundamental laws appeared to be quite independent of each other. By means of the theory of relativity they have been united into one law. (1920, pp. 45–6)

We have now discussed the basic postulates of special relativity – the invariance of physical laws, which entails the equivalence of all inertial observers and the limiting and invariant nature of c – together with the phenomena resulting from those principles, namely, the relativity of simultaneity, time dilation, length contraction and the relativity of the velocity-addition law. And we have seen how these phenomena, as well as others, are intimately related to one another and to the postulates. In other words, these phenomena are interdependent, in the sense that each of them must hold if one does.[29] This interconnection or unity is presented most completely in the transformation equations, which tell us how all space and time measurements are altered when there is a change in the reference frame. They provide a set of unified relationships that not only express the link between kinematical phenomena but also describe how relativity theory connects with the rest of physics. In that sense, the transformation equations perhaps compose one of the most powerful unifying structures that exists. However, there is another way to express the unity in special relativity, and that is through the concept of an invariant, the space-time interval, which originated in the work of Poincaré but was developed further by Minkowski in the formalization of the STR.

5.3. From Fields to Tensors: Minkowski Space-Time

The foundations of the Minkowski programme made special use of the Lorentz group. Recall that in discussing the velocity-addition law we saw that Einstein himself remarked that the parallel transformations between K and k necessarily formed a group. At around the same time that Einstein submitted the STR paper (1905), Poincaré developed a four-dimensional analysis of Lorentz's transformation equations. The importance of that work lay in the discovery of Lorentz invariance for certain quantities that remained the same in all frames of reference. The essence of Minkowski's 1908 work was an extension of those ideas involving the theory of invariants in a four-dimensional vector space. What he did was provide a complete geometric interpretation of the transformation equations that included Einstein's account of the relativity of time. That was achieved by showing how the relativity equations fit into a four-dimensional space-time structure; in fact, the Lorentz transformations can be seen as an immediate consequence of integrating space with time. As Einstein (1949) himself emphasized, the importance of Minkowski's work was not its four-dimensional analysis; even classical physics had been based on a four-dimensional continuum of space and time. The difference was that in the classical continuum the subspaces that had constant time values had an absolute reality independent of a reference frame. As a result, the four-dimensional continuum naturally split into a three-dimensional space and one-dimensional time. In the Lorentz transformations of the STR we can see that the x and t coordinates are inextricably linked; there is a formal dependence entailed in the ways in which these coordinates function in physical laws. Using Minkowski's formalism, it was possible to guarantee invariance for a particular law under the Lorentz transformations simply on the basis of its mathematical form. That formalism, a four-dimensional tensor calculus, became an especially powerful tool in the mathematical development of relativity. In addition, Minkowski showed that the Lorentz transformations could be understood as simply a rotation of the coordinate system in the four-dimensional space.

Without going into the details of Minkowski's graphical representation of space-time, let us briefly review the way in which his geometrical account provides a unification for relativity.[30] The second postulate of relativity tells us that the speed of light is a constant, c; in other words, it has a finite maximum speed. What this entails is that the additivity principle does not apply to speeds going in any given direction, even though they may approximate the principle at low speeds. Hence speeds, like gradients, are not additive in the way that, say, angles are. How, then, can we use additivity when considering moving objects? There exists a "regraduation" function that can transform magnitudes that combine according to some combination rule into magnitudes that are additive. In fact, given certain conditions, it is possible to construct a differentiable function in which the combination rule is just simple addition; that is, $f(u)$ is such that

$$f(u \oplus v) = f(u) + f(v) \qquad (5.95)$$

thereby allowing us to remeasure speeds using addition. This notion of speed, however, is not distance covered divided by time, but involves a different magnitude known as a "rapidity", where infinite rapidity corresponds to c. The move from speeds to rapidities involves a move from ordinary to hyperbolic trigonometry. In the way that gradients are ordinary tangents of angles, speeds are hyperbolic tangents of rapidities. Or, when we go from speeds to rapidities, we regraduate each speed by considering it a fraction of the universal speed and then taking the inverse hyperbolic tangent. Rapidities are also called pseudo-angles. Just as directions in sub-spaces of space-time subtend angles, directions that have both space-like and time-like characteristics subtend pseudo-angles. So two objects moving with a uniform speed are oriented to each other at a pseudo-angle; speed becomes like orientation, and the first derivative of velocity, acceleration, is like the first derivative of orientation, angular velocity. Time, then, can be seen as analogous to space insofar as it gives rise to hyperbolic pseudo-angles rather than ordinary circular angles.

In cases in which universal speeds are involved, the combination rule must involve hyperbolic functions that are defined in terms of exponentials and differ from ordinary trigonometric functions by a factor i (the square root of -1), which occurs at different places in the expression. When we want to apply comparable units to time and space (e.g., measuring distance in light-years) we simply multiply time by c (ct is proportional to the time, but has the dimensions of a distance) and add

i, which gives us

$$x_1 = x \qquad (5.96)$$
$$x_2 = y \qquad (5.97)$$
$$x_3 = z \qquad (5.98)$$
$$x_4 = ict \qquad (5.99)$$

This is a Euclidean four-dimensional space, which if translated back to x, y, z, t will yield the hyperbolic functions and pseudo-angles initially introduced to accommodate c. The time-like dimension of space-time can be seen to differ from the space-like dimension in the way that ict differs from x. This analogy between velocity and pseudo-angles allows us to understand the Lorentz transformations not just as a change from one reference frame to another moving at uniform translational velocity relative to it, but as a change from one set of coordinates to another set located at an angle to the first set. The inherent symmetry between space and time is further evident in the change from t to ct. We can rewrite the transformation equations to give

$$x' = \beta\left[x - \frac{v}{c}(ct)\right]$$
$$ct' = \beta\left[ct - \frac{v}{c}(x)\right] \qquad (5.100)$$

These equations are completely symmetrical between the space and time variables; if x and ct are everywhere interchanged, the equations will remain the same.

This alters our notions of space and time in that spatial and temporal separations are no longer seen as primary, but take on a "perspectival" nature; they are now regarded as aspects of the fundamental quantity: space-time separation. This quantity, s, sometimes referred to as the interval between two events, is given by

$$s^2 = x^2 + y^2 + z^2 - c^2 t^2 \qquad (5.101)$$

which also transforms in the following way:

$$s^2 = x'^2 + y'^2 + z'^2 - c^2 t'^2 \qquad (5.102)$$

Just as the distance r retains its form when we make a rotary transformation of axes in space, the interval s retains its form under a Lorentz transformation in space-time, and s^2 is simply interpreted physically as the square of the space-time distance from the origin O, with coordinates $(0, 0, 0, 0)$, to some point P, with coordinates (x, y, z, t). Despite the similarities to r^2 in Euclidean space, it is important to keep in mind that because of the minus sign, s^2 can take on negative, positive or zero values, with a slightly different interpretation in each case.[31] There are actually three invariants related to distance and time measurements in flat space-time, the functions S^2, ΔS^2 and the infinitesimal ds^2. These quantities characterize how clock measurements behave and how light travels in space-time, in addition to determining instantaneity and spatial distance measurements. S^2 characterizes these properties relative to the origin in flat space-time, and ΔS^2 does so for any two arbitrary points in space-time; ds^2 determines these properties for any two neighbouring points, which is necessary for building up the properties of curved space-time in general relativity. We can simply refer to this combination of space-time intervals as $(\Delta s)^2$, defined by the relation

$$(\Delta s)^2 = c^2(\Delta t)^2 - (\Delta x)^2 \qquad (5.103)$$

where the right side can be positive or negative, and the left side can be real or imaginary.

This geometrical approach of Minkowski is extremely powerful in its ability to mathematically unify the theoretical foundations of special relativity using just the geometrical structure of space-time. One can show how the Lorentz transformations emerged from the unification of space and time as a way of seeing velocities as analogous to angles in space-time. The transformations themselves can then be seen as the analogues to rotations in Euclidean space. But what does this unity of space and time consist in? At the beginning of his famous paper, Minkowski (1909) remarked that "henceforth space by itself and time by itself are doomed to fade away into mere shadows, and only a kind of union of the two will preserve an independent reality". That claim can be seen as having both positive and negative interpretations. Although time and space are not qualitatively identical with each other, they are inseparable parts of a larger unity. In that sense, space-time functions as a structure that synthesizes space and time but does not reduce them to the same entity. It is tempting, however, to claim that because we can treat the x and t coordinates as interchangeable for a space-like interval, we can treat space and time as having the same character. What we need to remember is that the formula for the invariant interval contains a minus sign that serves to differentiate space and time. Minkowski attempted to overcome the problem by introducing a new quantity w to measure time, where

$$w = (-1)^{1/2}t \quad \text{or} \quad \Delta w = (-1)^{1/2}\Delta t \qquad (5.104)$$

This has the effect of replacing the minus sign and giving the appearance of a Euclidean geometry in four dimensions instead of three. But because w is an imaginary number, rather than a real quantity, it cannot function as a way of identifying space with time. The latter always has the sign opposite to the distance term as defined by the interval. To that extent, space and time remain distinct but interdependent in a way that makes explicit their perspectival nature.

There is, however, another important unifying feature to Minkowski space-time, and that is the way in which the tensor calculus provides a mathematical unification of electromagnetism. In his 1905 paper, Einstein applied the transformation equations to the Maxwell field equations, demonstrating the relativity of electric and magnetic fields. Einstein defined E as the force that a source unit charge q_1 exerts on a unit charge q_2 located a short distance from the source; q_2 moves through the Coulomb field of q_1, but at any given time it can be taken as instantaneously at rest in some inertial frame k. E' is the force on q_2 due to q_1 as measured in k. On Lorentz's theory, the force on a unit charge q_2 is given by $E' = E + (v/c) \times B$, where $(v/c) \times B$ is the electromotive force. This describes what Einstein referred to as the "old manner of expression". On the relativistic interpretation, the force acting on q_2 can be considered from the point of view of the charge's instantaneous rest system. An observer in k describes q_2 as acted on only by an electric field, whereas an observer in K describes it as influenced by electric and magnetic fields. Only K's observer measures electromotive force, because K is not the instantaneous rest system for the charge q. As a result, it should be possible to reduce all electrodynamical problems to electrostatics. Lorentz had done that in the *Versuch* by interpreting the transformations as if the electron were at rest in a fictitious system that had the properties of being fixed in the aether. He then postulated a contraction effect for the moving electron in S_r. After completing the electrostatic calculations, he then transformed back to the inertial coordinates of S_r.

On the new interpretation, electromotive force plays only the part of an auxiliary concept, owing its introduction to the fact that electric and magnetic forces do not exist independently of the state of motion of the coordinate system. This electromotive force that acts on an electrical body moving in a magnetic field is nothing but the electric force considered from the point of view of the electrical body in its rest system.[32] Hence, electric and magnetic fields are relative quantities. This straightforwardly gives the solution to the asymmetries problem arising from the relative motion of the magnet and conductor. In proving that, Einstein drew the additional conclusion that questions concerning the "seat" of electrodynamic electromotive forces were essentially meaningless.

So how does Minkowski's work provide a unity for these fields? We think of space-time as the new formal structure for the STR, but the corresponding mathematical technique of a tensor calculus (or the theory of four-tensors) adapted specifically to this new structure makes it possible to show the intimate relationship between electricity and magnetism using the transformation properties of the electromagnetic-field tensor. Tensor equations can function as embodiments of physical laws because they have the property of being either true or false independently of the coordinate system. Or, to put it differently, tensors enable us to pick out intrinsic geometric and physical properties from those that are frame-dependent. This proves especially useful for relativity, because it states that the laws of physics must have the same form in all inertial frames.

Tensors are objects defined on a space V_N under all non-singular transformations of its coordinates or under a certain subgroup of admissible transformations. The latter is the important one for our purposes, because our concern is with objects defined on the four-space of events defined by the coordinates

$$x_1 = x, \qquad x_2 = y, \qquad x_3 = z, \qquad x_4 = ct \qquad (5.105)$$

These objects behave as tensors under the Poincaré group and can be called four-tensors.[33] Many of the spaces where tensors are applicable are metric spaces, which means that they possess a function ds that assigns distances to pairs of neighbouring points. The metric for the space of events is defined by an invariant of the squared differential interval mentioned earlier:

$$ds^2 = dx^2 + dy^2 + dz^2 - c^2 dt^2 \qquad (5.106)$$

This can be written in tensor notation as

$$ds^2 = g_{\mu\nu} dx^\mu dx^\nu \qquad (5.107)$$

where the metric tensor $g_{\mu\nu}$ is numerically constant under the admissible transformations

$$g_{\mu\nu} = \text{diag}(1, 1, 1, -1) \qquad (5.108)$$

The metric is used to define the scalar product of two four-vectors. A four-tensor equation is automatically, in virtue of its form, Poincaré-invariant. This allows us to determine whether or not it satisfies the principle of relativity, without applying the transformation equations.

In order to express the electric and magnetic fields in four-vector form, we need to assume that electric charge is invariant under the Lorentz transformations and is conserved over time. From that it follows that for any arbitrary volume V bounded by a surface S,

$$\int_V \frac{\partial \rho}{\partial t} dV + \int_S \mathbf{j} \cdot d\mathbf{S} = 0 \qquad (5.109)$$

where $\mathbf{j}$ is the electric current. By the Gauss theorem, the second term is equal to the change in electric charge throughout the whole volume, such that

$$\int_S \mathbf{j} \cdot d\mathbf{S} = \int_V \nabla \cdot \mathbf{j} dV \qquad (5.110)$$

Hence

$$\int_V \frac{\partial \rho}{\partial t} dV + \int_V \nabla \cdot j \, dV = 0 \qquad (5.111)$$

When the volume is arbitrarily small, we get the continuity or charge-conservation condition

$$\frac{\partial \rho}{\partial t} + \nabla \cdot j = 0 \qquad (5.112)$$

Because j is a three-vector, we need to find a four-vector in terms of (x, y, z, ict). The corresponding four-vector operator will be

$$\frac{\partial j_x}{\partial x}, \qquad \frac{\partial j_y}{\partial y}, \qquad \frac{\partial j_z}{\partial z}, \qquad \frac{\partial ic\rho}{\partial ict} \qquad (5.113)$$

Hence the continuity condition becomes

$$\frac{\partial j_x}{\partial x} + \frac{\partial j_y}{\partial y} + \frac{\partial j_z}{\partial z} + \frac{\partial ic\rho}{\partial ict} = 0 \qquad (5.114)$$

which can be expressed as the four-vector divergence of $j_x, j_y, j_z, ic\rho$ equal to zero. This four-vector can be written as J_μ, where μ represents the four components. J itself represents a combination of the electrodynamic current density in each direction and the electrostatic charge density. In fact, we can consider J as a wave of some other field A, so that for each $\mu = 1, 2, 3$ and 4,

$$\frac{\partial^2 A_\mu}{c^2 \partial t^2} - \frac{\partial^2 A_\mu}{\partial x^2} - \frac{\partial^2 A_\mu}{\partial y^2} - \frac{\partial^2 A_\mu}{\partial z^2} = J_\mu \qquad (5.115)$$

If we transform A such that

$$A' = A + \nabla \chi, \qquad V' = V - \frac{\partial \chi}{\partial t} \qquad (5.116)$$

where χ is an arbitrary potential field, then

$$\Box^2 A'_\mu = J_\mu \qquad (5.117)$$

We therefore add the gauge condition

$$\frac{\partial A_x}{\partial x} + \frac{\partial A_y}{\partial y} + \frac{\partial A_z}{\partial z} + \frac{\partial ic V}{\partial ict} = 0 \qquad (5.118)$$

If **A** is a real field, its curl is a second-order tensor, rather than a simple vector as in three-space. The tensor is a two-dimensional array Φ, where

$$\Phi_{\mu\nu} = \frac{\partial A_\nu}{\partial x_\mu} - \frac{\partial A_\mu}{\partial x_\nu} \tag{5.119}$$

This can also be expressed in matrix form as a 4×4 antisymmetric matrix. If the tensor is to represent the electromagnetic field, its elements must be functions of the components of the magnetic and electric fields. And if we identify the elements of the tensor with the components of the field in a particular way, then Maxwell's equations follow as a logical consequence.[34]

Without going into the details of the derivation, suffice it to say that the relationship between electricity and magnetism can be shown by the transformation properties of the electromagnetic-field tensor. If we begin with a field **E** due to a static charge distribution, with no magnetic field, and transform to another frame moving with a uniform velocity, the transformation equations show that there is a magnetic field in the moving frame even though there was none in the initial frame. Hence the magnetic field appears as an effect of the transformation from one frame of reference to another. The unity that was absent from Maxwell's initial formulation of the theory (the fact that the electric and magnetic fields were two entities combined by an angle of interaction) emerges from the Minkowski formalism in a relatively simple way. Here the electromagnetic field is one entity represented by one tensor; their separation is merely a frame-dependent phenomenon.

The question, however, is whether or not the physical implications of this picture are really as powerful as we are led to believe. In addition to Lorentz covariance, we need to assume invariance and conservation of charge, as well as Coulomb's law – that electrostatic attraction obeys an inverse-square law. In that sense it is incorrect to say that one can simply deduce Maxwell's equations from the Lorentz transformations. Moreover, the assumption of charge invariance is an empirical one, as is the assumption that Coulomb's law holds for this quantity. That is, we need a great deal more machinery than just the transformation equations to complete the picture; there are also non-trivial empirical assumptions lurking in the background that make the mathematical unity seem less grand than we might think. This is not, of course, intended to devalue the power of the tensor calculus in providing us a formal version of the unity implicit in special relativity; it is simply to heed the advice of Feynman, who warned that

whenever you see a sweeping statement that a tremendous amount can come from a very small number of assumptions, you always find that it is false. There are usually a large number of implied assumptions that are far from obvious if you think about them sufficiently carefully. (1963, vol. 2, p. 26-1)

The other question relevant to evaluating the impact of Minkowski space-time is whether or not the mathematics actually functions in an explanatory way with respect to the kinematics/dynamics of the STR. In the cases discussed in the earlier chapters, I mentioned the role of a theoretical parameter in the mathematical representation of the theory that functioned as the unifying mechanism. In Einstein's own formulation of the STR, the Lorentz transformations clearly provided the justification for claims about unity, with the relativistic parameter γ (or β) functioning as the component that allowed space and time to be transformed and integrated. When we move on to the Minkowski approach, we see that developing a geometric structure that integrates space and time together with the tensor calculus defined on that space-time allows us to represent the unity of the electric and magnetic fields by showing the transformation properties of the electromagnetic-field tensor. Again, we have a formal structure and a particular parameter playing a unifying role. Space and time, and electric and magnetic fields, remain physically distinct, but are united in a mathematical framework that integrates them in a seamless way.

The issue of explanatory power with respect to these frameworks is certainly not straightforward as in, say, the Maxwell/Lagrange case discussed in Chapter 3. There it was clear that the Lagrangian formalism unambiguously lacked explanatory power. The two postulates of the STR provide a unified theoretical core that involves explanatory claims about physical phenomena. But the real unifying power of the theory lies in its ability to transform physical systems in a way that unifies electricity and magnetism as well as electrodynamics and mechanics. However, the Lorentz transformations are not explanatory in and of themselves; they simply provide the tools for relating different reference frames, given the two postulates, the definition of simultaneity and the relativity of length and time. The transformation equations are in some sense the embodiment of the two postulates, but are not explanatory of the ways in which systems are constituted. That is to say, they don't provide a reduction of space and time, or of electricity and magnetism; instead they show how to integrate them so that physical phenomena and systems can be treated in a unified fashion. In the Minkowski case, geometry and physics are not integrated in the way they are in Einstein's general relativity. In the STR the geometry of space-time determines a possibility structure, rather than an explanation of how a particular system travels through space-time; that is, there is no geometric explanation of phenomena, in the way that curved space-time explains gravitational force. Consequently, the unifying power of the tensor calculus defined on that space-time does not provide a physical explanation of the integration of electric and magnetic fields that extends beyond the relativity described by Einstein in 1905.

The STR played a unifying role for physics proper by specifying formal constraints for its laws. That specification gave rise to a more localized unity of electricity and magnetism as well as space and time. Although the latter unity is properly characterized as synthetic rather than reductive, one can't help but feel that despite the similarities in synthetic character, it differs from the unity present in the electroweak case. The most obvious reason for this is that the parameter crucial for electroweak unification is, in fact, a free parameter. In that sense the STR exhibits a greater overall coherence and hence greater unity (at least for the time being).

6

⊲══════════════════════════════════════⊳

Darwin and Natural Selection: Unification versus Explanation

In the chapters on unification in physics we have seen how mathematics provided the kinds of structures that allowed us to bring together diverse phenomena under a common framework. Yet in both the electrodynamic case and the electroweak case, the additional parameters necessary for the unification (the displacement current and the Weinberg angle, respectively) provided nothing in the way of a substantive explanation of how and why unification processes take place. My point in those chapters has been to emphasize not only the differences between the ways in which theories can become *unified* but also how the unification process differs from the process of *explaining* specific phenomena.

Initially one might suppose that this situation would be limited to physics, especially in light of its rather abstract, mathematical form. However, in this chapter and the final chapter I want to show how a unification involving biological theories has many similarities to the unifications in physics, and that many parallel conclusions about the relationship between unification and explanation can be drawn in the two cases. More specifically, I want to show how the most successful unification in biology, the synthesis between evolutionary theory and genetics, was accomplished by applying particular kinds of mathematical structures that enabled early geneticists to bring together natural selection and Mendelism under a common framework. Although a theoretical unity was achieved, there was no agreed-upon explanatory model regarding the ways in which selection operated within that new synthesis. In fact, it is still an open question as to how the unification of evolutionary theory with Mendelism, under the structural constraints of population genetics, ought to be interpreted qualitatively. Ironically, the evolutionary synthesis succeeded in providing explanations only by introducing a disunity at the level of the models used in the application of population genetics to biological phenomena. That is to say, several diverse models have been required to account for the processes and effects encountered in this domain.

Before talking about the role of mathematics in unifying biological phenomena, I want first to examine Darwinian theory itself, especially because it often serves as a paradigm case for the connection between explanation and unification. If I am correct in thinking that the very project of unifying two theories is, for the most part, at odds with the procedures necessary for obtaining detailed explanatory knowledge, then it would significantly strengthen my case to show that this kind of opposition exists even when abstract mathematics has no role to play, which is seemingly the case in Darwin's theory.

As a way of addressing the question whether or not Darwin's theory exhibits both explanatory power and unifying power, I want to focus on Kitcher's account of explanation as unification. Kitcher (1981, 1989, 1993) suggests that Darwin's theory was accepted because it unified and explained a variety of phenomena. Two issues arise in connection with that strategy. The first is whether or not Kitcher's model is a good way of illustrating the process of unification in general. That is, if Darwin's theory does provide a unified account of biological phenomena, does Kitcher's model accurately capture the ways in which it does so? Second, is it the unifying aspects of the theory that function in an explanatory capacity?

In Chapter 7 we shall go on to discuss the evolutionary synthesis, particularly the work of Ronald Fisher and Sewall Wright and how their different mathematical approaches were crucial in the unification of evolutionary theory and genetics. As a result of their ground-breaking work, one can see how, especially in the case of Fisher's work, the mathematics became absorbed as part of the theory itself (in much the same way that it did in the physics cases). The immediate product of the synthesis was a unity that isolated natural selection as the main evolutionary mechanism in Mendelian populations, but one that provided no agreement on how selection actually operated. The theory that emerged as a result of the synthesis was one that retained a unity at the structural level (a unity of general principles) while exhibiting overwhelming disunity at the level of application via the models of population genetics.

6.1. On the Nature of Darwin's Theory

One of the most prominent debates in the philosophical literature on Darwinian evolution concerns the form of the argument presented in *The Origin of Species* (Darwin 1964) and in Darwin's notebooks (1960–67). Both Michael Ruse (e.g., 1979) and Michael Ghiselin (1969) interpreted the argument, albeit in different ways, as one involving hypothetico-deductivism (H-D), with natural selection providing the core from which diverse classes of facts were derived and explained. Although they had different accounts of how Darwin's argument fit that pattern, both claimed that H-D was not the only methodology used to support the theory. Moreover, Ruse argued that not all of Darwin's theory fit that model.

There are, however, many other interpretations, each of which provides its own particular strategy for reconstructing Darwin's theory. Paul Thagard (1978) claimed that Darwinian evolution represented a classic inference-to-the-best-explanation strategy. The theory was accepted because it provided a better explanation of biological phenomena than did creationist accounts. The three criteria of consilience, simplicity and analogy together accounted for the overall argument structure, with each of those criteria functioning as an important feature in what constituted a good explanation: Consilience was used in the classic Whewellian sense and referred to the number of different facts that were explained by the theory. Simplicity constrained the ways in which the facts were explained (i.e., no *ad hoc* hypotheses should be introduced). Analogy played a role insofar as it was responsible for the introduction and legitimacy of natural selection, on the basis of the demonstrated successes of artificial selection. According to Thagard, such features, and explanation in general, have to do with the way in which a particular theory is used; they involve a pragmatic dimension that is intimately connected with the historical and social context in which the theory is presented. Later in the discussion we shall take up the issue whether or not these conditions function, in Darwin's theory, in the way that Thagard suggests; but for now, let us move on to some of the other reconstructions, all of which ultimately bear some similarity to the way that Kitcher characterized Darwinian theory and his own account of explanation as unification.

Elizabeth Lloyd (1983) has offered an analysis of evolution by natural selection that is based on the semantic view of theories, a view advocated by van Fraassen (1980), Suppes (1961) and Suppe (1989).[1] The basic premise of the semantic view is that a theory consists of a family of models. So the models composing natural-selection theory belong to a specific group consisting of different model types provided by the theory to account for the empirical phenomena. The variables in those models are then specified by testing various hypotheses, and each model is evaluated by comparing its outcome with empirical observations. Lloyd gives the following example: A model type used to explain the presence of certain instincts would contain variables corresponding to the possible range or variation of a given instinct and its profitability in different circumstances, as well as standard natural-selection assumptions concerning the existence of fine gradations of instincts and their changes. A different model type might be used to explain the predominance of one species over another. Lloyd claims that the first half of the *Origin* is devoted to presenting these model types and providing empirical evidence for their basic assumptions. For instance, Darwin needed to be able to take for granted the existence of useful variation, because it would be a necessary variable in a natural-selection explanation. Consequently, he needed some evidence for the assumption that wild organisms spontaneously develop heritable variations that are advantageous to their reproduction and survival. The second half of the *Origin* is then understood as a collection of specific models (developed on the basis of the model types) that are used to explain various facts about the characteristics and behaviours of particular organisms, thereby establishing the empirical adequacy of natural selection.

Although Lloyd stresses the consilient nature of Darwin's theory of natural selection, she makes a point of separating that notion from the notion of explanatory power. Darwin himself frequently claimed that his theory could explain large classes of facts, but Lloyd maintains that his use of "explain" was decidedly non-technical and referred to the fact that the theory simply "accounted for" those phenomena. Her goal is to distance this latter notion of explanation from the pragmatic one defended by Thagard and to show that Darwin's main defence of his theory centred on claims regarding its semantic properties (i.e., its empirical adequacy), rather than on a pragmatic virtue like explanatory power. Consequently, she understands consilience as an index of the empirical adequacy of the theory (i.e., the more diverse facts the theory can "account for", the more empirically adequate it is). Hence, empirical adequacy is determined not only by the fact that natural selection allows us to construct several models of the phenomena but also by the fit between empirical data and the models. The foundation of Lloyd's criticism is the view that explanation *is* a pragmatic virtue of theories and therefore cannot be the main reason for a theory's acceptance. As a result, her disagreement with Thagard is simply that he uses explanatory power as *the* criterion for the success and acceptance of natural selection.[2]

Despite the fact that Darwin himself frequently referred to the theory's explanatory power as a condition of its acceptance, and did so in a context in which explanatory power seemed to refer to consilience (Darwin 1903, vol. 1, pp. 139–40), Lloyd downplays that use of explanatory power, without showing exactly why we should decouple consilience and explanation. She offers no evidence for the fact that the natural-selection explanations offered by Darwin were intended only to "account for" the phenomena, nor does she cite any reasons for thinking that consilience cannot be identified with explanatory power. Hence, nothing follows from Darwin's non-technical use of "explain" regarding the connection or lack there of between the two. Only if one assumes a pragmatic account of explanation does Lloyd's argument for their separation seem persuasive, and then only if one wishes to maintain that theories should not be and are not accepted on pragmatic grounds alone. Ultimately, of course, I agree with Lloyd on the separation of consilience and explanatory power; my point is simply that nothing in her argument would persuade one to adopt that view. In the physics cases, I claimed that there are good reasons for thinking that unifying power and explanation are at odds with each other. But that need not commit one to a pragmatic view of explanation. One can embrace the notion that explanations are "objective" or "epistemic" features of theories while claiming that the unifying mechanisms simply don't provide the kind of information necessary for a full explanation of the phenomena. This distinction between consilience and explanatory power will also be important in the discussion of Kitcher's views, but before moving on, let me briefly mention one other reconstruction of the *Origin*.

The final account I want to consider is one advanced by Doren Recker (1987), who sees the *Origin* as one long argument supporting evolution by natural selection. [Ernst Mayr (1964, 1991) also defends the one-long-argument view, but not in the way that Recker does.] Recker identifies three strategies for establishing the causal efficacy of natural selection and claims that different parts of the *Origin* are responsible for establishing each of the independent but related strategies: The first four chapters are intended to support the probability of the hypothesis that natural selection is causally efficacious by appealing to a *vera causa* strategy of the kind described by William Herschel.[3] This is done by showing how the effects brought about by natural selection are similar to those known to be caused by artificial selection. As a result, one is allowed to infer the existence of natural selection and its ability to produce effects of that kind in nature. The middle chapters are devoted to defending natural selection from possible objections, and the final chapters show the explanatory power of natural selection due to its applicability to large classes of facts.

There is a sense in which all of the approaches mentioned earlier address the question of unity in Darwin's theory, and yet there is an important sense in which the issue is simply bypassed. Recker sees the emphasis on one long argument as providing a unified "methodology" in the attempt to justify natural selection as a *vera causa*. The kind of Whewellian consilience emphasized by Thagard stresses the unity of the theory in virtue of its ability to "explain" broad classes of facts. Lloyd's semantic account can incorporate the unifying aspect of natural selection by showing how a few simple model types can be used to "account for" a variety of different phenomena. Each of those approaches is important, because they highlight the different ways in which unity can present itself in Darwinian theory, depending, of course, on how one reconstructs the argument of the *Origin*. Kitcher's account, as discussed next, embodies many of the ideas presented in those different approaches. He emphasizes unification as the key to understanding explanation. Both simplicity and unification are measured by the number and form of the argument patterns used in explanations; causality is construed as a relationship of explanatory dependency. The ideal of hypothetico-deductivism is upheld because explanation is thought to consist of derivation – in other words, explanations are, in fact, arguments. In order to analyse the merits of the claim that unification is explanation, let us look more closely at Kitcher's approach.

6.2. Kitcher's Argument Patterns: Explanation
as Derivation and Systematization

Kitcher's model exhibits both similarities to and differences from the traditional (Hempelian) D-N account of explanation. In addition to emphasizing the role of arguments in explanation, Kitcher stresses the importance of systematization. Explanations are not evaluated by considering them in isolation from each other, but rather by seeing how they form part of a "systematic picture of the order of nature" (1989, p. 430). To be successful, an explanation must belong to a set of explanations called the explanatory store of a particular theory or science. This set contains the derivations that collectively provide the best systematization of one's beliefs. The philosophical task of providing a theory of explanation is to specify the conditions on the explanatory store. The idea, then, is that we rank explanations with respect to their roles in a broader, systematic group of derivations supplied by a particular theory. Hence, it isn't sufficient that a conclusion simply be derived from certain premises; rather, one must show *how* the premises yield the conclusion, and that is done by showing how the argument forms part of the explanatory store.

The explanatory store $E(K)$ is the set of arguments that best unifies K, where K is the set of statements endorsed by the scientific community. The goal is to reduce the types of facts that one has to accept as brute by showing how to derive descriptions of many phenomena using the same patterns of derivation again and again. These argument patterns consist of a number of elements. First, there is the schematic sentence, which is an expression replacing some, but not necessarily all, of the non-logical expressions with dummy letters. For example, the statement "organisms homozygous for the sickling allele develop sickle-cell anaemia" is replaced by "organisms homozygous for A develop P, so for all x, if x is O and x is A, then x is P". Next, there are "filling" instructions for replacing dummy letters; for instance, one might specify that A be replaced by the name of an allele, and P by a phenotypic trait. A schematic argument is a sequence of schematic sentences, and the classification of a schematic argument consists of a set of statements describing the inferential characteristics of the argument, that is, which terms in the sequence are to be regarded as premises, which are inferred from which inference rules and so forth. Finally, a general argument pattern is a triple consisting of a schematic argument, a set of sets of filling instructions and a classification for a schematic argument.

A sequence of sentences instantiates the general argument pattern if and only if it meets certain conditions: First, the sequence must have the same number of terms as the schematic argument of the general argument pattern. Second, each sentence in the sequence must be obtained from the corresponding schematic sentence in accordance with the appropriate filling instructions. Third, it must be possible to construct a chain of reasoning that assigns to each sentence the status accorded to the corresponding schematic sentence by the classification. Kitcher gives the following example to illustrate the point. We can represent the schematic argument for a Newtonian one-body problem by specifying the following items:

1. the force on α is β
2. the acceleration of α is β
3. force $=$ mass $\times$ acceleration
4. (mass of α) $\times$ (γ) $= \beta$
5. $\delta = \theta$

The filling instructions tell us that all occurrences of α should be replaced by an expression referring to the appropriate body; β is replaced by an algebraic expression referring to a function of the variable coordinates of time; γ is replaced by an expression that gives the acceleration of the body as a function of its coordinates and time derivatives; δ is replaced by an expression referring to the variable coordinates of the body; θ is replaced by an explicit function of time. The classification of the argument tells us that items 1–3 are premises, and item 4 is obtained from them by substituting identicals. Item 5 follows from item 4 by simple application of algebra and calculus (Kitcher 1981, p. 517).

In order for this notion of an argument pattern to function as a mechanism for unification there must be some constraints on the kinds of instantiations permitted by the patterns. In other words, the stringency of the argument pattern will determine how difficult its instantiation will be. The conditions imposed by the filling instructions for substitution of dummy letters, along with the logical structure as determined by the classification, will determine the degree of stringency. The goal is to have an explanatory store that will best represent the trade-off between minimizing the number of patterns of derivation required to provide a systematic representation of K and maximizing the number of conclusions generated by those arguments. Hence, the unifying power of $E(K)$ will vary directly with the stringency of the patterns and inversely with the number of patterns.

How does this format apply in the Darwin case? Kitcher cites Darwin's remark that the doctrine of natural selection must "sink or swim according as it groups and explains phenomena" (F. Darwin 1887, vol. 2, p. 155). However, because Darwin was unable to provide a *complete* derivation for any biological phenomenon, Kitcher sees the explanatory power of the theory as simply its ability to unify a host of biological phenomena (1981, pp. 514–15).[4] Hence, Darwin's principal achievement was to bring together problems in biogeography, issues concerning relationships among organisms (past and present) and questions about the prevalence of characteristics in species or higher taxa. That was followed by a demonstration in outline how they might be answered in a unified way. That kind of approach would make history central to an understanding of biological phenomena, and it is those Darwinian histories that, according to Kitcher, provide the basis for acts of explanation (1993, pp. 20–1). We explain the distribution of organisms in a particular group by tracing a history of descent with modification that charts the movements of organisms in the lineage, terminating with the group in which we are interested. Sometimes an explanation will give a causal account of the modifications in the lineage, whereas other explanations will simply record the modifications. The unification will then consist in derivations of descriptions of those biological phenomena, derivations that instantiate a common pattern. The fewer patterns that are used, the more unified the theory is.

In the case of evolutionary theory, Kitcher distinguishes two patterns: one that deals with homologous characteristics in related groups (questions of the form "Why do G and G^* share the common property P?"), and one that deals with the primary idea of natural selection. The first homology example is as follows:

1. G and G^* descended from a common ancestral species S.
2. Almost all organisms in S had property P.
3. P was stable in the lineage leading from S to G; that is, if S was ancestral to S_n, and S_n immediately ancestral to S_{n+1}, and S_{n+1} ancestral to G, then if P was prevalent in S_n, almost all members of S_{n+1} were offspring of parents both of whom had P.
4. P was stable in the lineage leading from S to G^*.

5. P is heritable; that is, almost all offspring of parents both of whom have P will have P.
6. Almost all members of G have P, and almost all members of G^* have P.

The filling instructions for the patterns tell us to replace P with the name of a trait, G and G^* with the names of groups of organisms (populations, species, genera etc.) and S with the name of a species. The classification specifies that items 1–5 are premises, and item 6 is derived from items 1–5 by using mathematical induction on the lineages.

The simple-selection case is slightly more complicated and is concerned with questions like "Why do almost all organisms in G have P?" It exhibits the following pattern:

1. The organisms in G are descendants of the members of an ancestral population G^* that inhabited an environment E.
2. Among the members of G^* there was variation with respect to T: Some members of G^* had P, and others had $P\#, P\#\#, \ldots$.
3. Having P enables an organism in E to obtain a complex of benefits and disadvantages C, making an expected contribution to its reproductive success $w(C)$; having $P\#$ enables an organism to obtain a complex of benefits and disadvantages $C\#$, making an expected contribution of $w(C\#), \ldots$.
4. For any properties P_1, P_2 and so forth, if $w(P_1) < w(P_2)$, then the average number of offspring of organisms with P_1 that survive to maturity will be greater than the average number of offspring of organisms with P_2 that survive to maturity.
5. All the properties $P, P\#, P\#\#, \ldots$ are heritable.
6. No new variants of T arise in the lineage leading from G^* to G. All organisms in this lineage live in E.
7. In each generation of the lineage leading from G^* to G the relative frequency of organisms with P increases.
8. The number of generations in the lineage from G^* to G is sufficiently large for the increases in the relative frequency of P to accumulate to a total relative frequency of 1.
9. All members of G have P.

In this second example, the filling instructions require that T be replaced by the name of a determinable trait, the P terms by names of determinate forms of that trait, G^* by the name of an ancestral species, E by a characterization of the environment where members of G^* lived, the C terms by specifications of sets of traits and the $w(C)$ terms by non-negative numbers. The classification states that items 1–6 and 8 are premises, that item 7 is derived from items 1–6 and that item 9 is derived from items 7 and 8.

In the first example, Kitcher allows that common traits can result from circumstances other than homologies (i.e., parallelism or convergence), but he claims that Darwin himself proposed that many questions about common traits could be answered by discovering true premises that would instantiate the homology pattern. For the case of simple selection, Kitcher claims that the pattern is *implicit* in the explanations of the prevalence of traits that are found in the *Origin* and other Darwinian texts (Kitcher 1989). Variants of that pattern can be obtained by relaxing item 6 to obtain what Kitcher terms "directional selection", which allows for a more successful trait to arise in the population at the same time that an advantageous trait is increasing in frequency. Other cases may involve "correlated selection", where the increase in frequency of a characteristic P is explained by using one of the selectionist patterns to show that some other trait Q will increase in frequency, together with a premise asserting the correlation of P and Q, to conclude that P will increase in frequency. The idea is to show that the simple-selection pattern, with some slight modifications, is used again and again to derive a variety of conclusions. Kitcher presents similar results for the genetics case by providing examples of explanatory schemata that have been employed in genetics, beginning with the simple Mendelian case limited to one locus and two alleles, with complete dominance. He then goes on to show how refinements of this pattern can be introduced to deal with more complex cases, giving the impression of a cumulative and unified development of classic genetics.[5]

This way of reconstructing Darwinian evolution bears certain similarities to the view presented by Lloyd, who, as we have seen, emphasized the role of models, rather than argument patterns; the simpler, more unified theory is the one with fewer model types, as opposed to fewer patterns. In fact, Kitcher himself (1993, p. 45, n. 63) has pointed out that for some purposes the notion of a model type can be substituted for his use of general schemas:[6] If we characterize unification as simply an issue of *how many* model types or argument patterns are required to generate a number of different conclusions, unification becomes a wholly comparative notion. As a result, Kitcher says, theories we would not generally classify as unificatory may prove to be so when considered in comparison with their competitors. Undoubtedly that is in some sense right; most theories can unify at least some phenomena via their laws, without displaying the unifying power of, say, Newtonian mechanics or Maxwell's electrodynamics. But surely we want and perhaps need more from an account of unification. A unified theoretical core represented by a specific parameter suggests that unification is the product of a specific theoretical construction that involves more than an enumeration of the argument patterns and models that the theory generates. In other words, there is something distinct about the structure of a truly unified theory that differentiates it from others. Undoubtedly the measure of a unified theory will depend on the number of free parameters the theory has, but that is a different issue. Comparing argument schemas and model types on the basis of similarity relations is much more difficult than counting free parameters – it isn't simply the comparative nature of the task, but the difficulty in evaluating how the comparisons should be made.

If we want to draw an analogy between the physics cases and Darwinian theory, we can do so by isolating natural selection as the mechanism that serves to unify various biological facts under a single theoretical framework. Although that may indeed serve to unify the phenomena because of its broad applicability, it is not immediately clear that selection can also function as the source of explanatory power. One of the difficulties is that selection is by no means the only mechanism at work in Darwinian evolution; there are also the effects of use and disuse, spontaneous variation and also directed variation, where the tendency to vary, rather than the actual variation, is transmitted. None of these involves selection, and consequently they could lead to maladaptive differentiation in local populations. Hence, although natural selection is both sufficient and necessary for unification, it is neither with respect to explanation, in every case. That is to say, there is a difference between the unifying role of natural selection and its function as an explanatory hypothesis. This calls into question the suggestion that most of the Darwinian explanations are modifications of the selectionist schema.[7]

Before elaborating on the explanatory issue, I need make good on my claim about the unifying power of natural selection. To do that, I need to show, in more explicit terms, what it is about selection that unifies the phenomena (i.e., how and why it produces the unification). Once that framework is in place, we can see more clearly the extent to which selection is or is not capable of offering explanations. In other words, I do not want to *deny* that natural selection can have explanatory power; rather, my claim is that its explanatory power cannot be understood in terms of its unifying power. Explanations of biological phenomena are not necessary consequences of unification. Put another way, the unifying power of natural selection stands independently of the ways in which selection figures in explanations.

In the examples of unification we considered in the earlier chapters, mechanisms like the displacement current lacked explanatory power, because no account was given by the theory, other than a formal derivation, of how field-theoretic processes were possible. Similarly in the electroweak theory, there was an explanation of spontaneous symmetry-breaking that allowed for the introduction of other essential theoretical constraints (like the boson masses), but no causal explanation of the mixing of the weak and electromagnetic fields. What I want to show next is that the case of natural selection is somewhat different; although not always explanatory in and of itself, it can be made so with the addition of other theoretical hypotheses and assumptions. Although one could argue that this might also be true of the electrodynamic and electroweak examples, the important difference is that in the Darwinian case those assumptions were integrated into the theory in a way that was not true of the physics examples. Although it is certainly the case that those additional hypotheses often were problematic in and of themselves, they nevertheless could provide, with the aid of natural selection, explanations of evolutionary phenomena. But because of their rather questionable status, I want to claim that those explanations should not be seen to contribute to the evidential basis of the theory. Nevertheless, there emerges a degree of explanatory power that was not present in the physics cases. Although Maxwell introduced illustrative concepts into the theory, they were not intended as explanatory hypotheses; the core of the theory was simply that provided by the abstract dynamics of Lagrange. In the final analysis, regardless of the explanatory capacity of Darwinian theory, it is selection simpliciter that plays the unifying role.

In both Maxwell's theory and the electroweak theory, the kind of explanations we wanted were those that would give detailed causal stories about how and why specific processes took place. On Kitcher's account, the explanatory power provided by natural selection is structural insofar as it allows for the derivation of evolutionary facts. But it is also supposedly causal, in that selection is seen as the factor responsible for producing certain kinds of phenomena. If we were to characterize explanation simply as derivability, then of course we would be forced to change our minds about the theories that I claimed lacked explanatory power. Without getting too far afield in the debate over what counts as an acceptable explanation, it is reasonable to assume that we want explanations to provide some kind of theoretical understanding of how processes like the propagation of electromagnetic waves take place. Although derivation explanations are common in the natural sciences (and sometimes the social sciences), usually they are not without some theoretical background to which one can appeal to understand how the phenomenon in question came about. Derivations typically take the form of mathematical deductions from abstractly formulated laws. These are accepted as explanatory because they invariably are situated in larger theoretical contexts that can in turn provide explanations for why the derivations work. However, if one were interested only in mathematical derivations, then science would have no demand for anything other than phenomenological theories. It is because we are interested in the how and why of natural processes that we must acknowledge the importance of and demand for explanations that go beyond deduction. Without that we have what Lloyd characterized as an "accounting for", rather than an explanation. Thermodynamics was able to account for many phenomena, but one had to invoke the kinetic theory in order to provide explanations.

The unifying power of natural selection is what makes it extendable to a variety of phenomena. To understand how that situation arose, we need to look at the methodological context in which Darwin was working.

6.3. Methods and Causes: History and Influence

Some commentators, including Recker (1987) and Hodge (1977), as well as Darwin himself, have emphasized the way in which natural selection can be thought of as a *vera causa*. As noted earlier, this principle originated in Newton's rules for philosophizing, and although there was a span of almost 200 years between the publication of the *Principia* and the *Origin*, Newtonian ideas retained pride of place in British scientific theorizing. That was true not only with respect to the actual structure of mechanics but also with regard to methodological dictates concerning the roles of hypotheses and causes. The *vera causa* principle became an important constraint for both Herschel and Whewell, as well as Charles Lyell, all of whom had considerable influence on Darwin. The issue of how to establish a *vera causa* or how those nineteenth-century philosophers interpreted Newton's first rule is itself the subject of considerable philosophical debate. For my purposes, it will be sufficient to distinguish what I shall call the independent-existence constraint on establishing a *vera causa* from the "inferential" component.

Whewell was an advocate of the inferential approach, claiming that one could conclude that a cause was true if it led to a consilience of inductions. Because a theory achieves consilience by explaining a number of classes of facts, some of which were not contemplated in the formation of the theory, the argument is similar to a common-cause argument; that is, it would be extraordinary if a cause could explain such a diversity of phenomena and yet not be true.

We may provisorily assume such hypothetical causes as will account for any given class of natural phenomena, but that when two different classes of facts lead us to the same hypothesis we may hold it to be a true cause. (Whewell 1847, vol. 2, p. 286)

Hence, true causes are established in virtue of their ability to account for diverse facts. Darwin echoed that sentiment in the *Origin*:

It can hardly be supposed that a false theory would explain, in so satisfactory a manner as does the theory of natural selection, the several large classes of facts.... (1872, p. 784)

Again, in a letter to Hooker (February 14, 1860), he claimed that he always looked at the doctrine of natural selection as "an hypothesis, which, if it explained several large classes of facts, would deserve to be ranked as a theory deserving acceptance" (1903, vol. 1, pp. 139–40).

William Herschel, on the other hand, was concerned with initially establishing the *existence* of the cause in question and then showing how it could figure in the explanation of several different phenomena. The method involved basing hypotheses on causes that were already known to be capable of producing effects similar to those in question. Hence, one first looked for causes already known to exist and, failing that, justified any hypothetical causes in terms of their similarity to antecedently known causes. That was done by argument from analogy: One could invoke invisible causal mechanisms as long as they were linked analogically with experienced causal regularities. If the analogy between two phenomena A and B was very close and the cause of A was known, then one could not refuse to admit the action of an analogous cause in the case of B, thereby establishing a probable *vera causa*. Whewell was critical of that restriction to, or analogy with, known causes, on the ground that it hindered the discovery of new causes. However, both he and Herschel saw the search for *vera causae* as part of the search for general laws that would provide explanations of particulars.

It seems reasonable to assume that Darwin intended to formulate his account of evolution by natural selection along those lines, combining both versions of the *vera causa* principle and defending the theory on the basis of its consilience or unifying power. As I see it, the overall argument presented in the *Origin* consists of two parts: The first part itself consists of a two-part argument presented in the first four chapters, intended to establish the *existence* of natural selection (this is the first step in the *vera causa* argument). The second part, which occupies the remaining chapters, provides a presentation of the evidence for natural selection as a consilience of inductions (the Whewellian component of the *vera causa*). In that sense one can see Darwin as satisfying both Whewell's and Herschel's conditions for establishing a *vera causa*. That said, it is perhaps important to emphasize that Darwin did not subscribe to the idea that natural selection must be shown to be a "*vera causa* always in action" (1903, vol. 1, p. 140).

In the passages where Darwin first introduced the idea of natural selection (1964, p. 61), he did so on the basis of an analogy with artificial selection; yet that constituted only part of the existence argument. Although humans could, by the process of selection, produce variations in plants and animals, how could such a mechanism operate in nature without human intervention? In addition, he wanted to show that from the facts about variation in nature (which were not in dispute), together with the geometrical rate of increase in the population of each species (a fact that would lead to a struggle for existence), it would be, to use Darwin's words, "most extraordinary" if no variation useful to the welfare of an organic species had occurred, in the same way that variations useful to humans had occurred. In other words, we know that variations have occurred in nature, and we also know that useful variations have occurred in humans; given that all organic beings are involved in a struggle for existence, it is reasonable to assume that useful variations have occurred in them as well. Moreover, it would be those that exhibited the most useful traits that would have the best chance of being preserved in the struggle for life, and it was that preservation that was called natural selection (1964, pp. 80–1, 126–7). So the struggle for existence became the analogue in nature of man's selection in the domestic case, exerting a strong selective force on populations. The remainder of the *Origin* is devoted to establishing the second component of the *vera causa* argument – the role of natural selection in producing a consilient theory. "Whether or not natural selection has really thus acted in nature ... must be judged by the general tenor and balance of evidence presented in the following chapters" (1964, p. 127).

Given this reconstruction, can we characterize the unity of the Darwinian theory by simply pointing out the role of natural selection in bringing together geological, palaentological and other diverse phenomena? There can be no doubt that natural selection acts here in a unifying way; what we want to know, however, is how this is achieved, especially given my claim that unity cannot be understood as coincident with simply explaining these phenomena.

We have seen that part of the argument for natural selection as a *vera causa* was based on the geometrical rate of increase in each species, a fact that leads to a struggle for existence. It is this connection that constitutes the most difficult part of the argument, because it isn't immediately clear how natural selection follows from this principle of geometrical increase. The principle was originally formulated by Malthus (1798) in *An Essay on the Principle of Population*, and it states that a population, when unchecked, will increase at a geometrical rate, whereas the means of subsistence will increase at only an arithmetical rate. That principle supposedly follows from two main postulates or "fixed" laws of nature: (1) food is necessary for existence; (2) passion between the sexes is necessary and will remain nearly in its present state. As a result of the former, the effects of increases in population and subsistence (two unequal powers) need to be kept equal; but, given the difficulty of subsistence, that implies some strong and continually operating check on the population increase. That perpetual *tendency* for the increase in population to out-strip the means of subsistence is a general law of animated nature. The restraints that function in order to keep the ratio at an acceptable level are resolvable into moral restraint, vice and misery, qualities that create a general struggle for human beings, a struggle whose outcome is directly connected with their ability to survive and/or produce offspring.

There has been some debate in the literature as to whether Darwin was already acquainted with the idea of a struggle for existence before reading Malthus (Ruse 1979) or whether he adopted it from Malthus (Vorzimmer 1969), but regardless of the issue of historical priority, one thing seems clear: Malthus's principle was the key to unlocking the power and persuasiveness of natural selection. Prior to reading Malthus, Darwin claimed that it was "a mystery" how selection could be applied to organisms living in a state of nature. The missing link was a causal mechanism for understanding the selective process, the context for which was supplied by Malthus's principle.

The geometric–arithmetic ratio describing the tendency to reproduce to the limit of the available means of subsistence involved a system of checks to keep the ratio at a stable level. Those checks functioned as a system of survival pressures that affected entire populations. Natural selection could then be seen both as part of the system of checks that kept the ratio stable (it acted as a filter by limiting the survival chances of those beings that were poorly adapted) and as a response to the struggle, by ultimately producing species with traits well suited for survival and reproduction. Moreover, it could be seen to operate both at the level of individuals and at the level of populations or species. In the *Variations of Plants and Animals under Domestication* we find Darwin claiming that as soon as he had acquired a "just" idea of the power of selection, he saw, on reading Malthus, "that natural selection was the inevitable result of the rapid increase of all organic beings; for I was prepared to appreciate the struggle for existence having long studied the habits of animals" (1868, vol.1, p. 10). Under the circumstances described by Malthus, favourable variations would tend to be preserved, and unfavourable ones destroyed. And as Darwin remarked, it was on reading Malthus that he at once saw how to "apply the principle of selection – he at last had a theory by which to work" (F. Darwin 1887, vol. 2, p. 68).

Not only did Malthus's work provide a context in which natural selection could be seen to play a causal role in evolution, but also it provided Darwin a *method* for presenting his hypothesis of natural selection in a unified and systematic way.

The mathematical formulation of Malthus's principle allowed Darwin to see it as a kind of universal law that applied across populations, and as noted earlier, it enabled him to argue from the existence of useful variations in humans (an uncontroversial fact) to the existence of useful variations in other organic beings as a result of natural selection. Insofar as the struggle for existence could be seen to apply across all of nature, so could natural selection. Hence, it was not simply the idea of natural selection itself that functioned in a unifying role, but natural selection seen as part of the quantitative structure provided by Malthus. Not only did the Malthusian framework yield an empirical lawful basis from which one could view selection as a causal mechanism, but also its quantitative scheme furnished the kind of generalizability that allowed it to be applied across populations of all organic beings. It was in that sense that natural selection played a unifying role for Darwin as a mechanism that operated in the context of a larger quantitative framework, one that both supplied the foundation and illustrated the method for the general applicability of selection. To that extent one can see the influence of Newtonian methods extending well beyond the domain of mechanics.

6.4. Does Unity Produce Explanation?

Given the broad applicability of selection as a unifying principle/mechanism, can one then simply extend its role to include explanatory power? Although natural selection is not the only cause of evolutionary change (and therefore doesn't function in all explanations), that in itself does not constitute a reason for denying its explanatory power. The difficulty seems to have been that when Darwin presented the explanations that supposedly provided evidence that natural selection really had acted in nature, it became clear that several additional requirements would have to be met for the selectionist strategy to work. Darwin wanted to show how the continuing evolutionary effects of natural selection could be used to solve problems in geology, palaeontology, geographical distribution, morphology, embryology and so forth. If we consider his discussions of geographical distributions and his explanation regarding the inhabitants of archipelagos, we find, in addition to selectionist claims, the introduction of crucial assumptions about the methods and possibilities for transport across vast distances. He attempted to justify the assumptions by arguing that those methods were not simply possible, but were actually to be expected. For instance, seeds could be transported for many miles over oceans if they were embedded in driftwood. In addition, birds blown by gales across the ocean would be highly effective transporters of seeds (1964, ch. 11).

A similar situation obtained in his discussion of embryology, where Darwin wanted to explain why embryos frequently were radically different from adult forms. Part of the explanation involved the fact that the selective pressures experienced by embryos and adults often were very different. But because that did not by itself explain differences in structures, Darwin was forced to appeal to laws of heredity, arguing that there was no reason to suppose that new characteristics always appeared in embryos in the same form as in adults. In fact, some characteristics appeared only in adults, and unless there was selective pressure forcing them to appear at earlier stages of development, there was no reason why they would appear in embryos. On that basis he concluded that because selection varied throughout an organism's development, its structure would vary as well. The problem, of course, was the lack of an explanation or justification for the assumptions about variations and their effects on an organism's structure. In an attempt to overcome that difficulty, Darwin used cases of domestic organisms under artificial selection to justify claims made about variations in a natural environment (1964, pp. 444–6).

So in order for natural selection to be explanatory in diverse areas, several additional assumptions needed to be added to the theory, depending on the situation at hand – assumptions that would have a status quite different from the initial conditions responsible for specifying the problem situation. The claims about heredity and variation were especially important, because no suitable account of those mechanisms could be given, and yet they were necessary to complete the explanations. Once again, my claim is not that natural selection had no explanatory power; rather, it could function in an explanatory way only in conjunction with other quite specific assumptions, some of which lacked independent justification and were not well understood. In other words, selection could contribute to an explanation only if those other conditions were satisfied. And in that sense, although selection is a common feature in many explanations of evolutionary processes, selectionist arguments alone cannot ground the explanatory power of the Darwinian evolutionary theory. They are necessary, but not sufficient.

Because the explanatory power of the theory arises as a result of the many *applications* of selection, plus special assumptions about *other* kinds of mechanisms, the semantic view seems best able to capture how the overall theoretical structure is configured. By focusing on different models as the explanatory vehicles, we can see how natural selection provides the basis for explanatory strategies across diverse circumstances. Each of the domains in which selection functions, whether it be biogeography, morphology or embryology, makes use of a specific selectionist model that incorporates different kinds of assumptions in order to function in an explanatory way. However, if we follow Lloyd and identify the degree of unity or consilience of the theory with the number of model types necessary to account for the phenomena, then the degree to which a theory is unified becomes difficult to assess, because it isn't immediately clear how we ought to differentiate these model types. Should it be done on the basis of structural similarities, similarity of fundamental assumptions and so forth? At what point do we say that two models are of the same type? And at what point do we cease to call a theory unified on the basis of the number of model types it introduces? In addition, when unity is defined in this way, it focuses the question on the number of parameters and assumptions, rather than on a specific mechanism or parameter responsible for integrating the phenomena. As noted earlier, the number of free parameters in a theory is an important issue for unification, because the degree of integration will be a function of how cohesive the theoretical structure is. In other words, the more free parameters, the less unified. That natural selection plays a unifying role is not in doubt, but if we equate unity and explanation, the additional assumptions required for explanatory power detract from that unity.

It also seems clear, then, why focusing on the argument patterns discussed by Kitcher doesn't expose the true nature of unity. By identifying both explanatory power and unity with arguments/models, the semantic and deductivist accounts embody a fundamental tension between the desire for explanation (which involves the proliferation of schemes and structures) and the demand for unity (which involves their limitation). As I argued earlier, it is possible to isolate the unifying power of natural selection without embracing the difficulties of arbitrarily counting model types and argument patterns. Even in the case of a fundamentally qualitative theory like Darwin's we can see how the unifying power of selection arose from its association with a broadly based structural feature of populations that Darwin adopted from Malthus. Its explanatory power, on the other hand, is achieved at the level of models or specific applications. Here disunity and variety are required for success. Structural constraints are contextualized and coupled with additional assumptions necessary for understanding particular situations. By contrast, the quantitative framework within which the Malthusian principle was cast enabled it to function as a kind of universal law that could ground the general *application* of natural selection.

This way of thinking about unification also sheds light on how the modern synthesis of Darwinian evolutionary theory and Mendelian genetics produced a general unification in biology. In fact, what the synthesis shows is that it is possible, as in the case of the electroweak theory, to have both unity and disunity within the framework of one theory: a unity at the level of basic structural principles, coupled with disunity at the level of application and explanation. Again, crucial in all of this is the role of mathematics. In fact, my argument in the next chapter suggests that the synthesis in biology was possible only to the extent that the early geneticists, R. A. Fisher, Sewall Wright and J. B. S. Haldane, were able to mathematize certain kinds of biological phenomena. Just as Malthus's quantitative law provided a kind of mathematical basis for natural selection, the quantitative methods of Fisher and Wright showed that selection functioned in Mendelian populations.[8]

Now that we have seen how the traditional connection between unification and explanation can be severed even in non-mathematical theories, we can turn our attention to the way mathematics can function within biological contexts to produce unified theories. In the work of Fisher the mathematics not only provided a framework in which to express the unity of Mendelian genetics and Darwinian theory but also stood in the place of empirical work. By that I mean that it served as a kind of experimental foundation for the new biology. And to the extent that it furnished results that empirical research was unable to produce, it became a substantive part of biological theory. Insofar as the mathematics came to constitute a good deal of the theoretical edifice, it resembled quite markedly the way in which

mathematics functioned in the physics cases we considered, specifically the electroweak theory of Glashow, Weinberg and Salam. There we saw how a physical
dynamics emerged from within the constraints of a very powerful mathematical
framework – dictating the form that certain physical interactions must take. As
we shall see, a somewhat analogous case exists in the synthesis of evolutionary theory with Mendelian genetics. The mathematics tells us the effects of selection in
populations, something that could not be determined simply from the phenomena
themselves.

7

<⟹⟹⟹⟹⟹⟹⟹⟹⟹⟹⟹⟹⟹⟹⟹⟹⟹⟹⟹⟹⟹⟹⟹⟹⟹⟹⟹⟹⟹>

Structural Unity and the
Biological Synthesis

7.1. Synthesis and Unity

In Chapter 6, I claimed that one of the problems with selectionist explanations was a lack of understanding of the mechanisms of heredity and variation. That difficulty was partially alleviated through the work of Gregor Mendel in the late nineteenth century, but with the rise of Mendelian genetics came a supposed conflict with Darwinian evolution.[1] The debate was largely about the causal agents responsible for heredity and variation and about the nature of evolution itself. The controversy concerned the role of selection, as opposed to mutation, in producing changes in populations and whether evolution was gradual or discontinuous. Indeed, the issue of gradualism had a history that dated back to the publication of Darwin's *Origin of Species*.

Followers of Mendel saw the particulate nature of genetic factors as a basis for discontinuity in evolution and the hardness of inheritance (i.e., selection would have no effect on heritable traits). They argued that mutation pressures were the main agents of evolutionary change. Although selection was considered by Darwinians to be responsible for such change, some Darwinians claimed that individual variation was unimportant, that the causal factor in evolution was use or disuse, rather than selection. Consequently, the problems facing Darwinism included not just the origin of variation and the types of variations that were heritable but also the role of selection versus environmental conditions, which could give rise to use or disuse, the nature of species (whether they were real or abstract units) and the role of isolation in evolutionary change. One of the difficulties in trying to establish any compatibility between Darwinism and Mendelism was that many Darwinians were so embedded in the gradualism view that they were unwilling to seriously consider the possibility that Mendelism might have an explanatory role to play in evolution. Although discontinuous characteristics did obey the laws described by Mendelian theory, it was thought that those kinds of traits were of little importance in evolutionary change. With the rise of genetics studies came two different strands of research: Some geneticists were interested only in understanding the mechanics of inheritance, whereas others, who were more sympathetic to Darwinian theory, turned their attention to examining a possible genetic basis for evolution. Recognizing that evolution was a population phenomenon, researchers based their work on studies of gene frequencies in populations: Statistical populations were studied by mathematical geneticists, and natural populations were studied by ecological geneticists. Mathematical population genetics became further divided into two lines of research: The Mendelians (William Bateson in particular) were proponents of discontinuous variation, and the biometricians (Karl Pearson and later Ronald Fisher) were concerned with the compatibility of genetics and Darwinian theory.[2] The latter group stressed the role of continuous variation in selection and assumed that all inheritance was blending inheritance.[3] The Mendelians saw all evolutionary change as the result of genetic discontinuity produced by new mutations. As a result, there was no role for individual variation or selection, because neither was capable of producing anything new. Hence, the biometricians saw Mendelism as a threat, because it supported discontinuous evolution as the only possible option.

According to the Mendelians there were two fundamental problems that seemed to prevent any reconciliation between selection and the findings in genetics. First, the selective value of small variations was negligible, and thus it was extremely difficult to determine whether or not selection was actually operating. Second, the swamping effect in intercrossing seemed to obliterate the variation upon which selection acted. In other words, there was no explanation of how variations could persist in a population without being swamped. Interestingly enough, Bateson, who introduced Mendelian theory into British science, was also a proponent of evolution by selection. He suggested that if selection were to act on large discontinuous variations, the difficulties would disappear, because such variations would have high selective value and would not be obliterated by intercrossing.[4] Pearson, on the other hand, believed that the *continuous* variations in a pure line were heritable and that continued selection in such a line should be effective. Because blending inheritance would eliminate much of the heritable variation in each generation, the supply of variation in a population would soon be depleted unless new variations were heritable.

In order to resolve the controversy, what was needed was a way of showing *how* Mendelian inheritance was the mechanism that could explain variation in evolutionary theory. The importance of Mendel's work was his emphasis on variation among offspring, from which he derived laws that not only were founded on but also had predictive power for the nature of variation itself. Darwinian theories of inheritance, such as that proposed by Francis Galton, were based on the degree of resemblance between parents and offspring. Offspring of extreme parents (e.g., one short and one tall) tended to regress back to the mean of the parental generation. Laws were derived from observations of the mean height for all offspring whose parents belonged to a specific height class, with no emphasis placed on variation between parental pairs whose offspring were the same height. The result was a regression of means on means. Because Mendelian characteristics could be very small and were not blended away by crossing, and because Mendelian recombination provided the new variability for selection, it seemed the obvious complement for Darwinian selection. But in order to determine the evolutionary consequences of Mendelian heredity, it was necessary to show (predict) how gene frequencies would change under selection pressures as well as particular environmental conditions. Necessary for that approach was a reduction of populations to their composite genes; in other words, populations of individuals (humans, animals, etc.) were reconceptualized as populations of genes. The contribution from the biometricians was a methodology that approached the questions of heredity by applying statistical techniques to large populations.

That episode in the history of biology is important for my story about the relationship between unification and explanation, as well as the importance of mathematics in forging a unified theory. I want to claim that the evolutionary synthesis could not have taken place without the use of mathematical techniques that enabled geneticists to show how factors considered to have separate and conflicting roles to play in evolution could in fact be linked together. Because that could not have been achieved by empirical research on natural populations, one must view the mathematics itself as an integral and substantive part of evolutionary theory, one that was responsible for the unification of Darwinian selection and Mendelian genetics. In other words, the mathematical techniques "created", in a sense, the phenomena or populations that formed the foundation for the new version of evolutionary theory.

The unification consisted in showing, in a definitive way, that small selection pressures could be effective in producing evolutionary change, thereby showing that Mendelian heredity, when combined with assumptions about selection, breeding structure and population structure, could give a consistent theory of the mechanism of evolution. A noteworthy feature of the synthesis was that the mathematical techniques used by Ronald Fisher and Sewall Wright, both of whom were involved in the synthesis, were radically different. But even more important was the fact that they held different explanatory hypotheses about the role of selection with respect to specific biological phenomena. Consequently they emphasized different aspects of the evolutionary process. Although they eventually came to an agreement about quantitative comparisons (they could calculate the same numerical answer given the same assumptions), their initial assumptions when dealing with a biological problem usually differed significantly. As a result, the structural unity achieved by the employment of mathematical methods, the fact that selection and Mendelian heredity could together form a consistent evolutionary theory, was accompanied by striking disunity at the levels of both methodology and interpretation of the way selection and other factors operated. Embedded in the different models and mathematical methods were different views about evolutionary processes themselves. So although Fisher and Wright were interested in showing quantitatively how natural selection could operate under certain conditions, the unification was not accompanied by one consistent *explanation* about how that process actually took place.

I take this lack of a unique explanation about selection to be a significant problem for attempts to unite unification and explanation. The evolutionary synthesis was the product of a synthetic unity similar to that displayed by the electroweak theory. Unlike that case, here we have two *different* explanations of the unifying mechanism and how it operates. Recall that part of the motivation for linking explanation with unification was a desire to subsequently link explanation and increased confirmation/truth. The explanation that unified the phenomena was the one most likely to be true. But if we have a unified theoretical structure stating that natural selection is the mechanism for producing evolutionary change, accompanied by two conflicting explanatory accounts of how selection operates to produce that change, then the explanatory hypotheses offer nothing to choose between them, because each is consistent with the unified framework. Hence, any confirmation that the theory enjoys as a result of its unifying power cannot be the result of, and cannot be extended to, its explanatory assumptions. The bare fact of the evolutionary synthesis, that Mendelism is compatible with evolution by natural selection, furnishes the unifying core; but because that core admits of at least two different explanations, it becomes impossible to identify unification and explanation as one and the same thing. Although there is agreement that the synthesis produced a unified theory, that fact is *independent* of explanations of how selection itself functions, a problem that contemporary biological theory has ultimately not solved.

Before moving on to the details of the argument, let me begin by quickly reviewing the applications of statistical thinking that grew out of Mendel's own work. This will provide a context for appreciating the ways in which biometry, particularly in the hands of R. A. Fisher, could be used to reach an agreement between Darwinisn and Mendelian genetics.

7.2. Mathematics and Theory

7.2.1. Mendelian Ratios

Mendel's studies indicated that an immediate consequence of segregation in self-fertilizing organisms like peas was an orderly and predictable decrease in the proportion of mixed forms (heterozygotes) in successive generations produced from a cross between two initial pure forms (homozygotes). Hence, inbreeding should result in gradual reversion of offspring populations to the original homozygotic parental types. That investigation resulted in a specification of breeding conditions: When non-random mating is the rule, the frequency of heterozygotes will be expected to decline. Starting with a population formed by the hybridization of races AA and aa, Mendel was able to formulate the following generalization based on trial and error: In the nth generation of self-fertilization, the ratio of genotypes will be

$$(2^n - 1)AA : 2Aa : (2^n - 1)aa$$

In other words, there will be an increase in reversion to parental types AA and aa.[5]

The next obvious line of inquiry involved a calculation of the effect of Mendelian inheritance for a single locus with two alleles in a random breeding population, instead of one that was self-fertilized.[6] That idea led to the hypothesis that if the proportion of heterozygotes declines during inbreeding, then it should remain constant if there is crossbreeding or random mating. The mathematization of that hypothesis yielded the famous Hardy-Weinberg law, which states that the Mendelian 1:2:1 ratio will be stable from one generation to the next under conditions of random breeding. That can be expressed as the expansion of the algebraic binomial $p^2 + 2pq + q^2 = 1.0$, where p is the frequency of the dominant allele A, q is the frequency of the recessive allele a in the population as a whole, $p^2 = AA$, $pq = Aa$ and $q^2 = aa$.[7] The law holds provided there is no mutation or selection (except for balanced rates, when there will be no net changes in the frequencies of A and a) and the population is large enough to rule out sampling errors. The general idea is that gene frequencies will remain constant of their own accord from one generation to the next unless acted upon by outside influences like selection or non-random mating. That law makes it possible to translate statements about gene frequencies into more concrete statements about the frequency of occurrence of particular genotypes, thereby allowing us to think of heredity in terms of populations rather than individual genes.

It is important to remember that the Hardy-Weinberg law is a specification of *ideal* requirements; in other words, we use certain idealizing assumptions not representative of real populations to derive the variation-preserving character of the heredity mechanism. Hence, it is possible to think of specific applications of the law and extended developments in the field as deviations from those ideal conditions. For instance, the effects of selection can be understood in the context of the Hardy-Weinberg law as involving a specification of the number of generations required to alter gene frequencies as related to the intensity of selection. This is not meant to characterize the Hardy-Weinberg law as the starting point for all developments in population genetics; rather, my claim is that it provides a framework or baseline from which to examine the deviations responsible for producing evolutionary change. In that sense the synthesis of Darwinian selection and Mendelianism can be seen as the result of an interaction between the Hardy-Weinberg law and violations of the ideal conditions under which it supposedly holds. This allows for a way of thinking about the connection between the Mendelian gene and Darwinian variation in terms of an interaction within the gene pool that mediates between mutation and selection.

With this framework in place, we can now look at how the synthesis of these two mechanisms actually took place. We shall begin with the work of R. A. Fisher and then go on to show how Sewall Wright's radically different approach was successful in achieving a similar result.

7.2.2. Fisher's Idealized Populations

Fisher's first published work on genetics (1918) used Mendelian principles of inheritance to investigate the statistical effects of a large number of genetic factors in a mixed population. Fisher was interested in problems concerning the variety of individual factors that might affect genetic variability. But he was also preoccupied with a methodological issue, namely, the possibility of producing a statistical result based on a suitable specification of the population. The important factors included the role of dominance, general possibilities of preferential mating and survival, environmental effects and so forth, as well as the relative magnitude of each factor. At the time, it was thought that one had to list all the possible factors ascribable to a population in order to get an accurate description from which one could calculate a statistical answer regarding variability. That methodological assumption gained prominence largely from the work of Pearson (1903), who assumed that all Mendelian factors were equally important, that the alleles of each occurred in equal numbers and that all dominant genes had similar effects. Fisher's goal was to find a simple quantitative law that would allow evolutionary phenomena to fall into place in a manner similar to that of the ideal-gas law.

Part of the significance of Fisher's work was not simply to show the compatibility of selection and Mendelism, a remarkable result in itself, but to show how the statistical methods used in biometry actually hindered the chances of producing a synthesis. Fisher wanted to prove (contra Pearson) that an exact specification of individual factors relevant to the population was unnecessary for arriving at a proper statistical analysis. Instead, when factors were sufficiently numerous, the most general assumptions with respect to individual peculiarities would lead to the same statistical results.[8] Fisher saw the method as analogous to that used in gas theory, where it was possible to "make the most varied assumptions as to the accidental circumstances, and even the essential nature of the individual molecules, and yet develop the general laws as to the behaviour of gases, leaving but a few fundamental constants to be determined by experiment" (Fisher 1922, pp. 321–2). That is, one needed only general statistical laws about the interactions among individuals, rather than specific knowledge of the individuals themselves, in order to determine the effects of evolutionary mechanisms.

Fisher's results were achieved by using gas theory as a type of methodological analogy; populations were treated in the same fashion as ideal gases. However, instead of producing an abstract description that bore little resemblance to real populations as in the case of gas theory, the method enabled Fisher to determine the way in which selection operated in real cases and to show its compatibility with Mendel's laws. The methodological benefit offered by the notion of an ideal gas was that it presented the simplest account of a many-body system in virtue of the assumption of no interaction between the members. It was that kind of methodological simplicity that enabled Fisher ultimately to determine the role of selection in Mendelian populations.

However, there was more to the analogy than that. In his 1922 paper "On the Dominance Ratio", Fisher used the velocity-distribution law for gases as the model for calculating the distribution of the frequency ratios for different Mendelian factors. But there again, the *foundation* for that law was the notion of a perfect gas. When Maxwell developed his billiard-ball model of gases, he also presented a series of propositions that led to a statistical formula expressing the distribution of velocities in a gas of uniform pressure. The idea was that one could calculate the observable properties of a gas by knowing only the various positions and velocities of an average number of molecules, rather than those of the individual molecules. By assuming that the average number of molecules was the same at different places in the gas, one could then formulate a velocity-distribution function $f(v)$ that would depend only on the magnitude of v. Maxwell had employed two basic assumptions to establish his result: (1) When a moving sphere randomly collides with another fixed sphere, all rebound directions and hence all directions of motion in the gas are equiprobable. (2) The probability distribution for each component of the velocity is independent of the values of the other components. In other words, the components in the y and z directions are probabilistically independent of the x component; they can be treated as independent random variables.

The distribution law was identical in form with the "normal distribution" introduced by Laplace and Gauss in the theory of errors. But it was the idea that a statistical function should be employed in the description of physical processes that was truly revolutionary. However, the notion that the three resolved velocity components could be treated as independent variables seemed problematic, because the implication was complete randomness of molecular motion, an assumption that was not physically grounded in the motion and collisions of elastic spheres. In a later paper, Maxwell formulated another account of the distribution law that was more closely connected with molecular encounters. Instead of assuming that the velocity components of a single molecule were statistically independent, he assumed that that property held for the velocities of two colliding molecules. Hence the joint distribution function $f(v_1v_2)$ for the probability that molecule 1 has velocity v_1 and molecule 2 has velocity v_2 is the product of the probabilities of the two separate events $f(v_1)$ and $f(v_2)$. Maxwell claimed that the velocity-distribution law represented a stable equilibrium, such that the number of collisions of two molecules with initial velocities v_1 and v_2 that rebounded with velocities v_1' and v_2' would be equal to the number of collisions of molecules with initial velocities v_1' and v_2' that rebounded with velocities v_1 and v_2. The law was generalized and extended by Boltzmann into the basic principle used today. Although that account of the distribution law involved molecular collisions, the basic picture was still that of ideal gases; no attractive forces between the molecules were assumed, nor was any account taken of molecular size.

Why did the approach used in the kinetic theory, especially the distribution law, prove to be so valuable for Fisher? There are at least two possible answers to this question. The first, mentioned earlier, concerns advantages that derive from the simplicity of the method itself, and the second stems from the kinds of systems with which Fisher was dealing. To see how those two issues relate to each other, let us go back to Fisher's first genetics paper (1918).

Fisher's aim in that paper was to determine the correlation among relatives on the supposition of Mendelian inheritance. He found that by means of fraternal correlation it was possible to ascertain the dominance ratio (the expression of the effect of the dominance in the individual factors) and hence to distinguish dominance from all non-genetic (arbitrary) causes that could tend to lower correlations (e.g., environment). So in order to determine the total variance, it was necessary to separate environmental factors from genetic factors. Fisher first assumed that not all Mendelian factors were of equal importance and that different phases of each could occur in any proportions consistent with the conditions of mating. Also, the heterozygote was assumed to have any value between dominant and recessive. However, in order to move from a simple case to complex analysis of the population, several methodological assumptions were made, assumptions about the nature of a population that bore important similarities to the population of molecules in a perfect gas. They included complete independence of the different factors and the assumption that the factors were sufficiently numerous to neglect certain small quantities and also random mating (rather than collisions). No mention of gas theory was made in the 1918 paper. But in his 1922 paper, where he further discussed the dominance ratio, he specifically mentioned that his earlier analysis could be compared with the analytical treatment of the theory of gases. As mentioned earlier, the methodological analogy was founded on the fact that in the kinetic theory it was possible to make a number of different assumptions about the essential nature of molecules and the circumstances surrounding their motions and still develop general laws about the behaviour of gases, leaving only a few fundamental constants to be determined by experiment. That is, the kinetic theory had shown that knowledge of particular individuals was not required in order to formulate general laws governing the behaviour of a population. Similarly, in gas theory, as well as in Mendelian populations, when we have large numbers of individuals we can predict outcomes with a great deal of accuracy and reliability. Fisher wanted to prove that when factors were sufficiently numerous, the most general assumptions as to their individual peculiarities would lead to the same statistical results.

Although that methodological analogy was first introduced in 1918, its importance became clear in the 1922 paper, where Fisher was interested in determining the effects of selection and mutation on variance. In order to show the role of Mendelian inheritance in producing variation in evolutionary theory, it was necessary to show that selection did in fact operate in Mendelian populations. And in order to ascertain the evolutionary consequences of heredity, one needed to demonstrate (predict) how gene frequencies would change under selection pressures and particular environmental conditions. Necessary for that approach was the reduction of populations to their composite genes. Although important aspects of Fisher's method were opposed to those used by the biometricians, he did adopt their statistical way of approaching the heredity problems using large populations. But more importantly, Fisher's investigation required that he be able to isolate selection pressures from all other types of factors in order to show not only the compatibility of evolutionary theory and Mendelian inheritance but also the way in which selection operated in those populations. The basis for that separation was, again, the characterization of an ideal gas, which assumes that there is no interaction (force) between the molecules, that each behaves in an independent way.

In one way the significance of that methodological approach is obvious: How could one determine the effects of selection if it was impossible to separate it from other influences on the population? However, I want to make a stronger claim for the importance of the methodology provided by gas theory; that is, by using the model of an ideal gas, Fisher was able to "create" a population in which he could measure effects not measurable experimentally. Hence, he was able to show how and to what extent real processes operated in natural populations and to show the interaction between two supposedly incompatible phenomena, selection and Mendelian inheritance. It isn't simply that Fisher used the analogy with gas theory to *justify* his assumptions about population structure, nor did he use it to *explain* empirical results he had already arrived at. Instead, that mathematical technique provided the instrument for *investigating* the role of selection in human populations by replacing actual populations with idealized ones. What makes that even more significant is that it was only by means of those mathematical techniques that concrete results could be obtained, results that could then be compared with values determined using empirical measuring techniques. To see how that was done, let me briefly summarize the findings of the 1922 paper "On the Dominance Ratio", where the comparison with gas theory was made most explicit.

In that paper, Fisher argued that an equation representing the stochastic distribution of Mendelian determinants in a population over time was the key to an accurate and quantitative understanding of evolution in that population. He wanted to show that knowledge of individual factors relevant to the population was unnecessary for arriving at a proper statistical analysis.[9] That is, he needed only general statistical laws about the behaviour among individuals, rather than specific knowledge of the individuals themselves, in order to determine the effects of evolutionary mechanisms.

The paper began with a discussion of equilibrium under selection. He first demonstrated that the frequency ratio for the alleles of a Mendelian factor was a stable equilibrium only if selection favoured the heterozygotes. He then showed that the survival of an individual mutant gene depended on chance rather than selection. Only when large numbers of individuals were affected would the effect of selection override random survival, and even then only a small minority of the population would be affected. Fisher also examined the distribution of factors not acted on by selection, as well as the cases of gene extinction counterbalanced by mutation and extinction in the absence of mutation and selection, where one saw a steady decline in variation due to the effects of random survival (Hagedoorn effect). On the basis of his calculations of the number of genes exterminated in any one generation and the distribution of factors in successive generations, he was able to show that even in a population of roughly 10,000 random-breeding individuals without new mutations, the rate of gene extinction was extremely small.

Hence, the chance elimination of genes could not be considered more important than elimination by selection.

The other important component in the analogy with gas theory was that the distribution of the frequency ratio for different Mendelian factors was calculable from the fact that the distribution was stable in the absence of selection, random survival effects and so forth. Again, the source was the velocity distribution in gas theory. Just as the formulation of the velocity-distribution law assumed an independence of molecules in the gas, so, too, Fisher assumed the independence of various hereditary factors of each other and their independence from the effects of selection and random survival. The important difference was that he specified a population in which he could first calculate the distributions without selection, mutation and random survival, and then he used those results to go on and determine how the effects of selection operated in different contexts. To do that, Fisher considered the cases of uniform genetic selection in the absence of dominance and genotypic selection with complete dominance. From those distributions he was able to calculate the amount of mutation needed to maintain the variability given a specific amount of selection. To maintain variability in the case of equilibrium in the absence of selection, the rate of mutation had to be increased by a very large quantity. So the presence of even the slightest amount of selection in large populations had considerably more influence in keeping variability in check than did random survival. Consequently, the assumption of genotypic selection balanced by occasional mutations fit the facts deduced from the correlations of relatives in humans.

So, by making simplifying assumptions about the large size of the population and its high degree of genetic variability, Fisher was able to demonstrate how his stochastic distributions led to the conclusion that natural selection acting on single genes (rather than mutation, random extinction, epistasis etc.) was the primary determinant in the evolutionary process. He found that mutation rates significantly higher than any observed in nature could be balanced by very small selection rates. The distribution of the frequency ratios for different factors was calculated from the assumption that the distribution was stable. The methodological decision to isolate selection as an independent factor in producing variation was justified ultimately on empirical grounds, because it turned out, in Fisher's analysis, to be the only mechanism responsible for genetic variability. The kind of statistical independence that figured prominently in the velocity-distribution law was applicable to the effects of selection in Mendelian populations.

As more variables were added, the mathematics Fisher used became intractable, a situation that led to the development of his famous "fundamental theorem of natural selection", published in 1930 in *The Genetical Theory of Natural Selection*. The theorem states that the rate of increase in fitness for any organism at any time is equal to its genetic variance in fitness at that time (Fisher 1958, p. 35). The discussion of the theorem really began with the introduction of m, which Fisher designated as the Malthusian parameter of population increase, and v, the reproductive value. He then discussed the genetic element in variance and defined two variables: a, the excess over the average rate of increase of any selected group, and α. If m measured fitness by the fact of representation in future generations, then α was the average effect on m as a result of introducing the gene in question. That allowed him to state the contribution of each factor to genetic variance in fitness as

$$\sum{}' (2p\,a\,\alpha)$$

The frequency of the chosen gene was represented by p, and any increase dp in the frequency of that gene would be accompanied by an increase $2\alpha\,dp$ in the average fitness of the species. The definition of a required that $(d/dt)(\log p) = a$ or $dp = (p\,a)dt$. Hence

$$(2\alpha)\,dp = 2(p\,a\,\alpha)\,dt$$

which represented the rate of increase of average fitness due to a change in progress in the frequency of that one gene. Fisher focused on the rate of increase of a type because it measured fitness by the objective fact of representation in future generations. Summing for all allelomorphic genes, he had $dt \sum{}' (2p\,a\,\alpha)$, and taking all factors into consideration, the total increase in fitness was

$$\sum \alpha\,dp = dt \sum \sum{}' (2p\,a\,\alpha) = W\,dt$$

If the time element dt was positive, the total change in fitness $W\,dt$ was also positive. We see, then, that the rate of increase in fitness due to changes in the gene ratio was equal to the genetic variance of the fitness W that the population exhibited. The equation for W was defined in terms of two variables: a, the average excess of any factor, and α, which represented the average effect on the chosen measurement. So the effectiveness of selection depended on the total heritable variance available in the population at that time. For any specific gene constitution, an expected value X was built up of the average effects of the genes present in an individual. The variance of X was

$$W = \sum (2p\,q\,a\,\alpha)$$

and was arrived at by summing over all factors.

If fitness was a measure of the ability of a gene to survive and reproduce, then natural selection tended to increase the total fitness of the population. Although a change in fitness could be thought of as the result of two components, the effect of natural selection and the effect of environmental factors, what concerned Fisher was not the change in the measure of total fitness but the fraction that was due to natural selection. As a result, he regarded the effect of selection on m as limited to the additive effects of changes in gene frequencies. Other factors such as population pressure, climate, epistasis and interactions with other species were considered strictly environmental. What the theorem seems to state is that selection always acts to increase the fitness of a species to live under conditions that existed an instant earlier (Price 1972).

As Price (1972) has pointed out, it has long been a mystery what exactly Fisher meant by his fundamental theorem and how he derived it. For one thing, the word "fitness" was used in two quite different senses, and there were several mathematical additions that had to be made for consistency and coherence in the argument. I do not want to go into those intricacies here, because they do not really affect the nature of my claim about Fisher. But what I do want to stress is the role of the Malthusian parameter in the fundamental theorem. This is significant because of my claim in Chapter 6 regarding the importance of Malthus's law for Darwinian evolution. When Fisher introduced m, it was defined as a population measure, but by the time he fully stated the theorem (1958, pp. 37–9) it had become an individual or genotypic measure of fitness. Nevertheless, the question is where exactly m figures in the theorem. Because W is a measure of the genetic variance of fitness defined in terms of a and α, and because α represents the average effect of a specific gene on m, the connection between W and m is clear. A numerical determination of m can be found using the equation representing the aggregate contribution to the birthrate for all ages. When this equation is equal to unity, it supplies an equation for m for which only one real solution exists (1958, p. 26). In that sense, the number m is implicit in any given system of death and reproduction rates. Perhaps the most crucial feature of the theorem is the fact that m, the fitness (or mean fitness in a population), must remain near zero in every species most of the time to prevent extinction or overpopulation. Hence the change in fitness over time must also remain around zero. What is significant is that the natural-selection component of the change in fitness must always equal W, with m reflecting the dynamic conditions implied by Malthus's law of geometrical increase. Essentially there were two forces acting on a species at any given time: environmental changes, which, as Fisher showed, generally tended to decrease m, and the force of natural selection, which always tended to increase it. The balance remaining, after the rate of decrease due to deterioration in the environment is deducted from the rate of increase in the mean value of m produced by selection, results not in an increase in the average value of m (since this value cannot greatly exceed zero), but rather in a steady increase in the population.

The theorem was exact only for idealized populations in which fluctuations in genetic composition had been excluded. Nevertheless, it was important to have an estimate of the magnitude of the effect of those fluctuations; but, as Fisher was able to show, the random fluctuations in W (the genetic variance or rate of progress), when measured over a single generation, could be expected to be very small compared with the overall average rate of progress. At that point Fisher once again appealed to a physical model, namely, gas theory, as a way of grounding the regularity with which W could be expected to vary. He claimed that such regularity was guaranteed by the same circumstances that made a statistical assemblage of particles like a gas bubble obey the gas law (without appreciable deviation) (1958, pp. 34–40). He went on to compare his fundamental theorem to the second law of thermodynamics: Both are properties of populations that are true regardless of the nature of their individual units, and both are statistical laws, each of which requires continual increases in a measurable quantity (in one case entropy, and in the other fitness). And, he adds, "it is more than a little instructive that so similar a law should hold the supreme position among the biological sciences" (1958, p. 39).

The key for understanding Fisher's approach is, once again, the use of idealizing assumptions similar to those used in gas theory. The power of his model derived from elimination of parameters such as migration, isolation, genetic recombination and gene interaction. He was able to separate, in a way that could not be done in empirical studies, the key features in populational variation: genetic factors, environmental factors and sampling errors. Fisher's quantitative expression for genetic variance, $W = 2(pq\alpha)$, was used to determine the limits of each. Similarly, it enabled him to show the amount of variation any particular genetic trait could exhibit in a population that was distinct from environmentally caused fluctuations. In other words, a whole range of samples could, from statistical analyses of variance, show true genetic variability, as opposed to environmentally based variability. Consequently Fisher could deal exclusively with genetic variations, rather than with combined genetic and environmentally induced variations. He then related genetic variance to natural selection. Those techniques allowed him to determine the effects of selection on Mendelian variations and provided a model by which the effects of numerous variables (including the amount of variation or the degree of selection) could be quantitatively predicted.

The model consisted of two components: (1) variation due to mutation and (2) variation due to natural selection. In the case of mutations, Fisher pointed out that mutations from one form of gene to another occurred at specific rates, thereby providing a quantitative measure of how rapidly new variations were introduced into a population. With respect to natural selection, he was concerned with the factors in the environment that determined the possibility of a gene or group of genes being passed on, and in what quantities (i.e., the fitness of the gene). The two measures of fitness that could be applied to natural populations included one that was applicable to two or more species in comparison with each other. It was measured as a simple increase in the size of one population over the other. The fittest was the group that had the largest increase in population. The other measure applied to particular individuals or particular genes within a single group and was calculated as a change in gene frequency over successive generations.

Fisher's work produced a synthesis of evolutionary theory with Mendelian genetics by showing that the patterns of inheritance specified by Mendelian theory were consistent with gradual evolution based on the natural selection of small genetic differences. He also thought that he had shown that natural selection was the *only* mechanism of evolutionary change consistent with the new genetics. That unification resulted in changes in the theoretical foundations of both genetics and evolutionary theory. Fisher, building on Hardy's work, pointed out that in Mendelian inheritance there was no inherent tendency for variability to diminish over time. Alternative genes were conserved unless their proportions were changed by selection, chance or mutation; and as Fisher's calculations showed, selection was the factor most effective in changing gene frequencies. Mutation could no longer be thought of as the force of evolutionary change, for even very small selective effects could overpower it.

We have seen how the mathematical techniques used by Fisher enabled him to achieve the synthesis in a way that would not have been possible using only experimental results and methods. Implicit in that synthesis was the isolation of selection as the mechanism for producing evolutionary change. In that sense, selection can be seen as the foundation or factor that provides the basis for unity, insofar as it can be seen to operate in Mendelian populations. However, the sense in which selection became a "parameter", similar to the parameters I described in the physics cases, was made especially obvious with the publication of Fisher's fundamental theorem. The quantitative framework of Malthus, which proved so important for Darwin's account of selection, has similar implications for understanding Fisher's work. We know that the Malthusian parameter m, which represents fitness, has a natural-selection component that must always equal W, the genetic variance in fitness. So for Fisher, m did not simply measure changes in population numbers as it did for Malthus, but rather the change in the total reproductive value of population members; in other words, it measured the rate of population growth in terms of total reproductive value. Because Fisher was able to give a quantitative expression for W, it was possible to determine the effects of selection as a parameter in the overall determination of m. Hence, W "represents", in a way that builds on the foregoing account of the synthesis, the unification achieved between Darwinian theory and Mendelian theory.

I shall leave the discussion of Fisher's qualitative or explanatory version of selection until later, but now let us turn to a wholly different account of the synthesis given by Sewall Wright. Working at roughly the same time as Fisher, Wright (1921) also showed how selection operated in Mendelian populations. He did so not only by using a different quantitative method but also by incorporating an entirely different qualitative theory about the functioning of natural selection. What Fisher's work and Wright's work shared was the basic core of the synthesis, namely, the power of selection and the inability of mutation to produce changes in gene frequencies, the consequence of which was the hypothesis that evolution operated gradually as a result of the selective accumulation of small genetic differences. Hence, continuous variation was entirely consistent with discontinuity of genes.

Although there were several points of disagreement between Fisher and Wright (especially over Fisher's work on dominance), my focus here is their dispute over the mechanism of evolution (which involved some aspects of the controversy about dominance). That debate was important because it illustrated how the synthesis in biology failed to produce a consistent explanatory account of *how* selection actually operated as the agent of evolutionary change, while nevertheless producing a unified theory of evolutionary phenomena.

7.2.3. Wright's Path Coefficients

Wright was convinced by his experimental work with guinea pigs that it was the interaction systems of genes, rather than single genes, that were the important aspects of evolution. He also believed, contrary to Fisher, that natural selection operated most effectively in smaller populations where inbreeding was sufficiently intense to create new interaction systems as a result of random drift, but not intense enough to cause random non-adaptive fixation of genes. In those populations, natural selection could act on the newly produced interaction systems, resulting in a more rapidly changing population than that produced by mass selection of single genes (as described by Fisher's approach). Wright felt that there was sufficient evidence from animal breeders that mass selection was slow and somewhat unsure; hence he assumed that natural populations tended to become subdivided into partially isolated groups small enough to cause random drifting of genes. Thus, one of the key differences between Fisher and Wright in attempting to isolate the way selection worked concerned population structure. Fisher was convinced that it worked more efficiently in large groups in which there was mass selection and random breeding, whereas Wright's account favoured small inbred groups in which obvious variation would then serve as the basis for selection.

Wright's evolutionary models and mathematical techniques were intended to provide a way of focusing on a quantitative analysis of inbreeding. His use of the method of path coefficients (a system he devised) enabled him to determine the degree to which a given effect was produced by each of a number of different causes. In that analysis, the causes of factors that affected a trait (e.g., the general size of an organism) and factors that affected its separate parts could be treated as independent.

The idea behind path coefficients is as follows: We assume that there is a complex system of variables in which each one is represented as a linear function of others, as indicated by the arrows in Figure 7.1. In Figure 7.1, X and Y are two characters whose variations are determined in part by certain causes A, B and C, acting on both, and in part by causes that apply to either M or N, respectively. These causes are assumed to be independent of one another. Then a, b and c are the proportions of variation in X determined by these causes, and a', b' and c' are the corresponding proportions in Y. The extent to which a cause determines the variation in an effect is measured by the proportion of the squared standard deviation of the effect for which it is responsible. This is because the squared standard deviations due to single causes acting alone can be combined by addition to find the squared standard

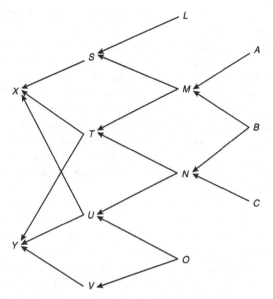

Figure 7.1. A schematic representation of Wright's path analysis showing how the causal pathways influence one another.

deviation for an array in which all causes are acting, provided, of course, that each cause is independent. The proportions of variation in a character determined by A, B and C and denoted by a, b and c are squares of the respective path coefficients. By making the sums of degrees of determination equal to unity,

$$a + b + c + d + \cdots + m = 1, \qquad a^1 + b^1 + c^1 + d^1 + \cdots + n^1 = 1$$

and expressing a known correlation between X and Y in terms of the unknown degrees of determination $r_{xy} = \pm\sqrt{aa'} + \sqrt{bb'} + \sqrt{cc'} + \cdots$, Wright was able to form a series of simultaneous equations that could be solved for the unknown effects of single causes.[10] In a later paper he defined the path coefficients as ratios of the standard deviations, rather than as the squares of the standard deviations. The plus signs and the square-root radicals in the equation for r_{xy} disappear, and the correlation can then be expressed as $r_{xy} = aa^1 + bb^1 + cc^1 + \cdots$. As a result, the coefficients of determination for the squared standard deviations will be just the squares of the path coefficients. A path coefficient can then be defined as the ratio of the variability of the effect that is found when all causes are kept constant except the one in question (the variability of which is kept unchanged) to the total variability (where variability is measured by the standard deviation).

The accumulations of paths and their interrelationships are the factors that give rise to the correlation. Wright wanted to emphasize the dangers of inferring causation from correlation, and he thought that the method could draw attention to the differences. The path coefficients were combined together in a particular sequential order, whereas correlation was a symmetrical relationship. The goal was to find a coefficient that would measure influence along each path and would be related in a simple way to the correlation. What we find, however, is that the product of the path coefficients along each path that connects two variables measures the contribution of that path to the correlation in such a way that the sum of all such correlations exactly equals the correlation between the two variables.[11] Each line of influence is measured by a coefficient that gives the proportion of the variation of the dependent variable for which the causal factor is directly responsible. The method is intended for use in establishing causal relationships where direct experience is impossible.

The path coefficient is identical with a partial-regression coefficient provided that it can be measured in standard-deviation units. However, Wright wanted to stress the difference between dealing with multiple regressions and dealing with his complex network of relations that involved hypothetical factors. As far as he was concerned, correlation and regression coefficients provided descriptive/predictive statistics from which scientists unjustifiably inferred the existence of causal relationships. By contrast, the method of path analysis was not used to *infer* a causal scheme or *deduce* particular causal relationships; instead, it required a specific causal hypothesis to be stated at the outset and was best applied in cases where causal pathways were known or could reasonably be assumed, but with their relative importance unknown.

Wright was able to extend his model to include cases in which causes were correlated instead of independent. He used the method to establish the relative importance of heredity and environment in determining piebald patterns in guinea pigs and found that in highly inbred stocks heredity was responsible for almost none of the variability; irregularity in development was practically the sole cause. He also applied the method to systems of mating. By looking at the effects of various systems of inbreeding on the genetic composition of the population, the effects of assortive mating, and the effects of selection, Wright was able to calculate the percentages of homozygosity in successive generations.[12] Like those of Fisher, Wright's mathematical methods allowed for precise quantitative determination of the effects of natural selection as the agent of evolutionary change and the basis for a unified theory.

It is important to note that nothing about the interpretive aspects of evolutionary theory follows from the method of path coefficients. It simply supplied a way to quantify causal relationships that were assumed by hypothesis – hypotheses that, in Wright's case, were formulated as a result of his experimental work on guinea pigs. As mentioned earlier, Wright believed, contra Fisher, that direct selection for single gene effects was significantly less effective than selection of interaction systems. He believed that such interaction systems composed an important part of the genetic makeup of organisms, and it was the random drift of genes caused by inbreeding that created new interaction systems. So although both he and Fisher believed in the role of selection in producing evolutionary change, they had differing explanations for the mechanisms leading to its successful operation. Fisher denied the importance of genetic interaction and random drift; he saw selection as most effective in large populations because of their greater genetic variance. Wright saw selection as most powerful in smaller inbred populations, where new interaction systems could be effectively produced.[13] Wright used those assumptions in the development of his "shifting-balance theory" of evolution, which, in combination with the mathematical models generated by path coefficients, provided the definitive account of Wright's view about the role of selection in Mendelian populations.[14]

7.3. Disunity and Explanation

We have seen how it was possible, using two different quantitative methods, to produce a unification of genetics with Darwinian theory by showing that selection, even in cases in which only small pressures were exerted, operated in Mendelian populations. Yet despite agreement on that basic unifying principle, there was significant disagreement at the explanatory level regarding how the evolutionary process actually took place. In his most recent discussion of explanation and argument schemas, Kitcher (1993) examines the work of Fisher, Haldane and Wright by showing how they provided an extension of Darwinian explanations through the instantiation of a pattern specifying genetic trajectories. The explanation is an answer to the following question: Given a population P with characteristics W and initial genetic distribution D_0, what is the genetic distribution in the nth generation, D_n? The answer or argument pattern consists of several steps: first, a specification of population traits like size and type of reproduction; second, a specification of the initial genetic distribution, including number of loci involved, linkage relations and fitness of the allelic combinations; third, a specification of further conditions such as the probability of mutation of alleles, frequency of migration and so forth; fourth, use of the principles of combinatorics and probability to derive the conclusion, namely, the probabilities that particular distributions will be found in the nth generation. Kitcher claims that Fisher, Haldane and Wright showed how important instantiations of that schema could be given, and insofar as simple selectionist arguments can be embedded in a pattern that contains these trajectories as a subpattern, we can see that the main structure of Darwin's selectionist patterns can be preserved.

The new pattern, "neo-Darwinian selection", asks this question: Why is the distribution of properties $P_1,\ldots,P_k$ in relative frequencies $r_1, \ldots, r_k$ ($r_i = 1$) found in group G? The answer involves the following:

1. the specification of a G_0-G sequence consisting of ancestors G and contemporaneous organisms G_0;

2. the variations at n loci among members of G_0, as well as the alternative alleles present in the sequence at the ith locus;

3. the probability of an organism having a particular trait p_j; and

4. an analysis of the ecological conditions and of the effects on their bearers of $P_1, \ldots, P_k$, which will show the fitness of the allelic combinations.

Finally, there is a specification of the genetic trajectories, followed by the number of discrete generations in the sequence, that will yield the expected distribution of phenotypic traits and the relative frequencies with which they occur. The results are given for both finite and infinite traits. Kitcher (1993, p. 46) maintains that this schema brings together in one pattern of explanation various factors that are relevant to evolutionary change and facilitates comparisons of their effects. In different instances one can emphasize certain factors and ignore others. For example, one may treat a large population as infinite and therefore ignore the effects of drift and sampling errors; or one may assume that mutation rates are negligible and treat loci as independent, thereby concentrating solely on fitness differences.

The idea behind Kitcher's account of explanation as unification is that one can determine the unity that a particular theory has by the number of explanatory patterns required to derive the phenomena it purports to explain. In the neo-Darwinian case, Kitcher is suggesting that this argument pattern can incorporate the approaches of Fisher and Wright and provide an explanatory schema that will unify Darwinian selection with genetics. We saw in the earlier discussion that Fisher and Wright had completely different and contradictory *explanatory* accounts regarding the operation of natural selection, yet they could arrive at quantitative agreement for particular results if they began with similar assumptions. But if Kitcher's schema can accommodate both explanatory stories as instantiations of the same pattern, then the situation becomes potentially problematic. Kitcher wants to understand explanation in terms of unification, as spelled out in the form of argument patterns. He also claims that explanation and unification provide reasons for theory acceptance. In this case, however, we have two competing theories/hypotheses about evolution and selection, each of which is compatible with a Darwinian and Mendelian story. But because both explanations are linked with unification and both are compatible with the evidence, there is no way of choosing between them. That is, we have no way of deciding which explanatory story to accept.

An alternative strategy would be simply to claim that the unification of Darwinian and Mendelian theories speaks only to the role played by selection, remaining neutral on the specific mechanisms associated with it. That would leave room for separating unity and explanation, because there is no acceptable explanation (one that is not underdetermined) of how selection actually functions. Consequently, fulfilling the constraints imposed by the argument pattern cannot provide an explanation of the selection process, because it is open to a variety of interpretations.

One might want to interpret Kitcher's strategy and aim as providing an argument pattern applicable to what has become known as the Wright-Fisher model in population genetics – a binomial distribution that can be used to determine changes in allele frequencies. What this provides is a quantitative model that represents the core of the unification. But again, that kind of unity does not entail a specific explanatory story. To summarize the problem: Theoretical agreement on unification of genetics with evolutionary theory could not have been achieved if the theorists had been required to supply explanatory details about how and why selection operated in particular types of populations. If Kitcher's argument pattern can reasonably accommodate both explanations of selection as instantiations of the same pattern, then one cannot simultaneously claim that explanation as unification provides a basis for theory acceptance, because it gives us no way of deciding between Fisher and Wright.

To claim, as Kitcher does, that Fisher and Wright "articulated a precise scheme of explanation that could be exemplified in different ways" (1993, p. 49) simplifies and perhaps even misrepresents several different issues regarding the biological synthesis. Although both Fisher and Wright produced precise quantitative schemes for deriving and predicting the occurrences and causal efficacies of particular phenomena, their methods were radically different. In Fisher's case, one could go on to argue that the mathematical account, in some sense, served to motivate the explanatory framework. The assumptions implicit in his statistical techniques about population size and structure were embedded in his views about how selection operated. Wright's method of path coefficients was completely neutral with respect to explanatory hypotheses or theories; those were provided by his shifting-balance theory of evolution and were motivated by his experimental work on guinea pigs. But again, what both showed, and the principle they both agreed upon, was the part that ultimately constituted the core of the synthesis, namely, that selection was a causal factor in evolutionary change consistent with genetics.

As Provine (1992) has pointed out, the debate between Fisher and Wright was not about quantitative methods, for in the final analysis their different methods led to almost the same numerical results regarding the quantitative effects of inbreeding, migration, selection and so forth on the statistical distribution of genes in a population. In other words, if one always starts with the same assumptions, the two methods will yield the same results. Hence, Provine concludes that differences in quantitative approaches did not *entail* significant differences in evolutionary theories. True. But there is no reason to assume that different quantitative approaches do not reflect different theoretical predispositions or assumptions about the nature of evolutionary systems. As we have seen, that was surely the case with Fisher's work. So although Provine's remark clearly is technically correct, to stop there would, I think, be to miss an interesting insight into the relationship between mathematics and the development of qualitative hypotheses that this case, in particular, exemplifies. Fisher was convinced that biological systems should be modelled on physical systems dealing with idealized populations having a certain structure. It was in that context that one could develop general laws that would focus on a single factor involved in the determination of specific outcomes. Because of his laboratory work, Wright thought that a proper analysis of evolutionary systems and organisms ought to focus on their multifaceted and complex nature (an approach directly at odds with Fisher's). That required that one be able to discern the relationships among several different variables that figured in the growth and reproduction of organisms. He maintained that gene interaction systems and genetic drift were important components in the production of variance for selection. Consequently, the kind of techniques he developed reflected that less idealized approach to determining the role of selection. The goal of Fisher's programme was to achieve a level of abstraction at which one could formulate general assumptions that would apply (approximately) to all situations. That kind of abstraction was anathema to Wright, whose method of path coefficients was a way of examining the roles of many different causal factors within a single population.

I want to claim that it was their differences regarding the structure of those populations in which selection operated most efficiently that *determined*, in each case, the appropriate methods for modelling biological systems and consequently determined the modelling assumptions (i.e., what variables needed to be included). Fisher's methods exemplified a top-down approach, where one starts with the view that biological systems ought to be like physical systems and proceeds by following a pattern of idealization that can produce biological analogues of general physical laws and principles. Wright adopted more of a bottom-up approach, with his methodology chosen on the basis of views about biological phenomena developed in response to experimental work. In that sense, differences in theoretical assumptions with respect to the organisms themselves undoubtedly influenced the choices of different methods and techniques. That becomes apparent if we look at Wright's addendum to Fisher's fundamental theorem, which states that "the rate of increase in fitness of any population at any time is equal to its genetic variance in fitness at that time *except as affected by mutation, migration, change of environment, and effects of random sampling*".[15]

Despite the fact that specific theoretical assumptions may have influenced the kind of mathematical techniques used by Fisher and Wright, that detail in no way limits the importance of the mathematics itself in producing a unified theory. What the different mathematical frameworks provided was a synthesis of two distinct hypotheses/theories that would not have been possible solely on the basis of experimental work. Beginning with the Hardy-Weinberg law, mathematics came to constitute the fabric of evolutionary genetics. There was very little evidence to support the view that evolution could proceed by small selection pressures, because observation of such effects in nature was almost impossible. Yet through mathematical analysis it was possible to show that natural selection could produce evolutionary changes in Mendelian populations. Consequently, the mathematics became part of the theory; biological concepts originally used in empirical investigation were mapped onto the mathematics in such a way that the models constructed by

Fisher and Wright became the determining force in structuring biological theory. The result was a unification at the level of a general principle or structure, but without any unity of explanatory detail. It is perhaps important to note that such disunity was not simply the result of historical contingency. Not only has the rise of "neutral models" in population genetics created a further discontinuity in the kind of explanatory progress that Kitcher's account attempts to maintain, but even models that are not part of the "neutral theory of evolution" differ in significant respects from those of Fisher and Wright. For example, one common and important assumption present in the work of Wright and Fisher was that generations did not overlap. That was challenged in a model due to Moran (1958), where the assumption of overlapping generations was introduced. And although some mathematical equivalences among these various new models can be shown, the qualitative understanding of population structure is very different in each. Insofar as unity is achieved, for the most part, through the use of certain kinds of mathematical structures, we must look elsewhere for a detailed understanding of how the phenomena behave. The explanatory power simply does not reside in the unifying structure of the theory.[16] Although unity operates at the level of general principles, explanation is better seen as an exercise in providing specific details that can function in the application of theories and models to real cases.

Conclusions

One of the themes I would like the reader to take away from this discussion is that unity, either in the broader context of science itself or in the more localized settings of specific theories, cannot be uniquely characterized. More specifically, no single account of theory unification can be given. A philosophical consequence of that claim is that unity should not be linked to truth or increased likelihood of truth; unification cannot function as an inference ticket. What exactly is the connection here? Although I have tried to explicate the features necessary for distinguishing a truly unified theory from a mere combination or conjunction, even within that framework different kinds of unity can emerge. Not only can unification exhibit a variety of ontological patterns (e.g., reductive and synthetic), but the way in which the unification is achieved can have a significant impact on whether or not there is an accompanying theoretical story, an account that can be claimed to provide a plausible representation of the phenomena. For instance, in neither the early version nor the late version of Maxwell's theory was there a viable physical interpretation of the electromagnetic field; hence, any affirmation of truth with respect to the theory would need to be severely limited in its content. Even if we consider Hertz's famous claim that Maxwell's theory was Maxwell's equations, prior to 1888 there was no guarantee that those equations were descriptively accurate, because there was no proof that electromagnetic waves existed. In what sense, then, would we be justified in claiming that Maxwell's theory, in virtue of its unifying power, was more likely to be true than not, or more likely than its rivals?

The second significant issue is that of empirical support. Intuitively one might want to claim that a unified theory has greater empirical support than two disparate theories simply because it can account for a larger number of phenomena with supposedly fewer assumptions. But here again we need to be mindful of the way in which the unification has been achieved. Obviously we want to minimize the number of free parameters in the theory, or, to use Weinberg's phrase, we want to increase the theory's rigidity. The goal is to have a tightly constrained unifying core applicable across a wide range of different phenomena. The electroweak theory has only one free parameter, the Weinberg angle, and although that parameter is representative of the theory's unity, its value is not derivable from the theoretical structure itself in the way that, for example, the value for the velocity of electromagnetic waves is a consequence of Maxwell's theory. Nevertheless, taken by itself, the electroweak theory certainly qualifies as a unified theory in the sense I have described, unlike, say, the standard model, which has no fewer than 18 arbitrary parameters, including particle masses, the number of charged leptons and so forth. These parameters need to be inserted into the calculations at the appropriate moments in order for the theory to account for the various particle interactions and to determine the forces between the various quarks and leptons. But these parameters are more than just calculational tools; they provide the foundation on which current particle physics is based. In fact, like the Weinberg angle, it is these parameters that function as the unifying core of the standard model; yet because they are "free" parameters, there is no apparent or discernible connection between them. And therein lies the problem.

In order for the theory to be truly unified, some mechanism linking these parameters must exist. Moreover, as long as there is no understanding of why the parameters have the values they do, the theory cannot explain at a fundamental level even why atoms and molecules exist. In other words, an explanation of the origin of these parameters is crucial for a basic understanding of the physical world. We see, then, that it isn't simply the number of parameters that creates a problem for unification; the mysterious nature of the parameters creates a corresponding difficulty for explanatory power. In the electroweak case we have only one arbitrary parameter, and, as we have seen, its status creates similar problems for the explanation of electroweak coupling, problems traceable to the mass of the Higgs boson. Although it and the standard model enjoy a great deal of empirical support, the unity provided by the standard model as a whole is surely suspect.

This issue of empirical support has another dimension that has consistently reappeared as a theme throughout the preceding chapters. In each of the examples I have discussed, a general and rather abstract mathematical framework has played an essential role in the unification of disparate phenomena. Even in Fisher's work his statistical analysis of genetics was based on a model from gas theory that incorporated only general assumptions about populations. The significance of these frameworks is that their generality allows one to ignore specific features of the systems/phenomena in question, thereby enhancing the potential for unification. The more general the structure, the more phenomena it is able to incorporate, and hence the greater the empirical support. But typically this is achieved at the expense of a detailed theoretical account that explains the mechanisms and processes associated with the phenomena. Hence the empirical support is limited to quantitative predictions. But philosophers of science know all too well the arguments against linking predictive power with increased likelihood of truth.

Given my arguments against the alliance of unification, truth and explanatory power, some might be tempted to associate my position with those of the "disunifiers", philosophers, historians and sociologists who argue against the unity of science on various grounds – metaphysical, epistemological, political. Although I am sympathetic toward arguments against a general unity of science, and I oppose arguments that attempt to infer a unity in nature on the basis of unified theories, obviously I do not want to deny that unity does exist within the confines of particular disciplines and theories. My goal has been to uncover the nature of that unity, the ways in which it is achieved and the kinds of consequences it can support with respect to the ontological status of theories. There can be no doubt that unity exists in science, that unified theories have been enormously successful and that unity is a goal pursued by many practicing scientists in a variety of fields. But nothing about a corresponding unity in nature follows from those facts; that is, we cannot, simply on the basis of the existence or success of unified theories, infer a unified world. In order for this negative thesis to be persuasive, one must first expose the foundations and structures of unified theories. Once that is done, it becomes possible to see the gap between the simple conceptual design of the theory and the complexity of the system it represents. The unification of electromagnetism with the weak and strong forces involves energies on a scale trillions of times beyond the reach of current instrumentation. A rather daunting problem indeed – one that led Sheldon Glashow and his colleague Paul Ginsparg to write that "for the first time since the dark ages we can see how our noble search may end, with faith replacing science once again". In that context the disunity arguments become extremely persuasive.

On the other hand, in contexts less ambitious than grand unification, where physics is somewhat less theological, it becomes difficult to dismiss undisputed facts such as the claim that light is a form of electromagnetic radiation. Although our understanding of electromagnetism has changed since the formulation of Maxwell's equations, it would be remiss to suggest that somehow the unity had not been preserved. However, as we saw in the preceding chapters, the unity produced by Maxwell's electrodynamics is in many respects unique. It effected a kind of ontological reduction that few, if any, subsequent theories have been able to match. However, ontological reduction in a case like Maxwell's theory needs to be distinguished from the kind of reduction that occurs when one theory supersedes another. It is common to argue that because newly constructed theories typically embody, as limiting cases, the things that were correct in their predecessors, we should see this as evidence for some form of convergence or unity in science. However, rather than contributing to scientific unity, such cases, on closer inspection, tend to reveal inconsistencies that actually speak against unity and convergence. Consider, for example, the replacement of Newtonian gravitation theory with general relativity. The fundamental ontology of the latter is curved space, rather than gravitational force; yet in the Newtonian limit the qualitative interpretation of the mathematics yields a flat space and gravitational forces. The problem is not simply that we want to retain the Newtonian ontology for particular purposes and ease of calculation, but rather that general relativity itself embodies two distinct ontologies that say different things about the constitution of the physical world. In that sense, then, the subsumption of old theories as limiting cases of new ones affirms the kind of disunity we expect from a science marked by periods of sweeping change. Nevertheless, the unity achieved in cases like Maxwell's theory, special relativity and evolutionary genetics requires us to take seriously the role of unification in scientific discourse and practice.

Hence, my position falls squarely between the two camps; arguments for unity and those for disunity have important lessons to teach us. However, viewed as a metaphysical thesis, the problem takes on a structure resembling a Kantian antinomy, where the truth of one of the disjuncts (unity or disunity) supposedly implies the falsity of the other. The resolution consists in showing that the thesis and antithesis (in this case, unity and disunity) are mutually compatible, because neither refers to some ultimate way in which the world is constructed. As empirical theses about the physical world, both are right, but as metaphysical theses, both are wrong, simply because the evidence, by its very nature, is inconclusive. To that extent the problem can be seen as presenting a false dichotomy. Neither position is adequate for providing a descriptive, explanatory or metaphysical account of nature; yet in certain contexts and for certain purposes, either may be sufficient to capture the phenomena at hand. As we saw in the discussion of specific theories, different kinds of unity were appropriate for different situations, and unity sometimes was coupled with an element of disunity.

My argument has focused on the rather narrowly defined issue of theory unification, but there are other forms of unity that may exist to greater or lesser degrees at the general level of scientific practice. The logical-empiricist goal of finding a common language that would bring unity to the sciences has not been and cannot be achieved; nevertheless, one might claim that there is a kind of unity that exists within specialized disciplines and sometimes even across disciplines – a unity produced by practices involving commonality of mathematical methods, instrumentation, measuring techniques and so forth. All of these function to produce a scientific community of the sort described by Galison (1998) in his metaphor of the "trading zone". Those unities are different in kind from the unity I have addressed in the preceding chapters. The issue of unity in nature need not arise in discussing unity of practice. And although such practical concerns are important for understanding the culture of science, they bypass the fundamental question that motivates much of the debate on scientific unity, specifically the ontological problem of whether or not nature is itself a unified whole. I have tried to address that particular issue by showing that the ontological question is, in most cases, distinct from the ability to construct a unified theory. In fact, my claim is that it is this lack of ontological unity that is partially responsible for the separation of unity and explanation at the theoretical level.

Throughout this book I have maintained that unification and explanation are often at odds with each other. Distinguishing between unity and explanation is important not only for a better understanding of the nature of theory unification but also to expose the fact that inferential practices grounded in the explanatory power of a theory often are simply inapplicable to the theory's unifying power. Initially that may seem rather odd, because it is usual to assume that the more phenomena that are unified within a theoretical framework, the more phenomena will be explained. We have seen, however, why that doesn't follow. In my critical discussion of the division between explanation and unification, little attention was paid to exactly what a scientific explanation should consist in. I spoke briefly about the desire to know the hows and whys of physical systems, but I did not specify a form that good explanations must instantiate. I have remained deliberately silent on the subject primarily because it is, in itself, the topic of a separate book and one that many philosophers of science have already addressed in detail. That said, there is a feature of explanation that everyone would agree on, namely, the desire to impart understanding. Yet in many of the theories I have discussed, the information required for understanding the fundamental processes that figured in the unification was simply unavailable. The electroweak theory provides a "how possibly" story that rests on the discovery of the Higgs boson. Minkowski space-time enhances the unifying power of special relativity, and yet it provides no understanding of its relativistic dynamics over and above what had already been explained by Einstein. But here again we have a theory that is itself capable of explaining and unifying diverse phenomena, and yet much is left unexplained at the level of its own fundamental postulates. In that sense, then, when speaking about explanatory power we must distinguish between the phenomena that the theory "accounts for" and a physical explanation of the theory's fundamental structure. Although all of the theories I have discussed followed similar patterns for producing unity, the unification manifested itself differently in each case; different levels of reduction, synthesis and integration yielded distinct ways in which phenomena could be united under a common theoretical framework. Accompanying those disparate representations of theoretic unity were diverse levels of explanation; yet what was most significant in each of the cases was that the processes associated with the piece of theoretical structure necessary for the unification were themselves left largely unexplained. To put it differently, the mechanism responsible for unity contributed nothing to the explanatory power of the theory.

It is important to understand that this is not simply an instance in which the request for explanation must terminate, as, for example, in the search for certain kinds of explanations of quantum systems. Nor is it a case of contextual opacity with respect to the information that is being requested or transmitted. In each of the examples I have discussed, a fundamental process has been left unexplained in the face of a remarkable feat of unification; consequently there is a basic lack of understanding as to how (or even whether or not) the unifying process actually takes place in nature. In some cases this speaks directly to the limits of science, imposed as a consequence of different levels of technological advance. We are simply unable to probe the fundamental building blocks of nature with the energies currently available in particle accelerators. However, in other cases the problem is not one of instrumentation, but rather lack of the theoretical knowledge needed to fill in the missing details. Natural selection and genetics provide a case in point; there are still debates as to whether or not and to what extent selection functions in certain contexts. Similarly, in Maxwell's electrodynamics there was a good deal of uncertainty about the nature of the field, a controversy that remains in the transition from classical theory to quantum field theory, and the relationships between fields and particles. In each of these cases the stumbling blocks to unity and understanding are empirical, grounded in science itself rather than in philosophical objections about the proper form for unification and questions about whether or not that has been achieved.

To conclude, let me summarize the connection between the philosophical and empirical parts of my argument. The motivating question is whether or not theory unification provides evidence for unity in nature. In order to answer that question, we first need to know what unification consists in. More specifically, is it connected with explanatory power, and if so, what is the link? Finally, how do these issues bear on the metaphysical thesis regarding a unified physical world? To put it slightly differently, how should we understand the unity that science has achieved? Has it achieved unity at all? And are there philosophical problems associated with this unity? These kinds of questions involve a philosophical dimension and an empirical dimension; in other words, we must begin by telling an empirical story about unity that is grounded in historical documentation before we can draw philosophical conclusions. My general conclusion is that it is a mistake to structure the unity/disunity debate in metaphysical terms; that is, one should not view success in constructing unified theories as evidence for a global metaphysics of unity or disunity. Whether the physical world is unified or disunified is an empirical question, the answer to which is both yes and no, depending on the kind of evidence we have and the type of phenomena we are dealing with. In some contexts there is evidence for unity, whereas in others there is not. But nothing about that evidence warrants any broad, sweeping statements about nature "in general" or how its ultimate constituents are assembled. The construction of theories that unify phenomena is a practical problem for which we only sometimes have the appropriate resources – the phenomena themselves may simply not be amenable to a unified treatment. Seeing unity in metaphysical terms shifts the focus away from practical issues relevant to localized contexts toward a more global problem – erecting a unified physics or science – one that in principle we may not be able to solve because nature may not wish to cooperate. The methodological point I want to stress is that eliminating metaphysics from the unity/disunity debate need not entail a corresponding dismissal of philosophical analysis. Philosophical methods and concepts are important for analysing the structure and implications of the evidence for both unity and disunity, but they should not be used as means for extending the domain of that evidence to claims about scientific practices and goals that are merely promissory. Debates about unity and disunity become productive when they take as their starting point an analysis of the accomplishments of empirical research. Science undoubtedly has achieved a certain level of unity, but in order to understand its limits, we need to know its nature. Uncovering the *character* of unity and disunity is a philosophical task, one that will contribute to the broader goal of better understanding the practice of science itself. And given the progress and methods of empirical science, there is perhaps nowhere that metaphysics is less helpful.

Notes

Chapter 1

1. Quoted by Westman (1972, p. 240).

2. That account maintained the central position of the sun in the universe, because if the forces between the planets and the sun were reciprocal, then the sun would be displaced from its position.

3. In other words, the forces of attraction or repulsion acting on the planet would be proportional to the cosine of the angle between the fibres and the radius vector from the sun. Or, because that angle would be equal to the true anomaly, the force causing libration and the magnitude of that force would be proportional to the cosine of the true anomaly.

4. For a detailed account, see Koyre (1973, pp. 446–7).

5. Although Keplerian astronomy is not usually taken as an example of a unified theory, it nevertheless displays the kind of interdependence that is characteristic of the unifying process.

6. Quoted by Butts (1968, p. 153).

7. There are also inductive methods that depend on the notion of resemblance; these are the law of continuity, the method of gradation and the method of natural classification. See William Whewell (1847, vol. 1; 1967, pp. 412–25).

8. For determinations of errors, see Scott Kleiner (1983, esp. p. 301, footnote).

9. See Butts (1973, pp. 53–85) and Harper (1989, pp. 115–52).

10. See Butts (1973, p. 81).

11. For Neurath, "physicalism" referred to the idea that the language of science ought to be a physical-thing language.

12. Einstein was never satisfied with quantum mechanics because he could not accept the kind of irreducible indeterminism that it entailed.

13. For example, in classical electrodynamics, if we investigate experimentally the effects of both a steady current and a variable current of electricity on the torsion of a longitudinally magnetized iron wire, we can deduce the effects of torsion and its variations in the wire.

14. The notion of a theory's "rigidity" is due to Steven Weinberg. Instead of thinking about the unifying power of a theory, he tends to see the issue as one involving the theory's ability to account for new phenomena without altering its structural properties. The more rigid the theory in achieving this goal, the more desirable it is. Instead of seeing this as an alternative to the notion of theoretical unity, I would instead claim that it is one of its defining features.

Chapter 2

1. Friedman uses the same argument to justify a realism about space-time structure. Although my discussion is limited to the kinetic-theory example, if the argument is unsuccessful in this rather straightforward case, then it would seem to face even more difficulty in the case of space-time structure, where the entities in question do not have any theory-neutral status.

2. I want to point out here that my objection is not directed specifically at the traditional conjunction objection that van Fraassen (1980) and others have addressed. What I want to emphasize is the incompatibility between Friedman's version of realism and the factors associated with unification, particularly as it was discussed by Whewell. See Friedman (1983, p. 242, n. 14), who likens his account to Whewell's consilience of inductions.

3. Van Fraassen (1980) exemplifies this approach, thereby presenting a more sophisticated version of anti-realism than the one that Friedman ascribes to him. Although Friedman is correct in seeing constructive empiricism as a generalization of van Fraassen's earlier views on space and time, he is mistaken in equating the position with traditional forms of instrumentalism. One obvious reason for Friedman's characterization would seem to be van Fraassen's use of the embedding approach rather than the sub-model approach, a move consistent with various forms of anti-realism. Although van Fraassen's literal interpretation renders our theories capable of being true or false, it doesn't entail the corresponding requirement that we believe them or that we can assign them a truth value. We present a theory by specifying its models and delineating certain parts of those models (the empirical substructures) as candidates for the direct representation of observable phenomena. Once our theory is accepted (at a minimum, what is actual and observable finds a place in some model of the theory), it guides our linguistic practice. As a result, the language receives its interpretation through the model(s) of the theory. Modal locutions as well as statements about theoretical structure and unobservable entities reflect the fact that our models specify many possible courses of events. Van Fraassen sees the literal interpretation as specifying the model as the locus of possibility, not a reality behind the model (1980, p. 220). Because language is interpreted through the model rather than by some mysterious hookup with reality, the constructive empiricist can advocate a literal interpretation of theories while remaining agnostic about metaphysical commitment to theoretical structure.

4. In an attempt to counter the Putnam-Boyd objections to anti-realism, van Fraassen argued that theory evolution could not proceed by simple conjunction; usually, if not always, substantial corrections were made to the individual theories that were brought together.

5. Richard Boyd suggests this in his "Lex Orenci est Lex Credendi" (1985).

6. For a more specific account of this case, see Khinchin (1949).

7. However, as Cartwright (1983, p. 60) points out, if a body is moving northeast, we cannot *physically* separate the northern and the eastern parts of that motion. The same (and perhaps even stronger) case can be made for the gas example.

8. I argued earlier that Friedman is committed to this stronger notion of a literally true identity for several reasons (e.g., the model/sub-model account seems to demand it, and if he were committed to merely a literal interpretation of theoretical structure, then there would be no reason to sanction realism about that structure, because no epistemological conclusions need follow from a purely semantic position).

9. In other words, the function $H(q_i, p_k)$ is an integral of the system described by the equations of motion.

10. See Appendix 2.1 to this chapter. For a brief discussion of Maxwell's dissatisfaction with the relationship of the van der Waals equation to the kinetic theory, see Maxwell (1965, vol. 2, pp. 407–8). For a detailed quantitative treatment of the relationship between the virial theorem and the van der Waals equation, see Segrè (1983, app. 14, pp. 281–3).

11. The critical temperature T_c specifies the point above which liquification is impossible, and P_c and V_c are the pressure and volume at which liquification first begins, the point when the substance is at a temperature just below T_c. So long as a gas is kept above critical temperature, no pressure, however great, can liquify it. A gas below critical temperature is usually described as a vapour. See Jeans (1917), as well as Tabor (1979).

12. See Jeans (1917, p. 94) for a table of comparisons.

13. It is interesting to note that the van der Waals explanation of critical temperature is not tied to the particular approximations used in deriving his equation. The existence of a critical temperature is usually assumed to depend upon the minimum in the potential-energy curve, but as yet there is no satisfactory detailed theory based on the molecular forces. For details, see Collie (1982, ch. 2).

14. That this happens can be seen from the "reduced equation of state". If a and b are eliminated from the equation in terms of critical constants, one obtains this "reduced" equation, which gives reduced values for pressure (where reduced $p = p/p_c$), temperature and volume. The reduced equation expresses the variables p, V and T as fractions of the critical pressure, critical volume and critical temperature. The equation supposedly is the same for all gases, because the quantities a and b, which vary from gas to gas, have disappeared. If this equation could be regarded as absolutely true, then whenever any two "reduced" quantities were known, the third could also be given, and similarly, when any two quantities were the same for two gases, then the third would also be the same (the law of corresponding states). This law holds only if the nature of the gas can be specified by two physical constants. But as in the case of the van der Waals law, the law of corresponding states is true only as a first approximation. For details, see Jeans (1917), Collie (1982), and Brush (1976, bk. 1, ch. 7).

15. A similar situation occurs in the case of nuclear models. For a discussion, see Morrison (1999b).

16. This is the issue raised by van Fraassen (1980).

17. One might suggest the following realist defence: There is a general molecular theory yet to be discovered, and in the meantime we can use the different incompatible models that have been formulated so long as we don't construe them literally. These models cannot be conjoined, so they lack unification and the boosts in confirmation that Friedman emphasizes. This line of defence seems unacceptable for several reasons. First of all, it implies that we understand the notion of a molecule as a mere mathematical representation that enjoys no confirmation, a result that seems to run counter to any realist defence of science. As pointed out earlier, if we opt for the embedding approach we *can* understand these models literally; we just don't interpret them as literally true. Second, and perhaps more important, the works of both Cartwright (1983) and Hacking (1983) have emphasized the variety of contexts in which different and incompatible models are used. If we construe each of these cases non-literally, then there is very little, if anything, that current theory tells us that can be literally understood.

18. A similar version of the semantic view is presented by Giere (1988). Cartwright (1983) has also presented persuasive arguments regarding the role of models in scientific theorizing, emphasizing the fact that a plurality of models can be used to account for a single

phenomenon. Although I am sympathetic to those accounts of modelling, I think they don't go far enough. There are many more methods of model construction that the semantic view, because of the importance it accords theory, cannot accommodate. Similarly, neither Giere nor Cartwright pays enough attention to the way that models function as independent sources of knowledge. For more on this, see Morrison (1998a,b, 1999a,b).

19. See Laudan (1971), Butts (1968, 1973, 1977), Hesse (1968), and Fisch (1985).

20. I would like to point out here that Forster (1988) does not argue in favour of the conjunction thesis.

21. Of course, this is not to downplay the achievement of universal gravitation, for it is certainly the case that Cartesian mechanics failed to furnish the kind of comprehensive theory that Newton provided. Nevertheless, one of the criteria that Whewell set forth as representative of a consilient theory was that it be able to explain phenomena in areas that it was not intended to cover, what we might now term "novel prediction". Without getting into the difficulties surrounding the confirmational status of novel predictions, Whewell seems to place a lesser value on a theory that is formulated in response to a predetermined idea about unified phenomena.

22. Levin (1983) makes this point in a different context.

23. It is interesting to note that one of the most celebrated cases of unification, Maxwell's theory of the electromagnetic field, did not depend for its viability on the ontological status of the aether. In discussing unification, a distinction should be made between theories that provide a mathematically unified account of the phenomena and those that have an ontological unity. Although Maxwell's original derivation of the electromagnetic-wave equation relied on a mechanical model of the aether, subsequent formulations disregarded that "unifying structure" as having any importance for the field equations. I shall discuss these issues in Chapter 3.

24. In the case of the synthesis of Mendelian genetics with Darwinian evolution there was an attempt to show a degree of compatibility. In fact, finding a synthesis was a motivating factor.

25. For a description of the derivation taking account of the sizes and forces of the molecules, see Tabor (1979, ch. 5).

26. For the original paper on this topic, see Clausius (1870, pp. 122–7). For additional discussion of the historical origins of the van der Waals law, see Garber (1978), Klein (1974), and Clarke (1976).

27. From equation (A2.2) it was possible to see that the pressure of a gas could arise from the motions of particles or from possible repulsive forces between them. Because experimental evidence (Laplace's theory of capillarity and surface tension of liquids) had already indicated that forces acted between the molecules of a gas, van der Waals assumed that those forces would affect the pressure term in the ideal-gas law. The size of the correction could be computed using the virial theorem. Because the form of $\phi(\mathbf{r}_{ij})$ (the virial of the intermolecular forces) was unknown, $\langle \frac{1}{2} \sum \mathbf{r}_{ij}\phi(\mathbf{r}_{ij}) \rangle$ could be replaced by an effective intermolecular-pressure term P', giving us

$$\left\langle \sum \frac{1}{2}m\,\mathbf{u}^2 \right\rangle = \frac{3}{2}(P + P')V$$

For gases, the molecular pressure P' is expected to be only a correction to the external pressure P, whereas for liquids the external pressure (in normal circumstances) is likely to be negligible compared with P'.

28. See Klein (1974, fn. 39) for details of the van der Waals method.

29. Because the mean free path supposedly is reduced for spherical molecules (by a factor of λ/λ_0, where λ_0 is the mean free path for point molecules), the pressure must be increased by the reciprocal of this ratio (pressure being proportional to the number of collisions per unit time). The equation of state can now be written in the form

$$\left\langle \sum \frac{1}{2} m \, \mathbf{u}^2 \right\rangle = \frac{3}{2}(P + P')(v - b)$$

where $b = 4N_0 v_m$, the force times the actual volume of the N_0 molecules in a mole. In short, $(V - b)$ is simply the volume of the container minus the sum of molecular volumes.

30. The number of interacting pairs was proportional to the square of the density of the fluid or inversely proportional to v^2, where v is the volume per mole. By the introduction of a proportionality factor a, the internal pressure a/v^2 gives us

$$\left\langle \sum \frac{1}{2} m \, \mathbf{u}^2 \right\rangle = \frac{3}{2}(P + a/v^2)(v - b)$$

31. Andrews (1869). See also Clarke (1976), Tabor (1979), and Brush (1976).

Chapter 3

1. In his first paper on electrostatic induction, in 1837, Faraday showed that the force between two charged bodies depended on the insulating medium that surrounded them and not merely on their shapes and positions. (Cavendish had demonstrated the same result long prior to Faraday, but his work remained unpublished.) According to Faraday's theory, induction was an action caused by contiguous particles that took place along curved lines that he called "lines of force". Faraday showed experimentally that different substances had different capacities for mediation of electrostatic forces. Those lines of force denoted the dispositions of individual particles to transmit an electric force from particle to particle. The particles of the medium (the dielectric) that transmitted the force were subject to a state of electrical tension that led to the propagation of electrostatic forces. Thus, the electrotonic state came to be identified with the actions of the particles and the lines of force.

Faraday used the term "lines of force" as a "temporary conventional mode" of expressing the direction of the power; they provided a kind of geometrical representation of lines of polarized particles of the dielectric that were subjected to electric tension. The polarization was not considered to be the result of mutual contact; instead, the particles acted by their associated polar forces. Although Faraday denied that electrostatic induction took place by action at a distance, he was at a loss as to how to explain the transmission of the tension from one polarized particle to another (Faraday 1839–55). There is a good secondary summary by Harman (1982b). See also Williams (1970).

2. In 1841 Thomson published a mathematical demonstration that the formulas of electricity deduced from laws of action at a distance were identical with the formulas of heat distribution deduced from the ideas of action between contiguous particles (an analogy drawn from Fourier's theory of heat). Again, in January 1845, Thomson, using Green's

notion of a potential function, demonstrated that Coulomb's electrical action at a distance and Faraday's notion of action by contiguous particles in the medium led to the same mathematical theory. Having undertaken a study of Faraday's electrostatics, Thomson extended the *mathematical* analogy between thermal phenomena and electrostatic phenomena to suggest a corresponding *physical* analogy. The physical model of propagation of heat from particle to particle suggested an analogous propagation of electrical forces by the action of the contiguous particles (molecular vortices) of an intervening medium. Although Faraday would later transform that analogy into his field conception of lines of force, Thomson denied any justification, based on the analogy, for its introduction as a true physical *hypothesis*. Instead, that mathematical correspondence merely indicated that a *possible model* of propagation of electrostatic forces could be based on a theory of action between contiguous particles.

Perhaps the most important mathematical analogy developed by Thomson, at least insofar as Faraday's final version of field theory was concerned, was the one outlined in his 1847 paper "Mechanical Representation of Electric, Magnetic and Galvanic Forces". The foremost difficulty with Faraday's early version of the field theory was its inability to explain the mechanism of molecular interaction between contiguous particles of the dielectric medium. Thomson's attempt at a solution was contained in what he termed a "sketch of a mathematical analogy" that dealt with the propagation of electric and magnetic forces in terms of the linear strain and rotational strain of an elastic solid. Thomson used the mathematical methods developed by Stokes to treat rotations and strains in a continuous aether. He showed that the distributions of the linear and rotational strains in an elastic solid in space would be analogous to the distributions of electric force (for point charges) and magnetic force (for magnetic dipoles and electric currents). Although that mechanical model had physical implications (Faraday's discovery of the rotational effect of magnetism on light was consistent with the representation of magnetic force by the rotational strain of an elastic solid), the relationship between mechanical strain and electrical and magnetic phenomena remained physically unclear. However, as Thomson noted, if a physical theory was to be discovered on the basis of that speculative analogy, it would, when taken in connection with the undulatory theory of light, most probably explain the effect of magnetism on polarized light and hence exhibit a connection with the luminiferous aether. See Thomson (1872) and Thompson (1910).

3. The chief proponent of action-at-a-distance theories was Wilhelm Weber, who in the 1840s developed a reduction of the laws of electricity and magnetism to consequences of the laws of mechanics. Weber's theory was based on the interaction of charged particles through distance forces that depended on their relative positions, velocities and accelerations. That, together with his Amperean theory of molecular currents, enabled him to provide a unified account of electricity, magnetism, electromagnetism and electromagnetic induction. Maxwell disliked Weber's theory because of its many startling assumptions; despite its many accurate results, Maxwell considered it only a mathematical speculation that ought to be compared with other accounts. See Maxwell's letters to Thomson, May 15 and September 13, 1855 (Larmour 1937). Maxwell believed that his analogy, though not nearly as comprehensive as Weber's theory, could nevertheless be justified on the basis of its mathematical equivalence to the action-at-a-distance formulation. "It is a good thing to have two ways of looking at a subject, and to admit that there are two ways of looking at it" (Maxwell 1965, vol. 1, pp. 207–8).

4. The model provided a kind of mathematical isomorphism between the equations of percolative streamline flow and the equations describing electric or magnetic lines of force. The

emphasis here and throughout Maxwell's work is on relations, rather than the objects that are related.

5. Maxwell Papers, Cambridge University, manuscript 7655, as quoted by Heimann (1970).

6. Heimann (1970, n. 84).

7. In a letter to Thomson, September 13, 1855, Maxwell remarked that he was setting down his theory for the sake of acquiring knowledge sufficient to guide him in devising experiments (Larmour 1937, p. 18).

8. Letter to Thomson, September 13, 1855 (Larmour 1937).

9. For details, see Williams (1970).

10. No tendencies for the magnetic lines to repel each other and contract along their lengths were derivable from the flow analogy; hence it could not explain magnetic forces. By assuming that the magnetic line of force represented the axis of a molecular vortex, the vortex theory was capable of explaining these phenomena without difficulty. It was possible to demonstrate that centrifugal forces caused by the rotational motion of the molecular vortices would tend to make each vortex tube expand in thickness, thereby tending to increase the spacing between magnetic lines. Moreover, because of the incompressibility of the fluid in the vortex tubes, they would tend to shrink in length, causing the magnetic lines to have a corresponding tendency to contract along their lengths. See Maxwell (1965, vol. 1, p. 467) and Seigel (1985, pp. 186–7).

11. An extension of the medium in the direction of the axes of the vortices, together with a contraction in all directions perpendicular to that, will cause an increase in velocity, whereas shortening of the axis and bulging of the sides will produce a decrease in velocity. See Maxwell (1965, vol. 1, p. 483; fig. 4, p. 489).

12. Daniel Seigel argues that Maxwell took a realistic attitude toward the vortex hypothesis. Later I shall claim that although the vortex hypothesis was more than simply a heuristic device, Maxwell interpreted it only as a *possible* physical hypothesis, unlike the idle wheels and the hydrodynamic analogy of "On Faraday's Lines". The problem was that no conclusive justification for the existence of vortices could be given, and in view of Maxwell's rather conservative methodology it seems reasonable to suppose that instead of seeing them as real physical entities, Maxwell saw the vortices as "candidates" for reality. Ole Knudsen (1976) argues that Maxwell was committed to a realistic interpretation of the phenomenon of rotation.

13. That Maxwell was less than fully convinced of the strength of his position is evident from the following remark:

Those who look in a different direction for the explanation of the facts may be able to compare this theory with that of the existence of currents flowing freely through bodies, and with that which supposes electricity to act at a distance.... The facts of electromagnetism are so complicated and various, that the explanation of any number of them by several different hypotheses must be interesting ... to all those who desire to understand how much evidence the explanation of the phenomena lends to the credibility of the theory. (1965, vol. 1, p. 488)

14. See Bromberg (1968), as well as Chalmers (1973). Seigel (1986) gives an excellent account of the origin of displacement, as well as a commentary on the debate surrounding the topic.

15. See Bromberg (1968) for a more detailed discussion.

16. See Bromberg (1968, p. 226). Earlier in the paper, Maxwell relates ρ to the magnetic properties of the medium by $\rho = \mu/\pi$, where μ is magnetic permeability. Hence $V = \sqrt{\mu/\pi} = \sqrt{E^2/\mu}$, and in air, $\mu = 1$ and $E = 310,740,000,000$ mm/sec.

17. In an earlier paper, written in 1857, Weber and Kohlrausch (1892–94) determined the ratio between the electrostatic and electrodynamic units of charge, a ratio having the dimensions of the velocity of propagation of electric action. Weber's constant referred to electrodynamic units rather than electromagnetic units, and so his ratio came out as $\sqrt{2}$ times the velocity of light. In a letter to Thomson, December 10, 1861, Maxwell stated what he had previously written to Faraday: that he had been unaware of Weber's result and that he had made out the equations before he had "any suspicion of the nearness between the two values of the velocity of propagation of magnetic effects and that of light". In addition to the fact that Maxwell supposedly had been unaware of Weber's result and hence of the similarity between the velocity of propagation of magnetic effects and that of light, further evidence for the unexpectedness of that unification stems from the way in which the wave equation was derived from the model. An examination of Maxwell's calculations reveals that the value for V was not in fact a direct consequence of the model. Because of the way in which Maxwell defines the value of m (the coefficient of rigidity), the correct expression for V is $\sqrt{m/2}$. Hence, Maxwell's model yields a velocity of propagation equal to the ratio of the electrical units divided by $\sqrt{2}$. Duhem (1902, pp. 211–2) was the first to point that out. Maxwell's definition of m is as follows: Let ξ, η and ζ be the displacements of any particle of the sphere in the directions x, y and z. Let P_{xx}, P_{yy} and P_{zz} be the stresses normal to planes perpendicular to the three axes, and let P_{yz}, P_{zx} and P_{xy} be the stresses of the distortion in the planes yz, zx and xy. Then m can be defined as $P_{xx} - P_{yy} = m \, (d\xi/dx - d\eta/dy)$ (Maxwell 1965, vol. 1, p. 493). Duhem concluded that Maxwell was guilty of deliberately falsifying one of the fundamental formulas of elasticity, implying that Maxwell had constructed his model in order to justify an electromagnetic theory of light that he already knew to be in error. However, as Chalmers (1973) points out, Maxwell cannot justly be accused of defining a coefficient of rigidity that differed from the customary one in order to conceal a faulty derivation, because he had defined the constant in precisely the same way 12 years earlier. See Maxwell's "On the Equilibrium of Elastic Solids" (1965, vol. 1, pp. 30–73).

18. See Whittaker (1951, ch. 8). For a more detailed mathematical treatment of the problems with the displacement current, see O'Rahilly (1965, ch. 3).

19. My illustration borrows from Seigel (1986).

20. Numerous possible reasons for that move are discussed in the literature. See, for example, Bromberg (1968), Bork (1963) and Seigel (1975).

21. According to earlier theories, a current employed in charging a condenser was not closed; instead, it terminated at the coatings of the condenser, where the charge accumulated.

22. My discussion follows Bromberg (1968). An alternative account is offered by Chalmers (1973), who argues that Bromberg is wrong in attributing an inconsistency to Maxwell. Chalmers claims that Bromberg's interpretation arises from failure to appreciate the distinction Maxwell makes between electromotive force and electric tension. However, for what seem to me to be obvious reasons, namely, the difficulty involved in giving an *electromechanical* theory of light, Maxwell seems guilty of interpreting $R = -4\pi E^2 h$ as both an electrical equation and an equation representing elasticity. Once the mechanical apparatus of Maxwell's model was abandoned (in "A Dynamical Theory" and the *Treatise on Electricity*

and Magnetism), the negative sign in the equation disappeared. As a result, the charge of inconsistency is perhaps slightly unfair, because the dual interpretation is a problem that arises directly from the model.

23. See Maxwell (1965, vol. 1, p. 486, vol. 2, p. 380). Seigel (1981) also makes this point.

24. The experimental facts concerned the induction of electric currents by increases or decreases in neighbouring currents, the distribution of magnetic intensity according to variations of a magnetic potential and the induction of statical electricity through dielectrics (Maxwell 1965, vol. 1, p. 564).

25. That generalized dynamical approach was also used in a somewhat different way by Thomson and Tait (1867), who integrated the energy approach with the Lagrangian formalism.

26. See Whittaker (1951) for discussion.

27. It is important to distinguish between the illustrative roles played by the idle-wheel model and the flow analogy (both of which were considered fictional models) and the more substantive interpretation given to the vortex hypothesis. Although the vortices performed an illustrative function, they also were considered a genuine physical possibility, unlike the former.

28. That liberation from mechanical influences allowed the law to figure nicely with the generalized form of Ampere's law. For an excellent discussion of the dynamical analogy in "A Dynamical Theory", see Everitt (1975).

29. Not only did Maxwell wish to identify charge with a discontinuity in displacement in a dielectric medium, but also he insisted that displacement involved a movement of electricity "in the same sense as the transference of a definite quantity of electricity through a wire is a movement of electricity".

30. An important dimension of "A Dynamical Theory" was its revival of several of Maxwell's earlier ideas about the geometry of lines of force. The intention was to extend the electromagnetic theory to explain gravitation, but unfortunately he was unsuccessful in that endeavour because of inability to represent the physical nature of lines of gravitational force. In "A Dynamical Theory" he pointed out that gravitation, being an attractive force, had the consequence that the energy of any gravitational field of a material constitution was less wherever there was a resultant gravitational force. Hence, those parts of space in which there was no resultant force would possess enormous energy. However, as Maxwell himself admitted, "I am unable to understand in what way a medium can possess such properties" (1965, vol. 1, p. 571).

31. See Maxwell (1965, vol. 2, p. 462) and Helmholtz (1867).

32. In the Newtonian formulation, quantities expressing relations between parts of a mechanical system appear in the equations of motion for the component parts of the system. Lagrange's method eliminates these quantities by reducing them to what are called "generalized coordinates" or quantities. Although Newton's laws and Lagrange's equations are simply different methods of describing mechanical systems, the latter approach makes it possible to reduce Newton's three laws of motion (necessary for describing each individual particle in the system) to a number corresponding to the number of degrees of freedom for that system. These new generalized coordinates represented by q_i define the configuration of the system, replacing the Cartesian coordinates of the Newtonian system. Basically, Lagrange's equations are second-order differentials that relate the kinetic energy (T) of the system to the generalized coordinates and velocities (q_i, $\dot{q}_i$), the generalized forces (Q_i) and the time (t). There is one equation for each of the n degrees of freedom possessed by the

system. In order to calculate the equations of motion for the system, we need only know the kinetic energy and the work of forces (a function of q_i, $\dot{q}_i$). Any force that does no work is automatically eliminated. If T and V (kinetic and potential energies) are known, the equations of motion enable the time development of the system to be deduced from knowledge of generalized coordinates and velocities at a particular time t:

$$\frac{d}{dt}\frac{\partial T}{\partial \dot{q}_i} - \frac{\partial T}{\partial q_i} = Q_i \quad (i = 1, 2, 3, \ldots, n)$$

33. This remark can be found in the Preface to *Mécanique Analytique* (1788).

34. In a letter to Thomson, Tait wrote that he saw great advantages in using "dynamics instead of mechanics"; for example, "dynamics" really meant the science of force or power and was erroneously used as a contrast to statics, whereas in reality there was no such thing as statics, only dynamical equilibrium (Thomson and Tait (1867, vol. 1, pp. vi, 250). See also Thompson (1910, vol. 1, pp. 459–60).

35. Thomson and Tait (1867, pp. 265–74). As Moyer (1977) points out, those arguments were borrowed from Rankine's comments in his second axiom of the science of energetics.

36. Thomson and Tait (1867, pp. 288–91). For the original formulation of the theorem, see Thomson (1872, vol. 1, pp. 107–12).

37. We should keep in mind that conservation of energy was also important in the mechanical approach made popular by Laplace's reductionist programme. Energy conservation was the unifying principle for thermal and mechanical phenomena. Heat production was thought to explain energy loss in mechanical processes. See Harman (1982, ch. 6). Energy was superior to the other two conserved magnitudes (momentum and angular momentum) because it took into account potential motions as well as actual motions. In addition, those other magnitudes were purely mechanical, depending on the equality of action and reaction, and therefore could not connect mechanics with other branches of physics. Energy, on the other hand, was not limited to rational mechanics and hence could function as a truly unifying concept or principle. See Rankine (1859, pp. 250–3, 347–8, esp. 253, 348).

38. See Harman (1987, pp. 267–97, esp. 289–90; 1988, pp. 75–86, esp. 87–8).

39. I would like to thank Jed Buchwald for drawing my attention to this point. For a comprehensive discussion of these differences, see Buchwald (1985).

40. As quoted by Harman (1982, p. 140).

41. Nearly all physical laws can be derived from variational principles from which differential equations mathematically identical with Lagrange's equations are deduced. The most straightforward changes in representation are produced by coordinate (point) transformations $Q_i = Q_i(q_i, t)$. In Lagrangian formulations, every transformation of generalized coordinates and momenta $[Q_i = Q_i(q_i, p_i, t), P_i = P_i(q_i, p_i, t)]$ generates the corresponding representation. Conditions of invariance are also imposed (given certain specifications about physical equivalence and significance). The representation of a physical theory is really then a study of its transformation properties, with the various representations pictured as many different mappings of the same object. For a detailed discussion of this topic, see Bunge (1957).

42. P_1 and P_2 are points in a medium of known refractive index $n(x, y, z)$. Using Fermat's principle, we calculate $A = \int_{P_2}^{P_1} n(x_1)\,ds$. Using the method of Lagrange, we introduce a physically uninterpreted parameter t that figures in the formula for optical length, i.e., $A = \int_{t_2}^{t_1} L\,dt$, where $L(x_i, \dot{x}_i) = n(x_i) \cdot (\sum \dot{x}_i^2)^{1/2}$.

43. For an excellent discussion, see Smith (1978). Smith points out that the full transition to physical dynamics occurred as a result of the irreversible nature of many heat phenomena and the possibility of moving beyond purely mathematical theories. One could argue that it was that period (the late 1860s and early 1870s) that marked the transition, for Thomson, from his earlier agnosticism to his move toward a realistic interpretation of the aether and the inner mechanisms of matter. That realism was evident in his critical remarks in the "Baltimore lectures" on Maxwell's approach to dynamics. In his address to the British Association in 1884 he claimed that "there can be no permanent satisfaction to the mind in explaining heat, light, elasticity, diffusion, electricity and magnetism, in gases, liquids, and solids, and describing precisely the relations of these different numbers of atoms, when the properties of the atom itself are simply assumed. . . ." (Thomson 1891–94).

44. In the Newtonian case, inertia defined a fundamental aspect of matter, the difference between external causes that generated change and internal causes that resisted it. Consequently, inertia was crucial to interpretation of the first law of motion. Moreover, using the third rule of philosophizing and his idea of "deductions from phenomena", Newton listed inertia as one of the essential properties of matter.

45. Maxwell (1873/1954, vol. 2, p. 182). Harman (1982) has characterized the difference between Newton's and Maxwell's definitions of matter and substance as that of a substantial representation versus a functional representation.

46. Maxwell defined polarization in terms of equal and opposite charges at opposite ends of a particle. In keeping with "On Physical Lines", dielectric polarization was represented by electric displacement, with variations constituting electric currents. According to the account provided in the *Treatise*, particles of a magnet were polarized, and the theory of magnetic polarization was developed in a manner analogous to the theory of dielectric polarization. The difference between the accounts in "On Physical Lines" and in the *Treatise* was that in the latter context displacement was defined by the motion of electricity, a quantity of charge crossing a specified area, rather than in terms of the rolling-particle motion.

47. Those experiments were performed between 1864 and 1867, but as late as 1876 we find Maxwell remarking in a letter to Bishop Ellicott that the aether was a "most conjectural scientific hypothesis". See Campbell and Garnett (1969, p. 394).

48. See, for instance, the discussion by Maxwell (1881).

49. Maxwell's interest in knowledge of relations can be traced to Sir William Hamilton, his teacher at Edinburgh, who is famous for his claim that all human knowledge is knowledge of relations between objects and not of objects themselves. For a discussion of the impact of the Scottish common-sense tradition, see Olson (1975).

50. I shall discuss the difficulties with this comparative notion of unification in Chapter 6.

51. Hertz was unable to solve the difficulty of relating the aether to matter and instead simply accepted the Stokes drag hypothesis for lack of a better alternative.

Chapter 4

1. Obviously this is not to say that unified theories do not have empirical components that are confirmed by experiment. Rather, I want to point out that the mechanism that facilitates the unification in the first instance is a mathematical model that can serve as a structural constraint on how the theories fit together.

2. At that time there was no experimental evidence for the neutrino, which was found much later, in 1956, by Cowan and Reines, two physicists from Los Alamos.

3. This theory was proposed by Sundarshan and Marshak (1958) and Feynman and Gell-Mann (1958).

4. A cross section is expressed as an effective target area, for example, 1 cm^2, which can then be related to the interaction probability by multiplying by factors like the flux of particles entering the interaction region. The basic unit for nuclear physics is the *barn*, which is equal to 10^{-24} cm^2. Neutrino collision cross sections are usually about 10^{-39} cm^2. See Dodd (1986).

5. There are also neutral-current processes that we shall discuss later.

6. Particles that decay in weak interaction do so extremely slowly, thereby giving them very long lifetimes.

7. I say "designed" here because at that time there was no experimental evidence for the existence of the W particle; in fact, the W would turn out to be a crucial and long-awaited piece of evidence that would confirm predictions of the Glashow-Weinberg-Salam electroweak theory. It was finally found in 1983.

8. Roughly speaking, the photon propagator can be thought of as the Green's function for the electromagnetic field.

9. But this is not the reason that local charge conservation is part of electrodynamics; the requirement stems from Maxwell's own modification of Ampere's law and the introduction of the displacement current. The continuity equation states that the rate of decrease of charge in a volume V is due only to the flux of current out of its surface. No net charge can be created or destroyed in V. Because V can be made arbitrarily small, charge must be locally conserved.

10. The norm of the vector is $f(x + dx)$, and expanded to first order in dx it is $f(x + dx) = f(x) + \partial_\mu f dx^\mu$, where ∂_μ is just $\partial/\partial x^\mu$.

11. A second gauge change with a scale factor Λ will transform the connection in the following way: $\partial_\mu S \rightarrow \partial_\mu S + \partial_\mu \Lambda$, which is similar to the transformation of the potential in electromagnetism, $A_\mu \rightarrow A_\mu + \partial_\mu \Lambda$, which leaves the electric and magnetic fields unchanged.

12. For a detailed discussion, see Mills (1989). Again, we can consider this very simply by thinking in terms of the invariance of the Lagrangian. Because of the space dependence of the phase of the electron wave, the Lagrangian is changed by this gauge transformation. The presence of the photon interacting with the electron and the changes in the photon wave function cancel out the changes in the Lagrangian, thereby restoring its invariance. So the invariance of the Lagrangian under local gauge transformations requires the existence of the photon (i.e., the electromagnetic field). For discussion of the derivation of charge conservation in quantum field theory, see Gitterman and Halpern (1981). For a general treatment of gauge theory, see Moriyasu (1983).

13. The axioms of group theory include the associative law, the identity law I, which involves a simple rotation and corresponds essentially to the act of leaving the object unchanged, i.e.,

$$I \times R_1 = R_1 \times I = R$$

and an inverse operation that corresponds to rotating an object backward. The latter axiom has the effect of undoing the original rotation, i.e.,

$$R_1 \times R_1^{-1} = I = R_1^{-1} \times R_1$$

These axioms apply to several kinds of symmetry operations in multidimensional spaces (reflections as well as interchanges of objects, etc.).

14. These can also be thought of as phase transformations, where the phase is considered a matrix quantity. See Aitchinson (1982) for a discussion of this topic.

15. "Isospin" actually refers to similar kinds of particles considered as two states of the same particle in particular types of interactions. For example, the strong interactions between two protons and two neutrons are the same, which suggests that for strong interactions they may be thought of as two states of the same particle. So hadrons with similar masses, but differing in terms of charge, can be combined into groups called multiplets and regarded as different states of the same object. The mathematical treatment of this characteristic is identical with that used for spin (angular momentum). Hence the name *isospin*, even though it has nothing to do with angular momentum. The SU(2) group is the isospin group and is also the symmetry group of spatial rotations that give rise to angular momentum.

16. To do this, one first specifies the up component of an isotropic spin vector at a certain position x in an abstract isotropic space. One then determines how much that up state at x needs to be rotated in order to correspond to an up state at position y. In other words, the connection between the isospin states at different points acts like an isospin rotation. The particle in the up state is moved through the potential field to position y in such a way that the spin direction is rotated by the field so that it is pointing up at y. The set of these rotations forms a symmetry group, and hence the connection between internal-space directions at each different point must also act like a symmetry group. My discussion follows Moriyasu (1983).

17. Isospin space is the imaginary internal space, as opposed to the ordinary four-dimensional space-time.

18. The generators of a group are a fixed subset of independent elements of the group such that every member of the group can be expressed as the product of powers of the generators and their inverses. In general, the number of generators equals the number of independent parameters in the group.

19. The component L_+ would have to carry one unit of isospin charge. The number of components of the internal part of the potential field in isospin space is equal to the dimension of the symmetry group.

20. Leptons are light particles, such as the electron, its neutrino and the muon that are involved in weak decay. See Schwinger (1957).

21. These are only partial symmetries, because of the mass splittings of the vector mesons. See Glashow (1961).

22. Private correspondence, April 5, 1994.

23. Quoted from private correspondence of April 5, 1994. The paper is Glashow and Gell-Mann (1961, p. 457).

24. The gap was thought to be caused by the attractive phonon-mediated interactions between electrons that produced correlated pairs of electrons with opposite momenta and spin near the Fermi surface. It then requires a finite amount of energy to break this correlation. See Nambu and Jona-Lasinio (1961).

25. See Goldstone (1961). The result was generalized to all cases of spontaneous symmetry breaking by Goldstone, Salam and Weinberg (1962).

26. See Higgs (1964a,b, 1966) for further details.

27. Hypercharge is a quantum number associated with an elementary particle, and it characterizes an isospin multiplet. It is equal to twice the mean electric charge (in e units) of the particles that form the multiplet.

28. Personal correspondence.

29. See Galison (1983, 1987) and Pickering (1984).

30. The strangeness of a particle is the sum of the number of antiquarks s that have strangeness $+1$ minus the number of strange quarks s that have strangeness -1. Strangeness was postulated to explain the fact that some elementary particles like kaons had longer lifetimes than expected.

31. My discussion follows, in part, Moriyasu (1983).

32. The Pauli-Villars regularization, which is a commonly used form, involves cutting off the integrals by assuming the existence of a fictitious particle of mass M that behaves like a ghost. The propagator takes on a modified form $(1/p^2)$ that is enough to render all graphs finite. $M^2 \rightarrow \infty$ is then taken as the limit, so that the unphysical fermion decouples from the theory.

Dimensional regularization involves generalizing the action to arbitrary dimension d, where there are regions in complex d space in which the Feynman integrals are all finite. When d is analytically continued to four dimensions, the Feynman graphs pick up poles in d space, allowing for the absorption of the divergences into physical parameters.

Lattice regularization is most commonly used in quantum chromodynamics (QCD) for non-perturbative calculations. Space-time is assumed to be a set of discrete points arranged in some hypercubical array. The lattice spacing then serves as the cutoff for the space-time integrals.

33. The idea of scale invariance simply means that the structure of physical laws does not depend on the choice of units. For any equation that is written in terms of dimensional variables there are always implicit scales of mass, length and time. These are not specified because of the form invariance of the equations themselves. Renormalization in quantum field theory requires the introduction of a new hidden scale that functions analogously to the traditional scale parameters in any dimensional equation. Consequently, any equation in quantum field theory that represents a physical quantity cannot depend on the choice of a hidden scale. For more, see Aitchinson (1982).

Chapter 5

1. For a complete discussion, see Miller (1981). In my commentary on Lorentz I have relied on Miller's helpful and detailed summary of the technical aspects of Lorentz's papers.

2. See Lorentz (1895, pp. 89–92) for a discussion of the experiment (reprinted 1952 in *The Principle of Relativity*. New York: Dover). For other discussion, see Miller (1981) and Zahar (1989).

3. Quoted by Schaffner (1969).

4. Even after Einstein's formulation of the STR there was reluctance to give up the aether as the source and carrier of electromagnetic waves.

5. A similar account was arrived at independently by Sir George Francis FitzGerald. Lorentz heard of it in an 1893 paper by Oliver Lodge; it had not been published at the time Lorentz was working on those problems.

6. Or, put a bit differently, if all the forces transformed like the Lorentz force, then the body would have to be shortened if the body were to remain in equilibrium in the moving frame. Of course, a result of special relativity is that all forces obey the same transformation law no matter how they arise.

7. See especially Poincaré (1906), Holton (1969) and Miller (1981). Zahar (1989) is a notable exception.

8. See Lorentz (1895).

9. See, for example, Zahar (1989, pp. 58–62) and Miller (1981, p. 36), both of whom advocate the view that the theory of corresponding states is the physical interpretation of the transformation equations.

10. It was not until Einstein's derivation of the transformation equations from first principles that they could be said to have acquired a physical meaning. But the notion of a physical interpretation here is somewhat different.

11. I shall later discuss this in greater detail.

12. Here, $r_r = \sqrt{x_r^2 + y_r^2 + z_r^2}$.

13. Only later did Lorentz and then Poincaré apply the contraction hypothesis to the electron, which was taken as flattened because of its motion through the aether.

14. "A Simplified Theory of Electrical and Optical Phenomena in Moving Systems", *Koninkl. Akad. Wetenschap. Proc.* (English edition) 1:427 (1899).

15. t'' is, however, still uninterpreted.

16. For a detailed discussion, see Miller (1981, pp. 79–85).

17. As Elie Zahar pointed out to me, Lorentz could still legitimately make a distinction between c qua "absolute" but frame-dependent velocity and c qua invariant "measured" velocity. Lorentz was, in fact, in a position (though unaware of it) to prove that the measured speed of light is the same in all frames.

18. The linearity was meant to ensure that the principle of inertia remain true in the STR.

19. For details, see Miller (1981).

20. In modern notation, $\beta = \gamma$.

21. Ironically, according to the STR there can be no strictly rigid rods, because they would instantaneously transmit action at a distance.

22. As Zahar (1989) points out, most of the revolutionary insights attributed to Minkowski had been anticipated by Poincaré, and Poincaré had made maximal use of them in constructing his Lorentz covariant gravitational theory.

23. It is worth mentioning that Maxwell's equations were the focal point of Einstein's programme. They were kept wholly intact while altering the laws of mechanics.

24. Einstein (1989, p. 268).

25. Letter from Einstein (1989) to Conrad Habicht, end of June–end of September 1905.

26. Einstein published three papers on the equivalence of mass and energy (1906, 1907, 1952b).

27. There has been a great deal of controversy over whether or not Einstein's derivation of the mass–energy equivalence was in fact fallacious. The debate began with a paper by Ives (1952), which was endorsed by Jammer (1961), a view that was then supported by Miller (1981). Stachel and Torretti (1982) have claimed that the derivation was in fact logically correct and have shown how the argument rests on equation (5.89), which can be justified using a relativity argument. Very briefly, the idea is that one can move from a consideration of a single body with respect to two inertial frames to a comparison of the body in two states

of motion with respect to the same inertial frame. The equivalence of those two viewpoints is just what the relativity principle asserts. I have simply assumed that the argument is valid, because my intention here is only to show how the unifying power of the relativity principle can be extended through the further unification of theoretical concepts and laws.

28. See Mills (1994) for a full discussion.

29. Although this is an obvious point when one thinks about it, Ellis and Williams (1988) have drawn attention to it in a rather nice way by indicating that it is only the whole package that is consistent. Any one of the phenomena makes sense only if the others also operate. The whole set of relativistic kinematic effects must be taken into account if one is to get a consistent description of what is happening. What appears to be length contraction in one frame may appear as time dilation in another.

30. For an extensive discussion of Minkowski diagrams, see Galison (1979).

31. For a detailed discussion of this and other related points, see Ellis and Williams (1988).

32. The transformed coordinates for E and B are as follows:

$$E_\xi = E_x, \qquad E_\eta = \gamma[E_y - (v/c)B_z], \qquad E_\zeta = \gamma[E_z + (v/c)B_y]$$
$$B_\xi = B_x, \qquad B_\eta = \gamma[B_y + (v/c)E_z], \qquad B_\zeta = \gamma[B_z - (v/c)E_y]$$

33. The Poincaré group is another name for the inhomogeneous Lorentz group with the addition of space-time translations.

34. For a full derivation, see Lucas and Hodgson (1990).

Chapter 6

1. When referring to those authors, I have not necessarily referenced the sources where that view was first developed. Instead, I have chosen to cite works in which the view is best elaborated and explained, leaving it for those works to direct the reader to the earlier papers in which the semantic account was initially formulated.

2. Although I want ultimately to agree with Lloyd about the distinction between consilience and explanation, I do so for very different reasons, reasons that have to do with the features necessary for *producing* explanations.

3. The idea of a *vera causa* can be traced to Newton's first rule, which states that "we ought to admit no more causes of natural things, than such as are both true and sufficient to explain their appearance". Roughly, Herschel's formulation is that in explaining phenomena, one should invoke only causes whose existence and competence to produce such effects can be known independently of their supposed responsibility for those phenomena. M. J. S. Hodge (1977) has an excellent discussion of the role of the *vera causa* principle in Darwin's argument.

4. What Darwin actually said in the passages (1964, pp. 243–4) quoted by Kitcher was that although the theory was strengthened by a few facts concerning instincts, those facts, although not logical deductions, were more satisfactorily seen as small consequences of one general law. There, of course, he was referring to natural selection. Although only instincts were mentioned, there is evidence that Darwin saw natural selection as unifying a variety of *different kinds* of phenomena, not just different manifestations of the same kind (i.e., instincts).

5. Kitcher (1989, pp. 438–42). In his recent work (1993, pp. 46–7), Kitcher also discusses the ways in which R. A. Fisher, J. B. S. Haldane and Sewall Wright extended the explanations that Darwin offered by deriving statements that previously had been taken as premises in selectionist patterns from genetic trajectories (i.e., general argument patterns that provide explanations of the genetic distribution in a population of particular characteristics W in some future generation). I shall deal in more detail with this feature of Kitcher's account in the following chapter on the evolutionary synthesis.

6. This seems to imply that in some contexts the use of models would be less straightforward than his account, which focuses directly on explanation and strategies for responding to explanation-seeking questions.

7. Kitcher (1989) makes this claim, but see Provine (1985) and Kohn (1985, pp. 828–9).

8. J. B. S. Haldane was the third in the trio of important figures in the evolutionary synthesis. His work also focused on mathematical analyses and addressed such issues as the intensity of selection and how the intensity of competition affects the intensity of selection, in addition to providing equations for changes in gene ratio and gene frequency. Haldane's work did not constitute a particular "approach" in the way that Fisher's work and Wright's work did. However, one of the things for which he is most famous is his questioning of the conflicts between different levels of selection. Unfortunately, for lack of space, I must limit my discussion to the work of Fisher and Wright.

Chapter 7

1. I use the word "supposed" here because the theories were not actually inconsistent nor incompatible with each other. They were thought to be as a result of certain interpretations associated with each theory. In that sense the unification did not produce a significant conceptual change in the core of either Mendelian genetics or Darwinian evolution.

2. The issue of compatibility between the Darwinian biometricians and Mendelism has been the subject of much debate in the historical, philosophical and sociological literatures. Some, such as David Hull (1985), have claimed that the delay in achieving the synthesis was an inexplicable embarrassment in the history of biology. Others, including MacKenzie and Barnes (1979), have argued that we should apply something called the "symmetry thesis" to that episode in an effort to search out the social and historical conditions that blinded scientists to the apparent compatibility of biometry and Mendelism. For more on that debate, see Nordmann (1994). I shall not go into those apects of the synthesis here; rather, my aim is to show *how* the theories became unified and the importance of mathematics in that process. Undoubtedly the story of why they were thought to be incompatible is significant, but it is, in many ways, a story different from the one I want to tell.

3. Variation is continuous insofar as one end of the curve of variation is thought to grade imperceptibly into the other, provided that a sufficiently large population sample is available. This kind of variation usually covers traits like size, intensity of colouration and so forth. Discontinuous variation is seen in the occurrence of an occasional individual that falls outside the norm of variation in a population.

4. The fact that Bateson embraced selection is one of the reasons that make it difficult to understand why the evolutionary synthesis took so long to emerge.

5. "Genotype" refers to the total genetic constitution of an organism.

6. An allele is any of the alternative forms of a gene that may occupy a given chromosomal locus.

7. If the frequencies in a population are $p = 0.7$ and $q = (1 - p) = 0.3$, then the genotype frequencies will be $AA = 0.49$, $Aa = 0.42$ and $aa = 0.09$.

8. Although several constants were required for a complete analysis, those necessary to specify the statistical aggregate were relatively few.

9. This of course departed from the methods of Pearson and the other biometricians who required that every factor be specified in all analyses.

10. When a given cause A produces effects in the same direction in X and Y, the sign of the term aa^1 is positive. When the effects are in opposite directions, the sign is negative. See Wright (1918).

11. Although the method differs from that used for partial-regression analysis, it does furnish a convenient way of calculating multiple correlations and regressions.

12. One of the most impressive applications of the method was Wright's work on the correlation between hog and corn variables. He took account of a wide range of factors: acreage, yield per acre, crop and average farm prices, numbers slaughtered at large markets, average weight, pork production, average market price, average farm price. He tabulated data from east and west as well as from winter and summer packing seasons. From those data, together with a statistical account of long-period trends, he calculated correlations between the variables, each with each other over a five-year period, producing 42 variables and 510 correlations. He took the major variables and constructed a causal scheme that, when outfitted with path coefficients, yielded the observed correlations. Trial and error yielded a causal scheme with four major variables (corn price, summer hog price, winter hog price, and certain aspects of hog breeding) and 14 paths of influence that predicted the observed correlations. For more, see Provine (1986).

13. For a detailed discussion of the differences between Fisher and Wright, see Provine (1992) and Wright (1988).

14. Very briefly, the shifting-balance theory comprises three basic phases. The first involves the breakdown of a large population into a number of subgroups, within which a number of random changes produce enough genetic variability to allow for gene-frequency drift. Once that drift has occurred, selection will then produce a favourable gene combination within a subgroup. That completes phase two. Phase three consists in an increase in numbers of the fittest subgroup, with the individuals from that group emigrating and mating in surrounding subgroups. See Crow (1992) and Provine (1986, 1992).

15. Letter from Wright to Fisher, February 13, 1931, quoted by Provine (1992, p. 213).

16. Although I shall not discuss the different models and approaches common in population genetics, it is interesting to note how the unity in modern biology closely resembles that in physics, insofar as there is often a unification at a rather abstract or higher-order level of theory, accompanied by a disunity at the level of models, application and explanation.

References

Abraham, Max. 1905. *Theorie der Elektrizität: Elektromagnetische Theorie der Strahlung*. Leipzig: Teubner.

Achinstein, Peter. 1968. *Concepts of Science*. Baltimore: Johns Hopkins University Press.

Aitchinson, I. J. R. 1982. *An Informal Introduction to Gauge Field Theory*. Cambridge University Press.

Anderson, P. W. 1958. "Random Phase Approximation in the Theory of Superconductivity". *Physical Review* 112:1900–16.

Andrews, Thomas. 1869. *Philosophical Transactions of the Royal Society* 2:575–90.

Bardeen, J., Cooper, L. N., and Schrieffer, J. R. 1957. "Theory of Superconductivity". *Physical Review* 108:1175–204.

Bork, A. 1963. "Maxwell, Displacement Current and Symmetry". *Americal Journal of Physics* 31:854–9.

Boyd, Richard. 1985. "Lex Orendi est Lex Credendi". In: *Images of Science*, ed. Paul Churchland and C. A. Hooker, pp. 3–34. University of Chicago Press.

Bromberg, Joan. 1967. "Maxwell's Displacement Current and His Theory of Light". *Archive for the History of the Exact Sciences* 4:218–34.

1968. "Maxwell's Electrostatics". *American Journal of Physics* 36:142–51.

Brush, Stephen. 1976. *The Kind of Motion We Call Heat*, 2 vols. Amsterdam: North Holland.

1983. *Statistical Physics and the Atomic Theory of Matter*. Princeton University Press.

Buchwald, Jed. 1985. *From Maxwell to Microphysics*. University of Chicago Press.

Bunge, M. 1957. "Lagrangian Formalism and Mechanical Interpretation". *American Journal of Physics* 25:211–18.

Butts, R. E. 1968. *William Whewell's Theory of Scientific Method*. University of Pittsburgh Press.

1973. "Whewell's Logic of Induction". *Foundation of Scientific Method: Nineteenth Century*, ed. R. Giere and R. Westfall, pp. 53–85. Bloomington: Indiana University Press.

1977. "Consilience of Inductions and the Problem of Conceptual Change in Science". In: *Logic, Laws and Life*, ed. R. G. Colodny, pp. 71–88. University of Pittsburgh Press.

Campbell, L., and Garnett, W. 1969. *The Life of James Clerk Maxwell*. New York: Johnson Reprint Corp.

Cartwright, Nancy. 1983. *How the Laws of Physics Lie*. Oxford University Press.

Chalmers, A. F. 1973. "Maxwell's Methodology and His Application of It to Electromagnetism". *Studies in the History and Philosophy of Science* 4:129–31.

Churchland, Paul, and Hooker, C. A. (eds.). 1985. *Images of Science*. University of Chicago Press.

Clarke, Peter. 1976. "Thermodynamics vs. Atomism". In: *Method and Appraisal in the Physical Sciences*, ed. C. Howson, pp. 41–105. Cambridge University Press.

Clausius, R. 1870. "Über einen auf die Wärme anwendbaren mechanischen Satz". *Philosophical Magazine* 4(40):122–7.

Collie, C. H. 1982. *Kinetic Theory and Entropy*. London: Longman Group.

Cooper, N. G., and West, G. B. 1988. *Particle Physics: A Los Alamos Primer*. Cambridge University Press.

Crow, James F. 1992. "Sewall Wright's Place in Twentieth-Century Biology". In: *The Founders of Evolutionary Genetics*, ed. S. Sarkar, pp. 167–200. Dordrecht: Kluwer.

Darwin, Charles. 1868. *The Variations of Plants and Animals under Domestication*, 2 vols. London: Murray.

1872. *On the Origin of Species*, 6th ed. London: Murray.

1903. *More Letters of Charles Darwin*, ed. Francis Darwin and A. C. Seward. London: Murray.

1959. *The Origin of Species, A Variorum Text*, ed. Morse Peckham. Philadelphia: University of Pennsylvania Press.

1960–67. "Darwin's Notebooks on Transmutation of Species", ed. G. de Beer (notebooks B, C, D and E). *Bulletin of the British Museum (Natural History)* 2:27–200; 3:129–76.

1964. *The Origin of Species* (facsimile of first, 1859, edition), ed. Ernst Mayr. Cambridge, MA: Harvard University Press.

Darwin, Francis. 1887. *The Life and Letters of Charles Darwin*, 3 vols. London: Murray. Reprinted 1969, New York: Johnson Reprint Corp.

Demopoulos, William. 1982. Review of *The Scientific Image*, by Bas van Fraassen. *Philosophical Review* 91:603–7.

Dewey, John. 1971. "Unity of Science as a Social Problem". In: *International Encyclopedia of Unified Science*, ed. O. Neurath et al., vol. 1, no. 1, pp. 29–38. University of Chicago Press.

Dodd, J. E. 1986. *The Ideas of Particle Physics*. Cambridge University Press.

Doran, B. 1974. "Origins and Consolidation of Field Theory in Nineteenth-Century Britain". *Historical Studies in the Physical Sciences* 6:132–260.

Duhem, Pierre. 1902. *Les Théories Electriques de J. C. Maxwell*. Paris.

Dupré, John. 1993. *The Disorder of Things: Metaphysical Foundations of the Disunity of Science*. Cambridge, MA: Harvard University Press.

1996. "Metaphysical Disorder and Scientific Unity". In: *The Disunity of Science: Boundaries, Contexts and Power*, ed. P. Galison and D. Stump. Stanford University Press.

Einstein, Albert. 1906. "Das Prinzip von der Erhaltung der Schwerpunktsbewegung und die Trägheit der Energie" (The Principle of Conservation of Motion of the Centre of Gravity and the Inertia of Mass). *Annalen der Physik* 20:627–33. Reprinted 1989 in *The Collected Papers of Albert Einstein*, ed. J. Stachel, vol. II, pp. 360–6 (English translation pp. 200–6). Princeton University Press.

1907. "Über die vom Relativitätsprinzip geforderte Trägheit der Energie" (On the Inertia of Energy Required by the Relativity Principle). *Annalen der Physik* 23:371–84. Reprinted 1989 in *The Collected Papers of Albert Einstein*, ed. J. Stachel, vol. II, pp. 414–28 (English translation pp. 238–50). Princeton University Press.

1920. *Relativity*. London: Methuen. Reprinted 1979, New York: Crown.

1949. "Autobiographical Notes" and "Reply to Criticisms". In: *Albert Einstein:*

Philosopher-Scientist, ed. P. A. Schlipp, pp. 2–94, 665–88. Evanston, IL: The Library of Living Philosophers.

1952a. "On the Electrodynamics of Moving Bodies". In: *The Principle of Relativity: A Collection of Original Memoirs on the Special and General Theory of Relativity*, trans. W. Perrett and G. B. Jeffery, pp. 37–65. New York: Dover. (Originally published 1905.)

1952b. "Does the Inertia of a Body Depend on Its Energy Content?" In: *The Principle of Relativity: A Collection of Original Memoirs on the Special and General Theory of Relativity*, trans. W. Perrett and G. B. Jeffery, pp. 69–71. New York: Dover. (Originally published 1905.)

1989. "Bemerkung zur Notiz des Herrn P. Ehrenfest: Translation deformierbarer Elektronen und der Fläschensatz". In: *The Collected Papers of Albert Einstein*, vol. II, ed. John Stachel, pp. 236–7. Princeton University Press. (Originally published 1907.)

Ellis, G., and Williams, R. M. 1988. *Flat and Curved Space Times*. Oxford University Press.

Everitt, Francis. 1975. *James Clerk Maxwell: Physicist and Natural Philosopher*. New York: Scribner.

Faraday, Michael. 1839–55. *Experimental Researches in Electricity*, 3 vols. London.

Feyerabend, Paul. 1981. *Realism, Rationalism and Scientific Method: Philosophical Papers*, vol. I. Cambridge University Press.

Feynman, Richard. 1963. *Lectures on Physics*, 3 vols. New York: Addison-Wesley.

Feynman, Richard, and Gell-Mann, M. 1958. "Theory of the Fermi Interaction". *Physical Review* 109:193–8.

Fisch, M. 1985. "Whewell's Consilience of Inductions: An Evaluation". *Philosophy of Science* 52:239–55.

Fisher, R. A. 1918. "The Correlation between Relatives on the Supposition of Mendelian Inheritance". *Transactions of the Royal Society of Edinburgh* 52:399–433.

1922. "On the Dominance Ratio". *Proceedings of the Royal Society of Edinburgh* 42:321–41.

1958. *The Genetical Theory of Natural Selection*, 2nd ed. Oxford: Clarendon Press. (Originally published 1930.)

Fizeau, Hippolyte. 1851. "Sur les hypothèses relatives à l'éther lumineux, et sur une expérience qui paraît démonstrer que le mouvement des corps change la vitesse avec laquelle la lumière se propage dans leur intérieur". *Comptes Rendus de l'Academie des Sciences* 33:349–55.

Forster, Malcolm. 1988. "Unification, Explanation, and the Composition of Causes in Newtonian Mechanics". *Studies in the History and Philosophy of Science* 19:55–101.

Friedman, Michael. 1974. "Explanation and Scientific Understanding". *Journal of Philosophy* 71:5–19.

1983. *Foundations of Space-Time Theories*. Princeton University Press.

Galison, Peter. 1979. "Minkowski's Space-Time: From Visual Thought to the Absolute World". *Historical Studies in the Physical Sciences* 10:85–121.

1983. "How the First Neutral Current Experiments Ended". *Reviews of Modern Physics* 55:277–310.

1987. *How Experiments End*. University of Chicago Press.

1998. *Image and Logic*. University of Chicago Press.

Galison, Peter, and Stump, David (eds.). 1996. *The Disunity of Science: Boundaries, Contexts and Power*. Stanford University Press.

Garber, Elizabeth. 1978. "Molecular Science in Late-Nineteenth-Century Britain". *Historical Studies in the Physical Sciences* 9:265–97.

Ghiselin, Michael. 1969. *The Triumph of the Darwinian Method*. Berkeley: University of California Press.

Gibbs, Josiah Willard. 1902. *Elementary Principles in Statistical Mechanics*. New York: Scribner.

Giere, Ronald N. 1988. *Explaining Science: A Cognitive Approach*. University of Chicago Press.

Gitterman, M., and Halpern, V. 1981. *Quantitative Analysis of Physical Problems*. New York: Academic Press.

Glashow, S. L. 1961. "Partial Symmetries of Weak Interactions". *Nuclear Physics* 22:579–88.

Glashow, S., and Gell-Mann, M. 1961. "Gauge Theories of Vector Particles". *Annals of Physics* 15:437–60.

Glymour, Clark. 1980. "Explanation, Tests, Unity and Necessity". *Noûs* 14:31–50.

Goldstone, J. 1961. "Field Theories with 'Superconductor' Solutions". *Nuovo Cimento* 19:154–64.

Goldstone, J., Salam, A., and Weinberg, S. 1962. "Broken Symmetries". *Physical Review* 127:965–70.

Hacking, Ian. 1983. *Representing and Intervening*. Cambridge University Press.

——— 1996. "The Disunities of Science". In: *The Disunity of Science: Boundaries, Contexts and Power*, ed. P. Galison and D. Stump, pp. 37–74. Stanford University Press.

Haldane, J. B. S. 1932. *The Causes of Evolution*. New York: Longman Group.

Harman, Gilbert. 1965. "Inference to the Best Explanation". *Philosophical Review* 74:88–95.

Harman, P. M. 1982a. *Metaphysics and Natural Philosophy*. Sussex: Harvester Press.

——— 1982b. *Energy, Force and Matter*. Cambridge University Press.

——— 1987. "Mathematics and Reality in Maxwell's Dynamical Physics". In: *Kelvin's Baltimore Lectures and Modern Theoretical Physics*, ed. Robert Kargon and Peter Achinstein. Cambridge, MA: MIT Press.

——— 1988. "Newton to Maxwell: The *Principia* and British Physics". *Notes and Records of the Royal Society, London* 42:75–96.

Harper, William. 1989. "Consilience and Natural Kind Reasoning". In: *An Intimate Relation*, ed. J. Brown and J. Mittelstrass, pp. 115–52. Dordrecht: Kluwer.

Heimann, P. 1970. "Maxwell and the Modes of Consistent Representation". *Archive for the History of the Exact Sciences* 6:171–213.

Helmholtz, H. Von. 1867. "On Integrals of the Hydrodynamical Equations, which Express Vector-motion". *Philosophical Magazine* 33:486–512.

Hesse, Mary. 1968. "Consilience of Inductions". In: *The Problem of Inductive Logic*, ed. Imre Lakatos, pp. 232–46. Amsterdam: North Holland.

Higgs, P. W. 1964a. "Broken Symmetries, Massless Particles and Gauge Fields". *Physics Letters* 12:132–3.

——— 1964b. "Broken Symmetries and Masses of Gauge Bosons". *Physical Review Letters* 13:508–9.

——— 1966. "Spontaneous Symmetry Breaking Without Massless Bosons". *Physical Review* 145:1156–63.

Hodge, M. J. S. 1977. "The Structure and Strategy of Darwin's Long Argument". *British Journal for the History of Science* 10:237–45.

Holton, Gerald. 1969. "Einstein, Michelson and the 'Crucial' Experiment". *Isis* 60. Reprinted 1973 in *Thematic Origins of Scientific Thought: Kepler to Einstein*, pp. 261–352. Cambridge, MA: Harvard University Press.

Holton, Gerald, and Brush, Stephen. 1985. *Introduction to Concepts and Theories in Physical Science*. Princeton University Press.

Hooker, C. A. 1975. "Global Theories". *Philosophy of Science* 42:152–79.

Howson, Colin. 1976. *Method and Appraisal in the Physical Sciences*. Cambridge University Press.

Hull, David. 1973. *Darwin and His Critics*. Cambridge, MA: Harvard University Press.

1985. "Darwinism as a Historical Entity: A Historiographic Proposal". In: *The Darwinian Heritage*, ed. D. Kohn, pp. 773–812. Princeton University Press.

Ives, H. I. 1952. "Derivation of the Mass–Energy Relation". *Journal of the Optical Society of America* 42:520–43.

Jammer, Max. 1961. *Concepts of Mass in Classical and Modern Physics*. Cambridge, MA: Harvard University Press.

Jeans, J. 1917. *The Kinetic Theory of Gases*. Oxford University Press.

Kallen, Horace. 1948. "The Meanings of 'Unity' Among the Sciences Once More". *Philosophy and Phenomenological Research* 4:493–6.

Kant, Immanuel. 1933. *The Critique of Pure Reason*, 2nd ed., trans. Norman Kemp Smith. London: Macmillan.

Kepler, Johannes. 1938. *Gesammelte Werke*, ed. Walther von Dyck, Max Casper, and Franz Hammer. Munich.

Khinchin, A. I. 1949. *Mathematical Foundations of Statistical Mechanics*. New York: Dover.

Kitcher, Philip. 1976. "Explanation, Conjunction and Unification". *Journal of Philosophy* 71:5–19.

1981. "Explanatory Unification". *Philosophy of Science* 48:507–31.

1989. "Explanatory Unification and the Causal Structure of the World". In: *Scientific Explanation*, ed. P. Kitcher and W. Salmon, pp. 410–505. Minneapolis: University of Minnesota Press.

1993. *The Advancement of Science*. Oxford University Press.

Klein, Martin. 1974. "The Historical Origins of the van der Waals Equation". *Physica* 73:28–47.

Kleiner, Scott. 1983. "A New Look at Kepler and Abductive Argument". *Studies in the History and Philosophy of Science* 14:279–313.

Knudsen, Ole. 1976. "The Faraday Effect and Physical Theory". *Archive for the History of the Exact Sciences* 15:235–81.

Kohn, David (ed.). 1985. *The Darwinian Heritage*. Princeton University Press.

Koyre, A. 1973. *The Astronomical Revolution*. Ithaca, NY: Cornell University Press.

Lagrange, Joseph L. 1788. *Mécanique Analytique*. Paris.

Larmour, Joseph. 1907. *Memoir and Scientific Correspondence of the Late Sir George Gabriel Stokes*. Cambridge University Press.

1937. *The Origins of Clerk Maxwell's Electrical Ideas*. Cambridge University Press.

Laudan, Larry. 1971. "William Whewell on the Consilience of Inductions". *The Monist* 65:368–91.

Levin, Michael. 1983. "Truth and Explanation". In: *Scientific Realism*, ed. J. Leplin, pp. 124–39. Berkeley: University of California Press.

Lloyd, Elizabeth. 1983. "The Nature of Darwin's Support for the Theory of Natural Selection". *Philosophy of Science* 50:112–29.

Lorentz, Hendrik Antoon. 1886. "De l'influence du mouvement de la terre sur les phénomènes lumineux". Reprinted 1935–39 in *Collected Papers*, vol. 4, pp. 153–214. The Hague: Nijhoff.

———. 1892a. "La théorie électromagnétique de Maxwell et son application aux corps mouvants". Reprinted 1935–39 in *Collected Papers*, vol. 2, pp. 164–343. The Hague: Nijhoff.

———. 1892b. "The Relative Motion of the Earth and the Aether". Reprinted 1935–39 in *Collected Papers*, vol. 4, pp. 219–23. The Hague: Nijhoff.

———. 1895. *Versuch einer Theorie der elektrischen und optischen Erscheinungen in bewegten Körpern*. Reprinted 1935–39 in *Collected Papers*, vol. 5, pp. 1–137. The Hague: Nijhoff.

———. 1904. "Electromagnetic Phenomena in a System Moving with Any Velocity Less Than That of Light". *Proceedings of the Royal Academy of Amsterdam* 6:809. (Reprinted 1952 in *The Principle of Relativity*, trans. W. Perrett and G. B. Jeffery, pp. 9–34. New York: Dover.)

Lucas, J. R., and Hodgson, P. E. 1990. *Space-time and Electromagnetism*. Oxford: Clarendon Press.

MacKenzie, Donald, and Barnes, Barry. 1979. "Scientific Judgement: The Biometry-Mendelism Controversy". In *Natural Order: Historical Studies of Scientific Culture*, ed. B. Barnes and Steve Shapin, pp. 191–210. London: Sage.

Malthus, T. R. 1798. *An Essay on the Principle of Population as It Affects the Future Improvement of Society*. London: J. Johnson.

Maxwell, J. C. 1856. "On Faraday's Lines of Force". *Transactions of the Cambridge Philosophical Society*, 10:27–83.

———. 1861–62. "On Physical Lines of Force". *Philosophical Magazine* 21:165–75, 281–91, 338–48; 22:12–24, 85–95.

———. 1865. "A Dynamical Theory of the Electromagnetic Field". *Philosophical Transactions of the Royal Society* 155:459–512.

———. 1873. *Treatise on Electricity and Magnetism*, 2 vols. Oxford: Clarendon Press. Reprinted 1954, New York: Dover.

———. 1881. *Elementary Treatise on Electricity*, ed. W. Garnett. Oxford: Clarendon Press.

———. 1952. *Matter and Motion*. New York: Dover.

———. 1965. *The Scientific Papers of James Clerk Maxwell*, 2 vols., ed. W. D. Niven. New York: Dover.

Mayr, Ernst. 1964. *The Growth of Biological Thought: Diversity, Evolution and Inheritance*. Cambridge, MA: Harvard University Press.

———. 1991. *One Long Argument*. Cambridge, MA: Harvard University Press.

Miller, A. I. 1981. *Albert Einstein's Special Theory of Relativity: Emergence (1905) and Early Interpretation (1905–1911)*. Reading, MA: Addison-Wesley.

Mills, R. 1989. "Gauge Fields". *American Journal of Physics* 57:493–507.

———. 1994. *Space-time and Quanta*. San Francisco: Freeman.

Minkowski, Hermann. 1909. "Raum und Zeit". *Physikalische Zeitschrift* 20:104–11. (Reprinted 1952 in *The Principle of Relativity*, trans. W. Perrett and G. B. Jeffery, pp. 73–91. New York: Dover.)

Moran, P. A. P. 1958. "A General Theory of the Distribution of Gene Frequencies". *Proceedings of the Royal Society, ser. B* 149:102–16.

Moriyasu, K. 1983. *An Elementary Primer for Gauge Theory*. Singapore: World Scientific.

Morrison, Margaret. 1986. "More on the Relationship Between Technically Good and Conceptually Important Experiments". *British Journal for the Philosophy of Science* 37:101–22.

1989. "Methodological Rules in Kant's Philosophy of Science". *Kant-Studien* 80:155–72.

1990a. "A Study in Theory Unification: The Case of Maxwell's Electromagnetic Theory". *Studies in the History and Philosophy of Science* 23:103–45.

1990b. "Unification, Reduction and Realism". *British Journal for the Philosophy of Science* 41:305–32.

1994. "Unified Theories and Disparate Things". *Proceedings of the Philosophy of Science Association* 2:365–73.

1996. "Physical Models and Biological Contexts". *Proceedings of the Philosophy of Science Association* 64:315–24.

1997. "Whewell and the Ultimate Problem of Philosophy". *Studies in the History and Philosophy of Science* 28:417–37.

1998a. "Modelling Nature: Between Physics and the Physical World". *Philosophia Naturalis* 35:65–85.

1998b. "Models, Pragmatics and Heuristics". *Dialectic* 1997/1:13–26.

1999a. "Models as Autonomous Agents". In: *Models as Mediators*, ed. Mary Morgan and Margaret Morrison. Cambridge University Press.

1999b. "Models and Idealisations: Implications for Physical Theory". In: *Idealisations in Physics*, ed. N. Cartwright and M. Jones. Amsterdam: Rodopi.

Moyer, Donald. 1977. "Energy Dynamics and Hidden Machinery: Rankine, Thomson and Tait, Maxwell". *Studies in the History and Philosophy of Science* 8:251–68.

Nagel, Ernst. 1979. *The Structure of Science*. Indianapolis: Hackett.

Nambu, Y., and Jona-Lasinio, G. 1961. "Dynamical Model of Elementary Particles Based on an Analogy with Superconductivity". *Physical Review* 122:345–58; 124:246.

Neurath, Otto, Carnap, R., and Morris, C. W. (eds.). 1971. *Foundations of the Unity of Science: Toward an International Encyclopedia of Unified Science*. University of Chicago Press. (Original publication began in 1938.)

Nordmann, Alfred. 1994. "The Evolutionary Analysis: Apparent Error, Certified Belief, and the Defects of Asymmetry". *Perspectives on Science* 2:131–73.

Olson, R. G. 1975. *Scottish Philosophy and British Physics, 1750–1880*. Princeton University Press.

O'Rahilly, Alfred. 1965. *Electromagnetic Theory*. New York: Dover.

Pais, A. 1982. *Subtle Is the Lord*. Oxford University Press.

Pearson, K. 1903. "On a Generalized Theory of Alternative Inheritance, with Special Reference to Mendel's Laws". *Philosophical Transactions of the Royal Society of London, ser. A* 203:53–87.

Pickering, Andrew. 1984. *Constructing Quarks*. University of Chicago Press.

Poincaré, H. 1906. "Sur la dynamique de l'électron". *Rendiconti del Circolo Matematico di Palermo* 21:129–75. Reprinted 1934–53 in *Oeuvres de Henri Poincaré*, 11 vols., vol. 9, pp. 494–550. Paris: Gauthier-Villars.

Price, G. R. 1972. "Fisher's Fundamental Theorem Made Clear". *Annals of Human Genetics* 36:129–40.

Price, G. R., and Smith, C. 1972. "Fisher's Malthusian Parameter and Reproductive Value". *Annals of Human Genetics* 36:1–7.

Provine, William B. 1971. *The Origins of Theoretical Population Genetics*. University of Chicago Press.

———. 1985. "Adaptation and Mechanisms of Evolution after Darwin: A Study in Persistent Controversies". In: *The Darwinian Heritage*, ed. D. Kohn, pp. 825–66. Princeton University Press.

———. 1986. *Sewall Wright and Evolutionary Biology*. University of Chicago Press.

———. 1992. "The R. A. Fisher–Sewall Wright Controversy". In: *The Founders of Evolutionary Genetics*, ed. S. Sarkar, pp. 201–29. Dordrecht: Kluwer.

Putnam, Hilary. 1975. *Mind, Language and Reality: Philosophical Papers*, vol. 2. Cambridge University Press.

Rankine, William J. 1859a. "On the Conservation of Energy". *Philosophical Magazine* 17:250–3.

———. 1859b. "Note to a Letter on the Conservation of Energy". *Philosophical Magazine* 17:347–8.

Recker, Doren. 1987. "Causal Efficacy: The Structure of Darwin's Argument Strategy in the Origin of Species". *Philosophy of Science* 54:147–75.

Ruse, Michael. 1979. *The Darwinian Revolution*. University of Chicago Press.

Salam, Abdus. 1968. "Weak and Electromagnetic Interactions". In: *Elementary Particle Theory: Relativistic Group and Analyticity. Proceedings of the Nobel Conference VIII*, ed. N. Svartholm, pp. 367–77. Stockholm: Almqvist & Wiksell.

Salam, A., and Ward, J. C. 1964. "Electromagnetic and Weak Interactions". *Physics Letters* 13:168–71.

Salmon, Wesley. 1984. *Scientific Explanation and the Causal Structure of the World*. Princeton University Press.

Sarkar, Sahotra (ed.). 1992. *The Founders of Evolutionary Genetics*. Dordrecht: Kluwer.

Schaffner, K. 1969. "The Lorentz Electron Theory of Relativity". *American Journal of Physics* 37:498–513.

Schwinger, J. 1957. "A Theory of Fundamental Interactions". *Annals of Physics* 2:407–34.

Segrè, E. 1983. *From Falling Bodies to Radio Waves*. San Francisco: Freeman.

Seigel, Daniel. 1975. "Completeness as a Goal in Maxwell's Electromagnetic Theory". *Isis* 66:361–8.

———. 1981. "Thomson, Maxwell and the Universal Aether in Victorian Physics". In: *Conceptions of the Aether*, ed. G. N. Cantor and M. J. S. Hodge, pp. 239–68. Cambridge University Press.

———. 1985. "Mechanical Image and Reality in Maxwell's Electromagnetic Theory". In: *Wranglers and Physicists*, ed. P. M. Harman. Manchester University Press.

———. 1986. "The Origin of the Displacement Current". *Historical Studies in the Physical Sciences* 17:99–146.

Smith, Crosbie. 1978. "A New Chart for British Natural Philosophy: The Development of Energy Physics in the Nineteenth Century". *History of Science* 16:231–79.

Stachel, John (ed.). 1989. *The Collected Papers of Albert Einstein*. Princeton University Press.

Stachel, John, and Torretti, Roberto. 1982. "Einstein's First Derivation of Mass–Energy Equivalence". *American Journal of Physics* 50:760–3.

Sundarshan, E., and Marshak, R. E. 1958. "Chirality Invariance and the Universal Fermi Interaction". *Physical Review* 109:1860.

Suppe, Fredrick. 1989. *The Semantic Conception of Theories and Scientific Realism*. Chicago: University of Illinois Press.

Suppes, P. 1961. "A Comparison of the Meaning and Uses of Models in the Mathematical and the Empirical Sciences". In: *The Concept and the Role of the Model in Mathematics and in the Natural and Social Sciences*, ed. H. Freundenthal, pp. 163–77. Dordrecht: Reidel.

Tabor, David. 1979. *Gasses, Liquids and Solids*. Cambridge University Press.

Tait, Peter G. 1876. *Lectures on Some Recent Advances in Physical Science*, 2nd ed. London.

Thagard, Paul. 1978. "Inference to the Best Explanation: Criteria for Theory Choice". *Journal of Philosophy* 75:76–92.

Thompson, Silvanus P. 1910. *The Life of William Thomson, Baron Kelvin of Largs*, 2 vols. London.

Thomson, William. 1854. "Note on the Possible Density of the Luminiferous Medium, and on the Mechanical Value of a Cubic Mile of Sunlight". *Transactions of the Royal Society of Edinburgh* 21:57–61.

1856. "Dynamical Illustration of the Magnetic and Helicoidal Rotatory Effects of Transparent Bodies on Polarized Light". *Proceedings of the Royal Society* 8:150–8.

1872. "Mechanical Representation of Electric, Magnetic and Galvanic Forces". In: *Reprint of Papers on Electrostatics and Magnetism*. London: Macmillan. (Originally published 1847.)

1891–94. "Presidential Address at Edinburgh to the British Association for the Advancement of Science". In: *Popular Lectures and Addresses*, 3 vols., vol. 2, pp. 175–6. London. (Originally published 1871.)

Thomson, William, and Tait, Peter G. 1867. *Treatise on Natural Philosophy*. Oxford: Clarendon Press.

1879–93. *Treatise on Natural Philosophy*, 2nd ed., 2 vols. Cambridge University Press.

van Fraassen, Bas. 1980. *The Scientific Image*. Oxford: Clarendon Press.

Vorzimmer, Peter. 1969. "Darwin, Malthus and the Theory of Natural Selection". *Journal for the History of Ideas* 30:527–42.

Watkins, Peter. 1986. *The Story of the W and Z*. Cambridge University Press.

Weber, W. E., and Kohlrausch, F. W. G. 1892–94. "Electrodynamische Massbestimmungen insbesondere Zurückführung der Stromintensitäts-messungen auf mechanisches Mass". In: W. E. Weber, *Werke*, 6 vols., pp. 609–76. Berlin: Springer-Verlag. (Originally published 1857.)

Weinberg, Steven. 1967. "A Model of Leptons". *Physical Review Letters* 19:1264–6.

1974. "Recent Progress in Gauge Theories of Weak, Electromagnetic and Strong Interactions". *Reviews of Modern Physics* 46:255–77.

1980. "Conceptual Foundations of the Unified Theory of Weak and Electromagnetic Interactions". *Reviews of Modern Physics* 52:515–23.

Westman, Robert. 1972. "Kepler's Theory of Hypothesis and the 'Realist Dilemma'". *Studies in the History and Philosophy of Science* 3:233–64.

Whewell, W. 1847. *The Philosophy of the Inductive Sciences Founded on Their History*, 2 vols., 2nd ed. London: John W. Parker. Reprinted 1967, New York: Johnson Reprint Corp.

Whittaker, Sir Edmund. 1951. *A History of the Theories of the Aether and Electricity.* New York: Thomas Nelson & Sons.

Williams, Pearce L. 1970. *Origins of Field Theory.* New York: University Press of America.

Wright, Sewall. 1918. "On the Nature of Size Factors". *Genetics* 3:367–74.

——— 1921. "Systems of Mating". *Genetics* 6:111–78.

——— 1988. *Evolution: Selected Papers,* ed. Willian Provine. University of Chicago Press.

Yang, C. N., and Mills, Robert. 1954. "Conservation of Isotopic Spin and Isotopic Gauge Invariance". *Physical Review* 96:191.

Zahar, Elie. 1989. *Einstein's Revolution: A Study in Heuristic.* La Salle, IL: Open Court Press.

Index

Analogy, 97–8. *See also* Consilience (of inductions)
  Maxwell on, 65–6
Argument patterns, 197–8
  in Darwinian evolution, 198–200, 227–8

Beta decay, in electroweak theory, 110–14, 111f
Boyle-Charles law, 59–60
Boyle's law, 48–9

Carnap, Rudolph, 22
Certainty, 16–22
Clock synchronization, 165–9, 170–1
  in inertial system, 170
Confirmation, Friedman on, 42
Conjunction, 39–43
  correction and, 42–3
  difficulties with, 41–2
  semantic vs. epistemological realism in, 41–2
  in unification, 37
Conservation of energy, 104, 105
Consilience (of inductions), 16–22, 52–7, 203
  certainty and, 16–22
  conditions for, 58
  in Darwinian evolution, 193–5, 203–4
  explanation in, 20
  reinterpretation of laws and terms in, 53
  in unification of Kepler's and Galileo's laws, 53–4
  in unification of relativity theory and Maxwell's electromagnetism, 54–5
Contraction effect, 153
Coulomb's law, 75

Darwinian evolution and natural selection, 192–209

argument patterns in, 198–200
consilience in, 193–5
explanation in, as derivation and systematization, 193, 196–202
explanation in, disunity of, 227–31
explanatory power of, 195, 206–9
heredity and variation in, 207
history and method in, 202–6
as hypothetico-deductivism, 193
as inference to best explanation, 193
Kitcher's model of, 193, 196–202
mathematics and theory in, 213–27
  disunity and explanation, 227–31
  Fisher's idealized populations, 214–24 (*See also* Fisher, R. A.)
  Mendelian ratios, 213–14
  Wright's path coefficients, 224–7
nature of theory in, 193–6
as one long argument, 195–6
population increase in, geometrical, 204–5
semantic view in, 194
unity of, 201–2, 204
unity of, explanation and, 206–9, 231
unity of, vs. in physics, 200–2
as *vera causa*, 202–3, 204
vs. Mendelism, 210–12
Decomposition, system, methodological paradox of, 45
Deductive nomological (D-N) model, 2–3
  criticisms of, 3
Dewey, John, 23
Dieterici equation, 48–9
Displacement (current), 73–5, 77, 79–81, 80f
Disunity
  explanation and, 227–31
  Kant on, 14
  unity through, 24, 208
Disunity/unity debate, 1, 6, 139–40, 237

Dynamical explanation, realism and,
  99–105
Dynamical theory, 81–90, 91–3, 96

Einstein. See Special theory of relativity
Electromagnetic induction, 162–3
Electromagnetism and optics, Maxwell's
  unification of. See Maxwell's
  electrodynamics
Electroweak theory, 109–46, 232–3
  electroweak interactions in, 130–1, 131f
  Fermi and beta decay in, 110–14, 111f
  Higgs mechanism in, 123–4, 135, 138
  neutral currents in, 126, 130–2, 131f,
    133
  renormalization in, 125, 127–30, 129f
  symmetry and gauge theory in, 118–21
  symmetry breaking in, 124–5, 127–8,
    134, 136, 138
  synthesis vs. reduction in, 125–6, 134–8
  synthetic unity in, 135, 136
  't Hooft, 125, 128
  unification in, 130–4, 137–40
  unity and disunity in, 135–40
  unity and, through symmetry breaking,
    121–7
  unity and renormalizability, 110
  Weinberg angle in, 109, 126, 135–8
Embedding approach
  van Fraassen on, 51
  vs. model/sub-model approach, 50,
    51
Empiricism, logical, 22–5
Energy, conservation of, 104, 105
Events, defined in STR, 168–9
Explanation, 202, 235–6. See also under
  specific theories, e.g., Electroweak
  theory
  as derivation and systematization, 193,
    196–202
  disunity and, 227–31
  explanandum, 2–3
  mechanisms, 3–4, 26–9
  as unification, 2–4, 29–30, 193, 228,
    231
  vs. truth, 35
  vs. unification, 1–2, 5, 35–8, 63–4,
    106–7, 138–9, 191, 192–209,
    212–13, 218, 227–31, 235–7
  (See also specific theories, e.g.,
  Darwinian evolution and natural
  selection)

Explanation, unity as, 1–3, 25–34
  applicability of theories in, 33
  combining of phenomena in, 32–3
  conditions for connection in, 32–3
  in Darwinian evolution, 206–9
  in electroweak theory, 138
  mathematical identities and
    contingencies, 30–1
  in Maxwell's electrodynamics, 63–4
  mechanical explanation in, 26–9
  objectivity in, 25, 26–30
  patterns in, 31–2
  properties of, 25
  reduction in, 25–6, 28
Explanatory store, 196–7

Faraday effect, 68, 84
Faraday's lines of force, Maxwell on, 64–8
Fermi, Enrico, 110
  beta decay and, 110–14, 111f
Feynman diagrams, 143
  and renormalization, 143–6, 143f, 145f
Fisher, R. A., 214–24
  approach and goal of, 229–30
  on dominance ratio and variance, 217
  on equilibrium under selection, 218–19
  gas theory model of, 215–19, 221–2
  on genetic factors in mixed populations,
    214–15
  on natural selection and fitness, 219–21
Fisher's fundamental theorem of natural
    selection, 219–22
  idealizing assumptions in, 222
  Malthusian parameter in, 221
  unification in, 222–3
Fitness, natural selection and, 221
Force, lines of
  and electric current, 70–1, 71f
  Faraday's, 64–8
  physical, 68–79
  polarity of, 68–9
Friedman, Michael, 25–6, 28. See also
    Conjunction; Reduction
  unification model of, 37–9, 47
  basic strategy of, 52
  confirmation of results in, 57
  historical relativism and, 58–9
  persistence of structure in, 59
  on realism, 36–7
Fundamental theorem of natural selection.
  See Fisher's fundamental theorem of
  natural selection

Gauge invariance, 114–17
  and renormalization, 127
Gauge theory, 109–10
  symmetry and, 118–21
  as unifying tool, 137–8
GIM mechanism, 133
Glashow, S. L., 119–20, 125, 132–3
Glashow-Weinberg-Salam model, 124–6
  renormalization in, 127–30, 129f
Glymour, Clark, 29–32

Hagedoorn effect, 218
Hardy-Weinberg law, 214
Herschel, William, 203–4
Higgs, Peter, 123
Higgs field, 124, 136, 138
Higgs model, 123–4, 135, 138
Hypothetico-deductivism, 98–9
  in Darwinian evolution, 193

Inference. See also Unification, realism,
    inference and
  conjunctive, 39–43
  unification and, 35–61, 232 (See also
    Unification, realism, inference and)
Inference to the best explanation (IBE),
    35–6
  Darwinian evolution as, 193
  objections to, 36, 139
Integration, method of, 22–5
Interconnectedness, 33
Invariance, local gauge, 114–15

Kant, Immanuel, 12–16
  logical principle and maxims, 12–15
  "order of nature", 12
  reason, 12, 14
  on unity, 5, 15–16
Kepler, Johannes, 5–6, 7–12
  on elliptical orbit of Mars, 9
  empirical confirmation, 10
  on geometrical vs. astronomical
    hypotheses, 6–7
  on interconnectedness and truth, 10–11
  on libratory motion, 9–10
  mathematical harmonies/hypotheses, 7,
    20
  on simplicity and unity, 7–9
  the Trinity and, 7–8
Kitcher, Philip
  on Darwinian evolution, 193, 196–202
  on explanation as unification, 193,
    227–9

Lagrangian approach, 30, 83, 90, 92–7
Logical empiricism, 22–5
Logical principle, 12–16
London, Fritz, 115
Lorentz-FitzGerald contraction hypothesis,
    153–4, 161
Lorentz's electromagnetic theory, 149–62
  asymmetries problem in, 163–4
  Fresnel's coefficient of entrainment in,
    150
  fringe shift in, 149–50
  Lorentz-FitzGerald contraction
    hypothesis in, 153–4, 161
  Michelson-Morley experiment and, 150,
    152–3
  molecular forces hypothesis, 153
  problems in, 161–2, 164
  Lorentz's Versuch, 156–62
  theorem of corresponding states in,
    158–60
  transforming electrodynamic into
    electrostatic problems in, 156–7,
    161–2
Lorentz transformations, 147–8, 151–3,
    157–61. See also Transformation
    equations

Mach, Ernst, 27–8
Malthus
  in Fisher's work, 223
  on population growth, 204–6
Mass-energy relation, Einstein's, 179–83
Mathematical metaphysics, 7–11
Mathematical models
  vs. physical systems, 30–1
  vs. statistical mechanics, 45–6
Mathematical theory, vs. physical theory,
    66
Mathematics, theory and, 213–27
  Fisher's idealized populations in,
    214–24
  Wright's path coefficients in, 224–7
Maxwell, J. C.
  on conservation of energy, 104, 105
  on mathematical vs. physical theory, 66
  on matter, 101
  on physical reasoning, 100
Maxwell's electrodynamics
  abstract dynamics in, 94–5
  aether vs. material media in, 102–3
  analogy in, 65–6, 71, 97–8
  charge in, 86

Maxwell's electrodynamics (*cont.*)
  consilience of, 77
  Coulomb's law in, 75
  dielectric constant in, 75–6, 77
  displacement in, 73–5, 77, 79–81, 80f,
    90
  dynamical theory in, 91–3, 96, 104
  early development of, 64–8
  elasticity equations in, 89
  electromagnetic field equations in,
    74–7, 86–7
  electromagnetic medium in, 73–4
  electromotive force equations in, 89
  electrostatics in, 65
  electrotonic state in, 66–7
  energy in, 83
  engineering analogy in, 71
  explanatory power of, 63–4
  history of, 62–3
  idle-wheel hypothesis in, 69, 71, 72
  important features of, 62–3
  Lagrangian approach in, 83, 90, 92–7
  late developments of
    1861–62, 68–79
    1865: dynamical theory, 81–90
    1973: electricity and magnetism,
      90–9
  magnetoelectric medium in, 73
  mathematical structure of, 4–5
  mathematical structure of, vs. physical
    concepts, 99–100
  mental representation in, 94–5
  motion of medium in, 83–4
  nature vs. its representation in, 96
  philosophical conclusions on, 105–8
  physical analogy in, 65–6
  polarity of lines of force in, 68–9
  problems with, theoretical, 79–81, 80f
  realism and dynamical explanation in,
    99–105
  reductive unity in, 63, 78
  stress and strain in, 102
  unification in, 56–7, 104–7, 135, 234
  unification in, realism and, 79–81
  unification in, vs. explanation, 106–7
  unifying power of, 2, 20
  vector potential in, 66
  vortex hypothesis in, 68–73, 71f,
    79–80, 91
  wave equation in, 87–8
Mechanical explanation, 26–9
Mendelian ratios, 213–14

Mendelism, vs. Darwinism, 210–12
Michelson-Morley experiment, 150,
    152–3
Minkowski space-time, 149, 183–91
Model/sub-model approach, 37–9
  literal vs. representational account in, 47
  phenomenological vs. theoretical
    structures in, 47
  in van der Waals law, 47–9
  vs. embedding approach, 50, 51
Molecular forces hypothesis, 153. *See also*
    Lorentz's electromagnetic theory
Molecular vortices, 68–73, 71f, 79–80, 91

Natural selection, 192–209. *See also*
    Darwinian evolution and natural
    selection
  Fisher's fundamental theorem of,
    219–22 (*See also* Fisher's
    fundamental theorem of natural
    selection)
  fitness of species and, 221
  Neo-Darwinian selection, 227–8
Neurath, Otto, 23–4
Newtonian mechanics, unifying power of,
    2–3
Newton's *Principia*, 4

Objectivity, 25, 26–30

Path coefficients, 224–7, 225f
Phase transformations, QED, 116
Physical theory, vs. mathematical theory,
    66
Populations
  Fisher's idealized, 214–24 (*See also*
    Fisher, R. A.)
  geometrical increase of, 204–5
  Malthus on growth of, 204–6
Principia, 4
Pseudo-angles, 184
Putnam, H., 41, 43

Quarks, 132–3

Realism, 36–7
  dynamical explanation and, 99–105
  molecular structure vs. properties in, 51
  unification and, 35–61 (*See also*
    Unification, realism, inference and)
  unification and, in Maxwell's theory,
    79–81

Reduction
  in electroweak theory, 125–6, 134–8
  and theory, 25–6, 28
  vs. representation, 43–52
  viability of, 43–7
Reductive unity, 5, 63
  in Maxwell's electrodynamics, 63, 78
Relativism, historical, 58–9
Relativity theory. See Special theory of
    relativity
Renormalization
  in electroweak theory, 110, 125,
    127–30, 129f
  Feynman diagrams, 143–6, 143f, 145f
  gauge invariance, 127
  in quantum field theories, 133
Representationalist account, 39

Salam, Abdus, 120–1, 124–5
Schwinger, Julian, 114, 118–19, 125
Scientific theory, reasons for belief in,
    29–30
Semantic view, 194
  and Darwinian evolution, 194
  Lloyd, E., 194–5
  of van Fraassen, 51–2
Simplicity, unity and. See also Consilience
  Kepler on, 7–9
Simultaneity, 166, 168, 170
Space-time, Minkowski, 149, 183–91
Space-time interconnection, 169–76
Special theory of relativity, 147–91
  constancy of c in, 164–5
  Einstein on, 147–8
  electromagnetic induction in, 162–3
  event in, 168–9
  geometrical vs. kinematical shapes in,
    175
  Lorentz's electromagnetic theory in,
    149–62 (See also Lorentz's
    electromagnetic theory)
  Lorentz transformations in, 147–8
  mass-energy relation in, 179–83
  Minkowski space-time in, 149, 183–91
  physical basis for, 148
  postulates of, 148
  synchronizing clocks in, 165–9, 170–1
  synthetic unity in, 176–9
  time in, 164–8, 176, 177
  transformation equations of, 169–77
Spin, particle, 113
Substantial identification, 98

Symmetries, 114–16, 137
  gauge theory and, 118–21
  objects and laws in, 114–17
Symmetry breaking, 121–8, 134, 136, 142
Symmetry groups, 116
Synthetic unity, 5, 100
  in electroweak theory, 135, 136
  in special theory of relativity, 176–9

Theory of corresponding states, 156–62
Theory unification. See Unified theories;
    specific theories, e.g., Maxwell's
    electrodynamics
Thomson, William, 64–5
  vortex model of, 66–7
Thomson and Tait's dynamics, 91–3,
    99–100
Time
  common, 167
  Einstein on, 164–8, 177
  local, 166–7
  relativity of, 176
  simultaneity, 166, 168
Transformation equations
  Einstein's, derivation of, 169–76
  Lorentz's, 147–8, 156–7, 161–2
Truth, vs. explanation, 35

Unification, 1, 105. See also specific
    theories, e.g., Maxwell's
    electrodynamics
  and the biological synthesis, 210–13
  combining of phenomena in, 32–3
  as confirmation, 138
  in electroweak theory, 137–40
  evidence and, 31–2
  as evidential or epistemic virtue, 57–9
  explanation as, 29–30, 193, 228, 231
    (See also Explanation, unity as)
  free parameters and, 207–8
  inference and, 232
  of Kepler's and Galileo's laws, 53–4
  in Maxwell's electrodynamics, 107
  reductive, 5, 63, 78
  of relativity theory and Maxwell's
    electromagnetism, 54–5
  synthetic, 5, 100
  synthetic, in electroweak theory, 135,
    136
  synthetic, in special theory of relativity,
    176–9
  in theory acceptance, 105–6

Unification vs. explanation, 139, 235–7
  in electroweak theory, 138
  in Maxwell's electrodynamics, 106–7
Unification, realism, inference and, 35–61
  conjunction and, 39–43
  consilience of inductions in, 52–7 (See
    also Consilience (of inductions))
  evidential or epistemic virtue of, 57–9
  Friedman model and, 38–9
  inference to best explanation and, 35–6
  many models problem in, 47–52
  reduction in, viability of, 43–7
  reduction vs. representation and, 43–52
Unified theories, 1, 35, 105. See also
    specific theories, e.g., Special theory
    of relativity
  convergence of quantitative results in, 11
  disparate things and, 135–40
  empirical support for, 232–3
  and evidential support, 108
  explanatory power of, 1–3, 55–6, 206–9
  interconnectedness in, 11, 33
  mathematical constraints in, 109–10
  mathematical framework of, 233
  mathematical vs. dynamical/causal proof
    of, 11
  prediction in, 55–6
  realism and, 36–7, 79–81 (See also
    Unification, realism, inference and)
  rigidity vs. flexibility in, 33–4
  synthesis vs. reduction in, 125–6
  uncertainty in, 235–7
  vs. explanations, 4–5, 35
  vs. theoretical deduction, 107
  vs. unity in nature, 140, 234, 237
Unity, 1, 2
  across vs. within science, 55
  arguments against, 139–40
  characteristics of, 232
  as consilience and certainty, 16–22 (See
    also Whewell, William)
  as explanation, 1–3, 25–34 (See also
    Explanation, unity as)
  as heuristic and logical principle, 12–16
    (See also Kant, Immanuel)
  Kant on, 12–16 (See also Kant,
    Immanuel)

Kepler on, 7–11 (See also Kepler,
    Johannes)
  as mathematical metaphysics, 7–11 (See
    also Kepler, Johannes)
  mechanical explanation and, 26–9
  as method of integration, 22–5
  simplicity and, Kepler on, 7–9
  synthetic (See Synthetic unity)
  through disunity, 24, 208
  Whewell on, 16–22 (See also Whewell,
    William)
Unity/disunity debate, 1, 6, 139–40, 237

V-A theory, 112
van der Waals law, 47–9, 59–61
van Fraassen, Bas, 51–2, 138
Versuch, 156–62
  theorem of corresponding states in,
    158–60
  transforming electrodynamic into
    electrostatic problems in, 156–7,
    161–2
Vortex hypothesis, 67–73, 71f, 79–80,
    91

Ward, J. C., 120–1
Weinberg, Steven, 121–2, 124–5, 134
Weinberg angle, 109, 126, 135–8,
    232
Weinberg's lepton model, 140–2
Weyl, Hermann, 114–15
Whewell, William, 16–22
  consilience, 16–17, 21, 52–7, 203 (See
    also Consilience (of inductions))
  induction and unification, 17–19, 21
  on inductive truth, 55
  methods of curves, least squares and
    means, 18
  predictive success, 16
  simplicity, 16
  unity, 5, 16
  vera causa of, 21
Wright, Sewall, 224–7, 225f
  approach and goal of, 229–30
Wright-Fisher model, 229

Yang-Mills gauge theory, 116–17, 123